SUSTAINABLE AGRICULTURE AND BIODIVERSITY:

INNOVATIONS AND INSIGHTS

SUSTAINABLE AGRICULTURE AND BIODIVERSITY:

INNOVATIONS AND INSIGHTS

Bibhuti Bhusan Sahoo
Regional Research and Technology Transfer Station (OUAT)
Bhawanipatna, Kalahandi-766001, Odisha, India

Harvinder Kumar Singh
Department of Plant Pathology,
College of Agriculture, IGKV, Raipur-492012, Chhattisgarh, India

Niranjan Chinara
All India Coordinated Research Project on Fruits (OUAT)
Bhubaneswar-751003, Odisha, India

and

Amitabh Panda
Krushi Vigyan Kendra, Kalahandi
Bhawanipatna, Kalahandi-766001, Odisha, India

NPH

NARENDRA PUBLISHING HOUSE
DELHI (INDIA)

First Published in 2025

ISBN: 978-93-5651-549-9

Published by :

NARENDRA PUBLISHING HOUSE

Publisher and Distributor

C-21, Varun Apartment, Sector -9, Rohini

NEW DELHI - 110085

Phones: 91-011-45025794, +919717874875

E-mail: info@nphindia.com, nphindia@gmail.com

Website: www.nphindia.com

Printed in India

Laser Typeset by Amrit Graphics, Shahdara, Delhi-110032

Contents

भा.कृ.अनु.प.-राष्ट्रीय लीची अनुसंधान केन्द्र
ICAR-NATIONAL RESEARCH CENTRE ON LITCHI
मुशहरी, प्रक्षेत्र मुशहरी, मुजफ्फरपुर-842002 (बिहार)
Mushahari Farm, Mushahari, Muzaffarpur - 842002 (Bihar)
ईमेल Email: director.nrcl@icar.gov.in वेबसाइट Website: https://nrclitchi.icar.gov.in

डॉ. बिकाश दास
निदेशक, भाकृअनुप- रा.लीची अनु.केन्द्र
Dr. Bikash Das
Director, ICAR-NRC on Litchi

FOREWORD

India has achieved a significant stride in total food grain production of 232.22 MT with a per capita food availability of 207.6 kg per year during 2023-24. However the total degraded land has increased by 16% and erosion by water, soil and others by 27% over last decades. Sustainability in agricultural production will emphasize environmental health, economic profitability by blending traditional farming methods with modern techniques by reducing use of harmful chemicals and promoting soil health, conserving water and ensuring energy efficiency besides supporting rural livelihoods.

For sustainability in agriculture, agro-biodiversity holds the key. Biodiversity ensures resilience against insect-pest, diseases and climate change, safeguarding global food security.

The synergy between sustainable agriculture and biodiversity is vital for building resilient farming systems. Together they foster a sustainable balance that protects the ecosystems while supporting the growing demand to food, feed and fibre. Promoting sustainable agricultural production along with biodiversity conservation is essential for achieving global sustainable goals and preserving our blue planet for future generations.

I congratulate the editors B.B Sahoo, H.K Singh, N. Chinara, A. Panda for their sincere efforts to bring out the book entitled "Sustainable Agriculture and Biodiversity: Innovations and Insights". I hope this book will be useful for the faculties, scientists, researchers and students.

(Bikash Das)

January, 2025
Muzaffarpur

Preface

Sustainable agriculture is a farming approach designed to meet the essential needs of humanity in an environmentally responsible manner. It is founded on the principle that we should satisfy current needs without jeopardizing the ability of future generations to meet theirs. Vegetable production plays a crucial role within this framework. Key practices such as organic farming, vertical farming, agroforestry and integrated pest management (IPM) are vital techniques that promote biodiversity while minimizing resource use. Moreover, mushrooms are recognized as horticultural crops. Cultivating mushrooms not only aids in managing agricultural waste and reducing environmental pollution but also provides a nutritious food source that helps address malnutrition and poverty. Unfortunately, many farming communities prioritize increased production, which can negatively impact the global ecosystem. Therefore, embracing organic and natural farming practices is of utmost importance.

A wealth of literature exists on the subject of sustainable agriculture, yet this book stands out by focusing on advanced and innovative techniques in vegetable cultivation. It not only addresses production methods but also emphasizes the systematic management of diseases and pests. Additionally, the book delves into mushroom production as a sustainable practice, detailing how various edible mushrooms can be cultivated scientifically throughout the year with minimal investment. Fish farming, another lucrative venture, is discussed in comprehensive detail as well. Ultimately, this initiative aims to provide a succinct overview of cutting-edge technologies in vegetable and mushroom production, pest management and aquaculture.

Authors

About the Editors

Dr. Bibhuti Bhusan Sahoo (b.1977) working as Scientist (Horticulture) at Regional Research and Technology Transfer Station, OUAT, Kalahandi, Odisha. He has published 55 research papers in national and international journals, 78 seminars/ symposiums/conference abstracts, 08 book chapters, 17 popular articles, 08 extension literatures and 03 textbooks. He has 09 papers in h-index and i10-index, with 350 citations in Google Scholar. He has guided 03 M.Sc. (Ag.) students in their theses. He has delivered twelve radio and television talks. He is an editorial board member of many research journals and life member of eleven professional societies.

Dr. Harvinder Kumar Singh (b.1984) working as Assistant Professor in the Department of Plant Pathology, IGKV, Raipur, Chhattisgarh. He has guided 17 M. Sc. students and 04 Ph. D students. He has authored 05 technical bulletins, 02 laboratory manuals, 12 book chapters and 23 research papers. Further, he also served as Zonal Councillor of Central Zone IPS in 2014. He is also the lead developer of CI-01 (DMRO 1214), a variety of milky mushroom and Vv-05-19 (DMRO 1213) a variety of paddy straw mushroom released for commercial cultivation.

Dr. Niranjan Chinara (b.1978) is working as Plant Pathologist in AICRP on Fruits, OUAT, Bhubaneswar. He has 15 years of experience in teaching, research and extension in the field of plant pathology, cotton and mushroom biology. He has been involved in the release of cotton variety (BS 279) and mushroom strain paddy straw mushroom (ICAR-HS-DMR- Strain-2023-094). He published 20 research papers, 45 abstracts, 20 booklets, 52 popular articles and 04 book chapters. He has guided 19 M. Sc. students from different disciplines.

Dr. Amitabh Panda (b.1975), is currently the Senior Scientist and Head of Krishi Vigyan Kendra, OUAT. He has completed his B. Sc. (Ag) from OUAT, M.Sc. Ag (Vegetable Science) from G.B.P.U.A & T, and Ph. D. from Visva-Bharati. He is PI and Co-PI of several projects such as NICRA, Plant Health Clinic, Biotech-KISAN, Pulse Model Village and Centre of Excellence on FPO. He has authored more than 10 research papers, 08 conference papers, 08 popular articles, 20 booklets and 02 book chapters. He has delivered more than 07 radio and television talks.

About the Book

This book consists of 15 chapters that delve into topics such as vegetable science, fruit science, pest and disease management, mushroom production and pisciculture. One chapter is dedicated to vegetable grafting, an effective technique utilized by farmers and growers to combine the desirable traits of different plant varieties. This section discusses various commercial grafting methods, as well as their benefits and limitations, enabling farmers to apply these techniques in their fields. Vegetables, in particular, face challenges in reproduction during periods of water scarcity. Anti-transpirants can reduce transpiration rates under drought conditions, helping plants to survive. This book explores the role of anti-transpirants during such critical times. Organic farming fosters the sustainable production of agricultural products while preserving soil health, environmental integrity, and human well-being. Similarly, hydroponics allows for the cultivation of various agricultural products in both laboratories and fields without the use of soil, facilitating the growth of high-value organic crops. The book meticulously outlines different vegetable breeding methods aimed at developing new strains or varieties with desirable traits, such as enhanced production, productivity, and resistance to diseases and pests. Biological management serves as a fundamental component of Integrated Pest Management (IPM). This book briefly describes various fungi, viruses, and insects utilized to mitigate negative impacts on plants. Mushrooms are recognized as horticultural crops with significant potential to combat poverty, hunger, and malnutrition while contributing to nutritional security. The book offers insights into various wild edible mushrooms, highlighting their importance, nutritional and medicinal values, and marketability, benefiting mushroom growers, traders, and students nationwide. Lastly, pisciculture is a crucial aspect of sustainable agriculture, focusing on the scientific production of different fish species. The book also includes chapters dedicated to sustainable agricultural practices that not only enhance agricultural productivity but also protect the environment for future generations.

Sustainable Agriculture and Biodiversity: Innovations and Insights, Pages: 1–24
Edited by: Bibhuti Bhusan Sahoo, Harvinder Kumar Singh, Niranjan Chinara and Amitabh Panda

CHAPTER - 1

VEGETABLE GRAFTING: AN EXCELLENT TECHNIQUE TO COMBAT BOTH BIOTIC AND ABIOTIC STRESS ON PLANTS

Barsha Tripathy[1*], Meenakshi Badu[2] and Suchismita Mishra[3]

[1]*Department of Vegetable Science,*
Faculty of Agricultural Sciences, SOADU, Bhubaneswar-760001, Odisha
[2]*Faculty of Agriculture, Sri Sri University, Cuttack-754006, Odisha*
[3]*Department of Floriculture and Landscaping,*
Faculty of Agricultural Sciences, SOADU, Bhubaneswar-760001, Odisha
[*]*Corresponding email: barshatripathy@soa.ac.in*

INTRODUCTION

The process of grafting involves using tissue regeneration to fuse two plant parts—a rootstock and a scion—so that the combined pieces physically reunite and grow into a single plant (Janick, 1986). Although fruit tree grafting has been done for thousands of years, vegetable cultivation is predicted to be a relatively new field for this approach. Using resistant root stocks for commercial vegetable grafting is one of the finest methods for producing vegetables in a sustainable manner. It lessens the need for agrochemicals in the production of organic food (Rivard and Louws, 2008). The main goals of utilising grafted seedlings are to: (1) increase resistance to nematodes and common soil-borne illnesses; (2) improve yield and quality; and (3) modify plant physiology to make plants more tolerant of unfavourable situations. Vegetable grafting is becoming more and more common around the world for cucurbits, tomatoes, eggplants, and peppers using robust and disease-resistant rootstocks to guarantee sufficient yields in situations where productivity is limited by biotic and abiotic pressures (Buller *et al.*, 2013). Increases in yield of up to 80% have been achieved in the Solanaceae family and 60-90% in the Cucurbit family using vegetable grafting.

HISTORICAL BACKGROUND

The self-grafting method, which was used to harvest a large gourd fruit by increasing the root-to-shoot ratio through many graftings, was elaborated in a historical text written in Korea in the seventeenth century and China in the fifth century. To prevent soilborne illnesses, Japan and Korea invented scientific vegetable grafting in the late 1920s by grafting watermelon (*Citrullus lanatus*) onto pumpkin (*Cucurbita moschata*) rootstock (Leonardi, 2016). Farmers in Korea and Japan were first exposed to this technological innovation through agricultural extension workers. By grafting watermelon onto bottle gourds, Japan was the first country to use grafted transplants for commercial purposes in the early 1930s. Among the solanaceous crops, scarlet aubergine (*Solanum integrifolium* L.) was first grafted on to aubergine (*Solanum melongena* L.) in the 1950s (Oda, 1999). According to the research reports confirmed by Isbell (1994) in the year 1935 and 1943 grafting technique was performed in tomato, eggplant and sweet pepper on selected weed rootstock. Nevertheless, the combination of tomato (scion) and jimson weed (rootstock) was not introduced on commercial scale and seems to be disappeared, due to the transportation of small number of alkaloids to fruit of the tomato and labour-intensive propagation method. In 2013, grafted "Pomato" plant was introduced in UK by a private seed company "Thompson and Morgan" later they launched another wonder plant known as Egg and chips plant in 2015. Grafting in tomato was commercially launched in the year 1960s with the prompt extension of protected cultivation technologies (high tunnels and greenhouses).

In an effort to identify the ideal rootstocks for wet environments, Dr. R. M. Bhatt and his colleagues at IIHR, Bangalore, initiated the first vegetable grafting project in India in 2013. Utilising Solanum nigrum as the rootstock, TNAU, Coimbatore has conducted research on brinjal grafting. *Momordica cochinchinensis* cucurbit grafting was investigated at the NBPGR regional station in Thrissur, Kerala, with the goal of increasing productivity by the grafting of female plants onto male plants.In order to introduce bacterial wilt and nematode resistance, Dr. Pradeep Kumar of CSKHPKV, Palampur, began working on grafting and discovered more than 22 rootstocks in the categories of brinjal, chilli, tomato, potato, and cucurbit. Flooding, soil-borne infections, and root-knot nematodes can all be reduced by grafting tomatoes onto resilient tomato or aubergine rootstocks. Certain private players also participate in grafting. Among these is VNR Seed Private Limited in Chhattisgarh, which provides farmers with grafted brinjal seedlings that are resistant to bacterial wilt. The company's blue ocean strategy has produced a number of excellent outcomes, including the introduction of grafted vegetables and root stock seed.The other seed company in India that specialises in vegetable grafting and provides high-quality grafted seedlings is called "Takii Seed India Private Limited." (Kumar *et al.*, 2015). Two significant public institutions in India, the Department of Horticulture at Tamil Nadu Agricultural University (TNAU) and the CSK Himachal Pradesh Krishi Vishvavidyalaya, Palampur, have modernised their vegetable grafting method through the use of robotic machines (Automated grafting machine). As a result, up to 800 grafted seedlings can be produced per hour. TNAU purchased the machine from Korea, and with mechanical grafting, they have created almost fifty thousand grafted plants to date. Similarly, Prof. A.K. Sarial dedicates a robotic vegetable grafting system to the farmers of Himachal Pradesh: CSK H.P. Agriculture University. The use of robots or grafting machines is growing in popularity. The first robotic grafting device, a one-cotyledon grafting system, was created in the 1980s by the Japanese company "I am Brain," especially for cucurbits. Following many years of development, fully automated robotic grafting machines were developed in the Netherlands and Japan in the 1980s. These machines could perform 1000 and 750 grafts per hour, respectively. The world's top producers of grafting machines are Korea, Japan, the Netherlands, and Spine.

PURPOSE OF CULTIVATION OF GRAFTED VEGETABLES

While preventing soil-borne illnesses like bacterial wilt in Solanaceae (tomato, pepper, etc.) and fusarium wilt in Cucurbitaceae (cucumber, melon, etc.) are the main objectives of grafting,

It is advantageous and serves a multitude of functions .Some of these functions include (Lee *et al.*, 2010):

- Increase in yield
- Shoot growth promotion
- Disease and nematode tolerance/resistance
- Enhancement of nutrient uptake
- Enhancement of water uptake
- Tolerance to elevated salt levels
- Ability to endure moist soil
- Tolerance to organic pollutants and heavy metals
- Tolerance for variations in quality

- Prolonging the harvest season
- Allowing successive or recurrent cropping
- Easy generation of ornamental values and organic wastes for teaching and exhibition

BASIC PREREQUISITES OF GRAFTING

a) **Root stock**

b) **Scion**

c) **Compatibility**

d) **Grafting aids** like grafting clips, tubes, pins, blade, cutter etc.

e) **Screen house**: It is used for growing seedlings prior to grafting. There must be provision for double door system at the entrance of screen house. The cladding material for covering the screen house should be made of UV stabilized polyethylene film to prevent UV light penetration.

f) **Healing chamber/Grafting chamber**: The healing chamber is a covered structure with controlled humidity and low light. The purpose of healing chamber is to provide congenial temperature and humidity for better union of the grafts. Grafted plants can also be healed under in plastic tunnels maintaining almost near to optimum conditions for better healing. In general, optimum rage of temperature and relative humidity is 25-30 ^{0}C and 85- 90%, respectively with low light intensity.

g) **Acclimatization chamber**: It is used for hardening of grafts prior to transplanting. This chamber helps to prevent leaf burning and wilting of the just healed seedlings. Grafted seedlings take 7 to 10 days for acclimatization.

GRAFT UNION FORMATION

The first prepared graft portions, the stock and scion, are positioned closely together but do not move or grow in conjunction. In contrast, some unions are completed only by the cells that grow following the grafting procedure. In contrast, some unions are completed only by the cells that grow following the grafting procedure. Three primary steps occur for the establishment of a compatible graft: vascular differentiation across the graft interface, callus cell growth at the graft interface or a callus bridge, and rootstock and scion adhesion.

The specific steps in the formation of a graft union are:

1) **Lining up of vascular cambium of stock and scion:** The creation of a cambial layer between the stock and the scion is necessary for effective

grafting. The parenchyma cells from the scion and stock that are generated in this region should be able to interlock because the cambial region should be sufficiently near together. The callus bridge's ability to maintain its vascular connection depends on the cambium. In order to prevent the graft components from moving or becoming dislodged during the parenchyma cell interlocking process, it is crucial that they are securely fastened together using a technique such as tying, wrapping, stapling, nailing, etc.

2) **Wounding response:** The cell wall and substance of stock and scion cells combine to produce the isolation layer, also known as the necrotic layer. Cutting the scion and stock kills the cells at least a few cell layers deep. The majority of the components in the necrotic layer eventually vanish or may stay in the spaces between the callus that is subsequently generated as a result of parenchyma cells really dividing. The unharmed parenchyma cells divide swiftly to generate an undifferentiated callus tissue. The wound periderm here is originally formed by callus tissue.

3) **Formation of callus bridge:** The creation of the callus is necessary for the transplant to be successful. In one to seven days, a fresh batch of parenchymatous cells proliferates from both the stock and the scion. Through additional cell division of the unharmed parenchyma cells in the scion and stalk's outer layer, callus tissue continues to form. The newly formed parenchyma cells are inside and next to the necrotic layer; they quickly mix and become entangled, filling the space between the stock and scion. The cement or bridging substance that protrudes in a bead-like way from the callus surface cells of both grafting fragments aids in the adhesion between the scion and stock cells. Next, there is a widespread cell fusion. The mixture of pectinase, proteins, and carbohydrates is contained in the projection that resembles a bead. Cellular communication or recognition may be necessary for the creation of the plasmodesmata in the vascular parenchyma, pits and perforation plates in the xylem elements, and superimposed sieve plates or sieve areas in the phloem sieve elements. Pectin fragments during the adhesion process may function as a signalling molecule for the cell structure. At least in certain plants, the parenchyma cells beneath the necrotic layer show increased cytoplasmic activity. A highly noticeable dictyosome accumulation may occur along the graft interface. Furthermore, by vesicle migration to the plasmalemma, these dictyosomes appear to secrete materials into the cell wall gap between graft components, which speeds up parenchymatous cell attachment at the graft contact interface.

4) **Wound repair:** Phloem and xylem were typically distinct at first in both herbaceous and woody plants until the vascular cambium bridged over the callus. Usually, the differentiated tissue that spans the graft union is the wound repair xylem, and then the wound repair phloem tissue. The callus differentiates into these vascular elements first, immediately forming the xylem elements and phloem sieve tubes. Between the circulatory systems of the scion and stock, a layer of vascular cambium subsequently forms. For the graft union to successfully form, the vascular cambium must connect in order for secondary vascular development to take place. After two or three weeks of grafting, the parenchyma cells that touch the cambial cells of the scion and stock develop into the new cambium cells at the borders of the newly created callus mass. Until continuous cambium connections are made between the scion and stock, the cambium creation in the callus mass continues to progress inward from the original scion and stock cambium and through the callus bridge.

5) **Secondary xylem and phloem production:** The normal cambium activity in the callus bridge is initiated by the newly produced layer of cambium, which lies down the new secondary phloem towards the outside and the xylem towards the inside. Graft cell partners that are next to the cambium have an impact on the sort of cambium cells that arise during the generation of new vascular tissue after cambial continuity. Thus, vascular linkage between stock and scion is made possible by the production of new phloem and xylem. It is crucial that the stage be finished before the majority of the newly developed leaves emerge from the scion's buds. When scion shoots extend their leaf surfaces, they may get desiccated and eventually die off.

THE MOLECULAR MECHANISM AT GRAFTING SITE

Genetic reactions at the graft unions, which involve a number of phenotypic alterations that are handed down to the following generation, mediate the developmental control process. Via vascular absorption, DNA is transported from stock to scion. HGT, or horizontal gene transfer, takes place as a graft is developing. The role of HGT is provided by the production of plasmodesmata and their link to the vascular tissue. Chloroplast genomes are observed to move across interspecific grafts, which are rather stable and passed on to the following generation.

In order to create stable allopolyploids and viable plants, nuclear genome migration also takes place during graft formation. Small RNA controls the epigenetic changes that occur throughout the grafting process. It demonstrates that in a successful graft, the short RNA of 24 nucleotides migrates from the scion to the stock and is crucial for modifying DNA methylation, which in turn affects epigenetic changes in the stock.

Gene silencing is greatly aided by the signal that SiRNAs (small interfering RNAs) mediate, which can pass through the graft site. SiRNA originating from the rootstock can facilitate gene silence in the scion and vice versa for a successful transplant due to its mobility ability. Because it controls some of the functional proteins involved in the growth of healthy plant tissue, messenger RNA, which codes for a variety of proteins, is also essential for the establishment of graft unions.

Numerous studies have demonstrated that mRNA transfers between grafts. Gibberellic acid insensitive RNA in grafted plants is transported by phloem tissue and is responsible for the change in leaf phenotype. A number of proteins bind to mRNA to promote molecular transport and slow down mRNA decay. In Cucurbita maxima, a chaperone (CmPP16) facilitates the movement of RNA from the stock to the scion. Plasma membrane intrinsic proteins, or PIPs, are important in the grafting process, and aquaporins primarily control active cell growth by facilitating cellular water transport.

PLANT HORMONES IN GRAFTING

Auxin is essential to the grafting process. The polar auxin transporter, or PAT, contributes to the growth of xylem tissue and aids in the maintenance of auxin levels. The two primary protein families that regulate PAT and function as the efflux carriers in auxin transport are PIN FORMED auxin transport protein (PIN) and ATP-binding cassette subfamily B (ABCB). The graft formation response also modifies transcript levels, including ABCB and PIN. Auxin Response Factor (AUX) gene up-regulation during graft union formation regulates multiple biochemical pathways and other hormones involved in vascular connection between scion and stock. According to theory, auxins control the expression of the gene HIGH CAMBICAL ACTIVITY 2 (HCA2) at the grafting site, aiding in phloem reconnection. Auxin supplementation stimulates the growth of the graft union via controlling the metabolism of phenylpropanoid, cytochrome P450, and carbohydrates. Auxins also play a role in the development of lateral roots and control the growth of the cambium and xylem in plants. Genes such as ALF4 (ABERRANT LATERAL ROOT FORMATION) of auxin signalling pathways regulate the reconnection of phloem tissue.

Cell division, the development of the vascular system, the growth of the shoot apical meristem, root growth, shoot organogenesis, tissue patterning, and plant biology under unfavourable environmental conditions are all regulated by cytokinin. It also promotes the production of calluses at the grafting site. The cytokinin role during grafting is also favoured by increased zeatin riboside at the grafting site. Auxin and cytokinin together stimulate vascular differentiation and improve the ratio of phloem to xylem. It has been noted that the increased cytokinin in the xylem of the rootstock promotes auxin transport from the shoot and speeds up the establishment of the graft union. After successful grafting, the administration of external cytokines leads to enhanced formation, faster phloem regulation, improved nutrient transfer, root architecture, root meristematic activity, and bud production. Since cytokinin is a cable for long-distance transport, it can modify PAT, which is also regulated in the formation of root vascular systems. Additional cytokinin produced in the shoot can control the expression of the PIN protein and auxin signalling, both of which are necessary for the growth of the grafted plants. During the primary and secondary development of vascular bundles, cytokinin also aids in the creation of vascular tissue and cambial tissue. Cytokinin signalling controls the LONESOME HIGHWAY (LHW), which promotes stele cell proliferation and development as well as protoxylem production.

Gibberellins are essential for cell division, late embryogenesis, cambial activity, cell elongation, xylem expansion, xylem fibre differentiation, and secondary growth in plants. Gibberellins control the biological processes of plants under abiotic stress via the gibberellin-signalling pathway. Gibberellins have been observed to enhance the Xylo genesis process in grafting. Phloem formation is induced by decreasing IAA:Gibberellins ratios, whereas xylem formation is induced by increased IAA:Gibberellins ratios. Gibberellins induce PAT at the cambial area by upregulating PIN1, a crucial auxin transporter. When cotyledon seedlings were compared to seedlings lacking cotyledon, there was a greater endogenous gibberellin application in the former. It was discovered that giving gibberellin to cotyledon-less cucumber plants enhanced tissue union or the wound-healing process in the hypocotyl. Furthermore, ethylene and jasmonic acid control the mechanisms involved in vascular reunion.

GRAFTING TECHIQUES IN VEGETABLE CROPS

The rootstock and scion compatibility, the age and number of seedlings, the grafting techniques, the quality of the grafted section, and the post-grafting management are the main factors influencing the survival rate of the grafted plants. The selection of scion and rootstock cultivars, the selection of suitable grafting techniques, the healing of the graft union, and the acclimatisation of the grafted plant are the four basic processes in the grafting process.

1. **Selection of scion and stock cultivars-** They comprise the initial and fundamental stage of every grafting procedure. Based on factors such market demand, yield, quality, purity, and viability, scion cultivars are chosen. Compared to scion cultivars, rootstock cultivars have more vigour. The viability, purity, disease resistance, compatibility with scion cultivar, and adaptability to the local environment and soil conditions are all taken into consideration while choosing the rootstock cultivar.

Table 1: Rootstocks and grafting methods used in the vegetable crops

Scion plant	Rootstock	Grafting Methods
Tomato	*Lycopersicon pimpinellifolium*	Cleft grafting
	Lycopersicon hirsutum	Cleft grafting
	Lycopersicon esculentum	Cleft grafting
	Solanum nigrum	Tongue and cleft grafting
Eggplant	*Solanum torvum*	Tongue and cleft grafting
	Solanum sissymbrifolium	cleft grafting
	Solanum khasianum	Tongue and cleft grafting
	Solanum integrifolium	Tongue and cleft grafting
Cucumber	*Cucurbita moschata*	Tongue grafting and hole insertion grafting
	Cucurbita ficifolia	Tongue Approach grafting
	Cucurbita maxima	Tongue grafting
	Cucurbita maxima × *Cucurbita moschata*	Tongue Approach grafting
	Cucumis sativus	Tongue Approach grafting
	Sicyos angulatus	Tongue Approach grafting
Bitter gourd	*Cucurbita moschata*	Hole insertion grafting and tongue grafting
	Lagenaria siceraria	Hole insertion grafting
Water melon	*Benincasa hispida*	Hole insertion and cleft grafting
	Cucurbita moschata	Hole insertion and cleft grafting
	Cucumis melo	Cleft grafting
	Cucurbita maxima × *Cucurbita moschata*	Hole insertion grafting
	Sicyos angulatus	Tongue Approach grafting
	Lagenaria siceraria	Splice grafting
Bottle gourd	*Cucurbita moschata, Luffa* spp.	Hole insertion and tongue grafting
Melon	*Cucumis melo*	Tongue and Cleft grafting

2. **Grafting methods-** The techniques chosen for grafting a specific rootstock and scion (Table 1) vary depending on the necessary grafting number, the aim of the grafting, the accessibility of labour, the availability of infrastructure and machinery, etc. All things considered, there are two types of grafting techniques: mechanical and manual grafting. Even though a lot of grafting machines and robots have been developed, manual grafting is still a common procedure. A variety of techniques are employed in hand grafting, which are described below:

    - **Tongue approach grafting -** The scion and rootstock in this procedure need to be roughly the same diameter. Here, the scion has its first real leaf and cotyledons, while the rootstock has formed cotyledons. Halfway through the stem, below the cotyledons, a 45-degree cut is made downward; in the scion, an equal angle upward slice is created. To enable the scion to be positioned on the rootstock top, the cut's location and angle must be quite exact. The stems are brought together so that they overlap, and then they are clipped or wrapped with plastic wrap, foil, or parafilm to secure. The combined transplant is set inside a little tray or planter. As needed, give the plants some water. Cut off the piece of the scion below and the portion of the rootstock above after five and seven days, respectively. This technique is often used on crops like watermelon, cucumbers, and melons.
    - **Cleft grafting-** The methods of cleft grafting used in herbaceous plants are different from those used in woody plants. This technique, which places the grafting junction high on the hypocotyl, helps prevent infections spread by the soil. When the first genuine leaf and cotyledons appear, the rootstock is ready to be grafted in seven to ten days. After beheading the rootstock seedlings, a longitudinal downward cut of 1-1.5 cm in length and 3/4 depth is made. In order to create a tapered wedge, the lower part of the scion is chopped below a slant angle, leaving one to three true leaves on the scion. A clip is positioned over the cleft in the rootstock to ensure a tight hang when the scion is positioned there. Plants are stored in greenhouses until they recover enough to be moved. Although technique was first applied to cucurbits, cleft grafting is now also limited to solanaceous crops.
    - **Hole insertion grafting-** Because watermelon seedlings are smaller than their rootstock (squash or bottle gourd), hole insertion grafting is used on watermelon. The true leaves and the meristem tissue at the rootstock's growth tip are cut away in preparation for the grafting procedure. Next, a slit is cut from one side of the hypocotyl to the other

across the developing point of a cotyledon.The insertion tool is a shaved stick, such as a toothpick or bamboo barbecue skewers. The scion hypocotyl is sliced in a V shape, and the inserted stick is left at the growth point. After that, the stick is taken out and the scion is put into the hole. Because the grafted plants maximise the contacting surface area between the scion and rootstock, their quality is good. The plants are moved once they have healed.

- **One cotyledon grafting-** They go by the names slant or splice grafting as well. Here, the rootstock is initially cut at a 45-degree angle, removing the meristem tissue, a cotyledon, and the true leaves. After that, the scion's hypocotyl is chopped like the rootstock's. Here, a grafting clip is used to secure the scion to the rootstock. Since they function best when the scion and stock have the same hypocotyl diameter, they are more suited for melon.
- **Tube or Japanese grafting-** This procedure involves cutting the scion and the rootstock at a 45-degree diagonal angle. In this case, both cuts are performed below the cotyledon to lessen rootstock suckering after the grafts have healed. Now, the two sections are connected by means of a grafting clip or plastic parafilm. The grafted plants are placed in the healing room, which has a constant temperature of at least 18 degrees, high relative humidity, and little light. The plants are taken out of the chamber after about 7 days. This technique is used in the commercial world for tomato grafts.
- **Pin grafting-** With the exception of using specially made pins to secure the graft union in place of grafting clips, they are identical to those of a single cotyledon graft. In order to help align and secure the linked portions, the cotyledon of the scion and the rootstock are cut horizontally. A ceramic pin is then put into the cut surface. Additionally, the transplanted grafted plants undergo healing.
- **Slant grafting-** This grafting technique has gained favour recently. It was created with robotic grafting in mind. With this procedure, when a rootstock cotyledon is cut on a slant, it is imperative to remove the initial leaf and lateral buds.
- **Mechanized grafting-** Grafting is a laborious process, and efforts are being made to minimise the amount of labour needed. Since 1987, attempts have been made to automate the grafting process. The shape of the seedling, where the cut is made, how the seedling is gripped, the cutting technique, the fixing materials and tools, and other basic criteria are what

determine whether grafting by machine or robot is successful. Adhesive and grafting plants have been combined to create grafting robots for plugs. Eight tomato, eggplant or pepper plugs can be grafted at once with the help of this robot. Recently, fuzzy logic and neural networks were used to create a fully autonomous grafting system that clones seedling quality estimation. Additionally, a healing chamber with regulated air quality has been developed to improve graft survival.

3. **Healing of grafts-** The most important factor in creating an environment that will encourage the grafted union's callus to form in the healing chamber is healing. To encourage the production of callus at the union, the temperature should be between 28 and 29 degrees Celsius with a relative humidity of 95% for five to seven days in partially shaded areas with one to two days of darkness. By keeping the ideal temperature and high humidity while lowering light intensity and transpiration, it aids in the establishment of a better graft union.
4. **Acclimatization of grafted plants-** In order to acclimatise the plants and prevent wilting and leaf burning, they are placed under a mist chamber, greenhouse, or plastic cover after the wound has healed and formed a callus.

SOME INNOVATIONS OF VEGETABLE GRAFTING

A selection of the numerous modern advancements made possible by vegetable grafting are outlined below:

i) **Grafting Robots**: A fully automated model created in the Netherlands has the capacity to graft 1,000 tomato or aubergine seedlings in an hour. It also performs additional duties including automatically matching rootstock and scion seedlings—an essential step in raising the success rate. In 1993, Iseki & Co. Ltd., Matsuyama, Japan released the GR800 series, the first commercially available grafting robot model for cucurbits. Since then, a variety of semi- and fully automated grafting robots have been developed. The reports also noted of grafting robots developed in other countries (Table-2).

Table 2: Some robots developed for grafting vegetables

Robots	Model	Developed by	Vegetable crops
AG1000 robot	Fully automated	Yanmar Agricultural Equipment Co. (Osaka, Japan) 1994	Solanaceous
Arnabat S. A	Semi-automated	Barcelona, (Spain) 2000.	Cucurbits, Solanaceous
Iseki's GR800 & GR-600	Semi-automated	Helper Robotech Co., Gimhae, Korea (2004)	Cucurbits

ii) **Micro-grafting**: To eradicate the viruses from infected plants, in vitro grafting employs extremely tiny or micro explants (< 1/1000th mm3) derived from meristematic tissues. In herbaceous plants, micrografting has been utilised to assess grafting physiology and ascertain the molecular underpinnings of cell-to-cell interactions. Although costly, this approach allows for the quick propagation of virus-free plants.

iii) **Double grafted and single grafted tomato**: The tomato plant is a product of vegetable grafting. In this instance, cleft grafting was used to graft tomato scions onto potato rootstocks. Harvest more than 500 cherry tomatoes above ground with a Brix TSS above 100. There are tomato grafts that are single, such as Sun Sugar, Brandywine, and Indigo Rose. In the United States, Log House pioneered the process of creating double-grafted tomato plants using Big Beef or Geronimo rootstock as scions, yielding red and yellow pear tomatoes by 2010.

GRAFTING IN VEGETABLE CROPS

The primary purpose of vegetable grafting is to mitigate both biotic and abiotic stresses like as salinity, heat, humidity, drought, and other soil-borne infections. In addition, grafted veggies remain in the field longer than non-grafted ones. Several studies are been carried out for the above field are presented in the following tables (Tables 3-5).

CONSTRAINTS

Not all vegetable species can be grafted, despite the advantages of doing so. Grafting success is determined by a variety of parameters, including genetic background, growth characteristics, anatomy, and physiological and biochemical aspects. Grafting and the process of creating grafted seedlings are typically associated with a number of issues. The labour and methods needed for the grafting process, as well as how to handle the seedlings after the graft to allow for rapid healing for seven to ten days, are the most significant issues. A skilled grafter may graft up to 1200 seedlings a day, or 150 seedlings every hour; the exact amount depends on the grafting technique. In a similar vein, the grafting techniques largely determine the post-graft handling strategy. vegetables that can suffer a variety of illnesses, including worms, many insect pests, Fusarium wilt, Verticellium wilt, and Bacterial wilt.

Table 3: Impact of grafting in major vegetable crops

Crop	Rootstock	Scion	Impact of grafting
Tomato	*Solanum torvum, Solanum sisymbriifolium*	Tomato	Early flowering
	Heman and Primavera	Tomato cv. 'Big Red'	Early flowering
	Solanum sisymbriifolium	Tomato	Early flowering
	Beaufort and Arnold	Tomato cultivars Beril, Swanson and Yeni Talya	Highest fruit yield/plan
	Dinafort	Corbarino	Increased fresh and dry yield, harvest index, and fruit/clusters
	Silex	Corbarino	High levels of simple sugars, flavonoids, ascorbic acid and trans-lycopene were also observed
	Tomato cv. 'Beaufort'	Tomato genotype 'Catalena'	More fruit /plant
	African eggplant cultivars 'Aworoworo' and 'Green'	Tomato varieties 'Tropimech' and 'Roma	Early flowering, early harvest and total soluble solids highest average fruit weight
	Tomato cv. 'Unifort'	Tomato cv. 'Faridah'	Early fruiting, highest average fruit weight
	Maxifort	cv. Big Dena	Increase in range of yield under favourable saline conditions
	Physalis and Cocona	RVTC 20, 0224-53 and RVTC 57, *Solanum habrochaites*	Good compatibility
	Emperador and Groundforce	Siriana and Buzau 1600	Highest yield, more dry matter
	Brinjal rootstock VI034845	Tomato cv. Avatar	More fruits per plant, maximum pericarp thickness, highest yield per plant
	Solanum pimpinellifolium	Tomato cv. 'Ramellet'	More fruits per plant, highest average fruit weight
	Maxifort (*Solanum lycopersicum* × *Solanum habrochaites*)	Tomato	Productivity
	Tomato rootstock LS-89	Tomato	Early harvesting, highest fruit weight, maximum pericarp thickness, highest TSS
	Solanum sisymbrifolium	'BARI Tomato-4	More fruits per plant
	Brinjal var. Navkiran	Tomato var. Sona NTH 2829	Highest number of fruit/plants, low titratable acidity

[Table Contd.

Contd. Table]

Crop	Rootstock	Scion	Impact of grafting
Brinjal	*Solanum incanum* × *Solanum melongena*	Black Beauty	Highest yield
	Solanum torvum	Longo	Highest marketable yield
	Solanum xanthocarpum	Pusa Shyamala	Highest solasodine content
	Solanum torvum	Pusa Shyamala	Lowest infection of bacterial wilt
	L23B	cv. Classic F1	Highest marketable production/hectare
	Torpedo	Classic F1	Highest average fruit weight
	Beaufort	Brinjal	Fruit quality and yield, resistance to Verticillium wilt
	'Birgah' eggplant	*Solanum torvum*	increase Iodine-stress tolerance of eggplant, improved total anthocyanins concentration
	BARI Hybrid Begun-4	BARI Begun-8	Highest average fruit weight, highest yield/plant
	Solanum torvum	Scarlatti	Lowest glycoalkaloid
	Solanum torvum	Surati Ravaiya Purple	Maximum fruit set, highest marketable fruit yield/plant, highest marketable yield/hectare, minimum infestation of fruit and shoot borer
	Solanum grandiflorum x Solanum melongena	Madonna	Highest average fruit weight
	MWR rootstock	MCV2	Highest average fruit weight
Cucurbits	Xiuli, Nanzhen No-3, Nanzhen No-4	melon	Higher plant height, diameter of stem, and number of leaves per vine
	Cucumis and *Cucurbita*	Tainan No 1	Parallel grafting survival rate and best vegetative growth, significant effect on fruit quality was observed
	Squash	*Cucumis sativus*	Increased plant height, physical characters, early and total yield of cucumber fruits in both summer and winter seasons

[Table Contd.

Contd. Table]

Crop	Rootstock	Scion	Impact of grafting
	Pumpkin	Bitter gourd	Recorded with highest protein content
	Pumpkin	Palee F_1 hybrid	Highest ascorbic acid content
	RS-17 and RS-8	Sardes	Increased yield
	Lagenaria siceraria	cucumber cv. Kalaam	High plant survival rates (95%), better vegetative growth and maximum fruit yield
	Lagenaria siceraria	cucumber cv. Kalam F1	Highest fruit weigh, Total soluble solids, fruit dry matter, fruit yield
	(*Cucurbita maxima* (Duchesne) × *Cucurbita moschata* (Duchesne ex Poir)	Water melon	Quality fruit
	Lagenaria siceraria	Bottle gourd	Quality fruit
	Cucurbita pepo	Watermelon	length of roots, number of leaves were significant high
	Squash (*Cucurbita moschata)*	Cucumber	Wax-free fruit
	Carnivor	Cantaloupe and honeydew melons	Root stock 'Carnivor' improved ascorbic acid content in grafted cantaloupe and honeydew melons, Root stock 'Carnivor' improved β-carotene content in grafted cantaloupe melons.

Table 4: Grafting of vegetables for resistance/tolerance to abiotic stresses

CROP	ROOTSTOCK	TRAIT
Tomato	AR-9707 (*Solanum lycopersicum*)	Salinity
	Maxifort (*S. lycopersicum* × *S. habrochaites*)	Saline conditions tolerant
		Heavy metal toxicity
	Beaufort (*S. lycopersicum* × *S. habrochaites*)	Drought tolerence
	S. lycopersicum × *S. chessmanii*	Salt stress
	Cheong Gang and 602/Jjak Kkung	Drought stress
	IC-111056IC-354557	Flood tolerant
	He-man and Maxifort	Greater nutrient and water uptake and abiotic stress tolerant
	Black beauty (eggplant)	Heat stress
	Unifort cv	Increased uptake of Ca^{+2} and $K^{+}$and Salinity tolerant
	B-blocking	Thermal stress
	Solanum pennelli	Tolerance to drought
	Solanum chessmanii	Resistant to salt
	Solanum galapagense	Tolerance to salt
	Solanum habrochaites	Resistance to cold
	Solanum chilense	Resistance to drought
	Solanum lycopersicum var. *cerasiforme*	Tolerance to humidity,
	Solanum integrifolium x Solanum melongena	High temperature tolerance
Brinjal	*Solanum elaeagnifolium*	Resistant to drought
	Solanum macrocarpon	Tolerant to flooding
Sweet pepper	Creonte	Drought stress Tolerant
Potato	*Solanum cardensi*	Low temperature

[Table Contd.

Contd. Table]

CROP	ROOTSTOCK	TRAIT
Watermelon	Hongdun (*Citrulus lanatus* sp.)	Potassium Deficiency Tolerant
	Cucurbita pepo	Drought tolerant
	PS 1313 *(Cucurbita maxima* × *Cucurbita moschata)*	Water use efficiency
	Brava (*Cucurbita pepo*)	N metabolism
	Jinxinzhen (*Cucurbita moschata*)	K uptake and metabolism
	TZ 148 (*Cucurbita maxima* × *Cucurbita moschata*)	Salinity
Cucumber	Shintoza (*Cucurbita maxima* × *Cucurbita moschata*)	High temperature
	Shintoza (*Cucurbita maxima* × *Cucurbita moschata*)	Cold
	Shintoza-type rootstock	Copper toxicity tolerant
	Luffa (*Luffa cylindrica* Roem. cv. Xiangfei No. 236)	Drought conditions
	Fig leaf gourd	Low temperature
	Pumpkin (*Cucurbita maxima* × *Cucurbita moschata*)	Nutrient deficiency or toxicity
	'Chaojiquanwang' rootstock (*Cucurbita moschata*)	Salinity tolerant
	Pumpkin	Salt tolerance
	Three rootstock cultivars Kilameki, Tielizhen and Fig leaf gourd	Low temperature
	Luffa [*Luffa cylindrica* (L.) M. Roem] rootstock	Heat stress tolerance
	Luffa rootstock	Drought tolerance
	Cucurbita ficifolia	Low temperature tolerance
Watermelon	Hongdun (*Citrulus lanatus* sp.)	Potassium Deficiency Tolerant
	Cucurbita pepo	Drought tolerant
Melon	TZ 148 (*Cucurbita maxima* × *Cucurbita moschata*)	Salinity
	Shintoza (*Cucurbita maxima* × *Cucurbita moschata*)	Mineral and water uptake
	Cucurbita rootstock	Nutrient deficiency tolerant

Table 5: Grafting of vegetables for resistance/tolerance to biotic stresses

Crop	Identified resistant rootstock genotypes	Scion (s) cultivar/F1 hybrids	Resistance against
Tomato	RVTC-66 (*Solanum lycopersicum* var. *cerasiforme)*	Santa Cruz Kada	Bacterial wilt
	Big power, Beaufort and Maxifort	Cherokee Purple, German Johnson	Root knot nematode
	CRA 66 and Hawaii 7996 (*Solanum lycopersicum*)	German Johnson	Bacterial wilt
	Maxifort (*Solanum lycopersicum* x *Solanum habrochaitus*)	German Johnson	Fusarium wilt
	Beaufort	Tomato	Root-knot nematode
	Solanum habrochaites	Tomato	Resistance to cold as well as insects & diseases (TMV)
	Solanum chilense	Tomato	Resistance to drought and diseases (CMV, TYLCV)
	Solanum neorickii	Tomato	Resistant to bacterial diseases
	Solanum pimpinellifolium	Tomato	Colour, quality, resistance to disease
	Solanum pimpinellifolium	Tomato	Nematode resistance
	Solanum lycopersicum var. *cerasiforme*	Tomato	Tolerance to humidity, resistance to fungi and root rot
	Solanum peruvianum	Tomato	Resistance to tomato spotted wilt virus and root knot nematode (RKN)
Brinjal	*Solanum tovrum*	Pusa Shymala, Pusa hybrid-6	Bacterial wilt
	Solanum torvum and *Solanum sisymbrifolium*	Tsakoniki	Verticillium dahliae
	Haritha, Surya, SM 1, SM 2, SM 3, SM116	Green Long Hybrid	Bacterial wilt
	Solanum integrifolium	Brinjal	Root-knot nematode
	Solanum torvum	Brinjal	Verticilium wilt
	Solanum khasianum; Solanum viarum	Brinjal	Resistant to shoot and fruit borer (BSFB)
	Solanum xanthocarpum	Brinjal	Immune to phomopsis blight
	Solanum sisymbrifolium	Brinjal	Resistant to little leaf
	Solanum auriculatum	Brinjal	Immune to little leaf disease
	Solanum sisymbrifolium; Solanum indicum	Brinjal	Immune to RKN

[Table Contd.

Contd. Table]

Crop	Identified resistant rootstock genotypes	Scion (s) cultivar/F1 hybrids	Resistance against
Chilli	*Capsicum chinensis; Capsicum baccatum*	Chilli	Anthracnose resistant species
	Capsicum pubescens; Capsicum microcarpum	Chilli	Powdery mildew resistance species
Sweet Pepper	AR-96023 *(Capsicum annuum)*	Sweet Pepper	Root-knot nematode
	AF-2638 *(Capsicum annuum)*	Sweet Pepper	Phytophthora capsici
Hot and Sweet Pepper (Capsicum annuum)	Serrano type Criollo de Morelos-334 (CM334) PI-201232 Weishi and Buyeding	Baker, Beldi Indra hybrid Xinfeng 2	Phytopthora rot Bacterial wilt Fusarium wilt
Potato	Solanum desmissum	Potato	Resistant to late blight
Cucumber (Cucumis sativus)	*Benincasa hispida and Cucurbita maxima* x *Cucurbita moshata*	Sinai hybrid	Meloidogyne spp.
	Cucurbita maxima x *Cucurbita moshata*	Zena	Fungal wilt
	Cucurbita maxima	Caspian340	Fungal wilt
	Peto 42.91 F1, TS-1358 F1, and TZ- 148 F1	Brunex F1	Fusarium wilt
	Benincasa hispida	Cucumber	Tolerance to Fusarium wilt Tolerance to Verticillium wilt
	Cucumis hystrix	Cucumber	Resistant to downy mildew, gummy stem blight, virus and nematode
	Citrullus colocynthis, Cucumis metuliferus	Cucumber	Nematode tolerance
Bitter gourd	*Cucumis metuliferus*	Bitter gourd	Nematode tolerance
	Cucurbita moschata	Bitter gourd	Nematode tolerance
Watermelon	*Cucurbita maxima* x *Cucurbita moshata*	Tri x Palomar	Verticillium wilt

[Table Contd.

Contd. Table]

Crop	Identified resistant rootstock genotypes	Scion (s) cultivar/F1 hybrids	Resistance against
(*Citrullus lanatus*)	*Cucurbita maxima* x *Cucurbita moschata*	Crisp'n Sweet	Verticillium wilt
	Lagenaria siceraria	Thuy Loi and PhuDong	Fusarium wilt
	Lagenaria siceraria	Watermelon	Improved tolerance to Phytophthora capsici Tolerance to powdery mildew
	Shintoza *(Cucurbita maxima* x *Cucurbita moshata)*	Watermelon	Fusarium wilt
	Cucurbita spp.	Watermelon	Verticilium wilt
	Shintoza *(Cucurbita maxima* x *Cucurbita moshata)*	Watermelon	Root-knot nematode
Melon	*Cucumis melo* var. *flexuosus, varconomon, varagrestis,* Asian Landraces	Melon	Tolerance to nematodes,Tolerance to Monosphorascusvine decline Fusarium wilt tolerance
	TZ 148 (*Cucurbita maxima* x *Cucurbita moshata*)	Melon	Fusarium wilt
	Shintoza (*Cucurbita maxima* x *Cucurbita moshata*)	Melon	Root rot diseases
	Cucumis melo var. *momordica*	Musk melon	Resistant to DM and PM
	Cucumis trigonus	Musk melon	Resistant to fruit fly
	Cucumis anguria; Cucurbita ficifolia; Cucumis metuliferus	Musk melon	Resistant to nematode
Pumpkin	*Cucurbita lundelliana*	Pumpkin	Resistant to powdery mildew
Wax gourd	*Benincasa hispida*	Wax gourd	Resistance to Fusarium wilt
Melon, Watermelon, Cucumber	*Cucumis maxima* × *Cucurbita moschata*	Melon, Watermelon, Cucumber	Tolerance to Fusarium wilt and fruit quality, Controlling Verticillium, Adaptability to nematodes ,Higher productivity and bigger fruit
	Citrullus lanatus var. *citroides*	Melon, Watermelon	Tolerance to Fusarium wilt and nematodes
Cucumber	*Cucurbita moschata*	Cucumber	Tolerance to Fusarium wilt

Problems associated with grafting and cultivating grafted vegetables seedling

Factors	Category
Labour	Grafting operation and post-graft care
Techniques	Rootstocks
Management	Fertilizer application
Growth	Excessive vegetative growth and physiological disorders
Fruit quality	Size and shape, appearance, insipid taste, soluble solids, yellow band in flesh and internal decay
Expense	Rootstock seeds
Scion rooting	External rooting, internal or fused rooting

CONCLUSION

Given the wide range of uses for vegetable grafting that this technology has across the globe, it has the potential to assist India's vegetable industry in overcoming obstacles and boosting farmers' income through increased crop productivity and decreased costs associated with fertilisers, pesticides, and disease control products. An eco-friendly way to encourage the production of organic vegetables is through grafting. A thorough understanding of grafted plants would be very helpful in the development of some ideal rootstocks for sustainable farming. In-depth studies on the interaction between stock and scion as well as long-distance communication are required in order to improve the resistance of grafted plants. There are still documented incompatibility issues, which call for more research into cell recognition, the relationship between the connective tissues of the stock and scion, the presence of growth regulators and how those regulators interact with other metabolic enzymes or phenols. Even though autonomous grafting techniques are still being developed, more automated robots using less expensive approaches will need to be standardised soon. Vegetable grafting is a cutting-edge method for enhancing long-term vegetable production by reducing soil degradation and the loss of natural resources. Grafted transplant trading is growing quickly on the global market, and with the growth of the grafted vegetable industry in India, this option can be taken advantage of. The large-scale implementation of this method is something that needs more attention, especially from farmers, in order to encourage the long-term, economical, and year-round production of vegetables.

FUTURE THRUST

- Selecting plants for breeding via the use of particular physiological indicators is novel for future rootstock breeding, and selecting suitable rootstocks still requires trial and error.
- There are numerous and largely unidentified physiological and molecular processes that contribute to some stress-tolerant rootstocks' beneficial reaction.
- In order to provide a framework for the early detection of incompatible grafts for commercial utility, it is necessary to understand the functional physiology of grafting plants.
- To improve the use of this technology, a thorough understanding of its different genetic and physiological components is necessary.
- The main goal of future study should be to identify the physiologically significant root-derived processes that have a strong correlation with the desired rootstock characteristics.
- After the biomarkers have been established, they can be utilised as general instruments to create a successful strategy for choosing rootstocks that increase the ability of vegetable crops to adjust to abiotic stressors such as water or heat shock, organic contaminants, or other problems.
- In order to solve this problem, scientists need to focus on creating and making widely available infrastructure, tools, and grafting robots that will increase grafting productivity while reducing labour expenses.
- The creation of grafted vegetable databases, software, mobile applications, and crop models will help farming communities and nursery managers choose the best cultivars for scion and rootstock.

REFERENCES

Aminul Islam AKM, Mominul Islam AKM, Era FM and Hamim I (2021) Vegetable grafting: a century-old technique to improve organic production. *J. Hort. and Postharvest Res.,* 4(3):303-322.

Ashok Kumar B and Kumar (2017) Grafting of Vegetable Crops as a Tool to Improve Yield and Tolerance against Diseases- A Review. *Int J. Agri. Sci.,* 9(13):4050-4056.

Buller S, Inglis D and Miles C (2013) *Hort. Sci.*, 48(8):1003– 1009.

Ilakiya T, Parameswari E, Davamani V, Yazhiniand G and Singh S (2021) Grafting Mechanism in Vegetable Crops. *Res. J. Chem. Env. Sci.*, 9(4):18-26.

Janick J (1986) Horticultural Science, 4th edn. W.H. Freeman & Co., New York.

Kumar P, Lucini L, Rouphael Y, Cardarelli M, Kalunke RM and Colla G (2015) Insight into the role of grafting and arbuscular mycorrhiza on cadmium stress tolerance in tomato. *Frontiers in Pl. Sci.* 6:477.

Kumar R, Rajasree V, Sagar L, Ahuja A, Savithiri N, Karthick K, Mehta A and Saini R (2018) Vegetable Grafting: A Recent Advance in Olericulture: A Review. *Int. J. Curr. Microbiol. App. Sci.* 7(09):1877-1882

Lee JM, Kubota C, Tsao SJ, Bie Z, Echevarria PH, Morra L and Oda M (2010) Current status of vegetable grafting: Diffusion, grafting techniques, automation. *Scientia Hort.*, 127(2):93-105.

Leonardi C (2016) Vegetable grafting tour introduction. University of Catania, Sicily, Italy, 23.

Nirosha K, Ashwin Kumar B, Mamatha A and Sreenivas M (2023) Vegetable grafting: An emerging approach in vegetable production: A brief review. *The Pharma Innov. J.* SP-12(7): 133-138.

Oda M (1999) Grafting of vegetables to improve greenhouse production. Food and Fertilizer Technology Centre Extension Bulletin 480, 1–11.

Pradeepkumara N, Subhasmita S, Kowsalya KB, Jain S, Hongal D, Meher SS, Kiran B and Narayana Swamy G (2022) Grafting technique a novel advancement in vegetable breeding. *The Pharma Innov. J.,* 11(10):469-480.

Rivard CL and Louws FJ (2008) Grafting to Manage Soilborne Diseases in Heirloom Tomato Production. *Hort. Sci.*, 43:2008-2111.

Thakur D and Savita (2020) Role of grafting in vegetable crops: A review. *J. Pharmac. and Phytoche.*, 9(6): 1170-1174.

Sustainable Agriculture and Biodiversity: Innovations and Insights, Pages: 25–40
Edited by: Bibhuti Bhusan Sahoo, Harvinder Kumar Singh, Niranjan Chinara and Amitabh Panda

CHAPTER - 2

ANTI-TRANSPIRANTS IN VEGETABLE CROPS

Meenakshi Badu[1]*, Barsha Tripathy[2] and Suchismita Mishra[3]

[1]*Faculty of Agriculture, Sri Sri University, Cuttack, Odisha-754006, Odisha*
[2]*Department of Vegetable Science,*
Faculty of Agricultural Sciences, SOADU, Bhubaneswar-760001, Odisha
[3]*Department of Floriculture and Landscaping,*
Faculty of Agricultural Sciences, SOADU, Bhubaneswar-760001, Odisha
**Corresponding email: badumeenakshi@gmail.com*

INTRODUCTION

The most prevalent substance in nature, water also poses the greatest obstacle to agricultural productivity. Transpiration is the term used to describe the loss of water vapour from a plant's exposed portion. Transpiration causes a significant amount of water loss: 2 litres per day for sunflowers, 36–45 litres for apples and up to 1 tonne per day for elm trees. Transpiration causes a plant to lose between 98 and 99 percent of the water it absorbs. Only around 0.2% of the total is used for photosynthesis; the remainder is stored by the plant for growth.

Types of Transpiration

- The majority of transpiration happens through the leaf's surface, or foliar surface. The term for it is "foliar transpiration." More than 90% of transpiration is accounted for by foliar transpiration.
- Young stems, fruits, flowers and so forth also transpire often. Mature stems do not transpire much. "Cauline transpiration" refers to transpiration from stems.

Depending upon the plant surface transpiration is of the following four types-stomatal, cuticular, lenticular and bark.

i. Stomatal Transpiration:

- This kind of transpiration is the most significant. Roughly 50–97% of transpiration is made up of stomatal transpiration. It happens via the stomata. The majority of stomata are located on leaves. On the young stems, flowers and fruits, a few of them are present.

- The plant's stomata allow the atmosphere to seep into its moist interior. Water vapour consequently becomes saturated in the inside air. Other than right after a rainstorm, the outer air is rarely soaked with water.
- As a result, water vapour diffuses forth through stomata. To replenish the water vapour that is released, more water evapouration occurs from the interior cells. As long as the stomata remain open, stomatal transpiration will continue.

ii. Cuticular Transpiration:

- It happens through the leaf cuticle or epidermal cells, as well as other exposed plant components.
- Cuticular transpiration accounts for only 3–10% of total transpiration in common plants.
- Cuticle-thin herbaceous shade-loving plants can have up to 50% of their transpiration coming from cuticular transpiration.
- Cuticular transpiration happens all day and all night.

iii. Lenticular or Lenticellate Transpiration:

- It can only be found in the woody branches of trees that have lenticels.
- Less than 0.1% of transpiration occurs in the lenticular region.
- However, because lenticels lack a closure mechanism, it persists day and night.
- Through the intercellular gaps between the complimentary cells, the lenticels establish a connection between the ambient air and the cortical tissue of the stem.

iv. Bark Transpiration:

- This kind of transpiration happens when the stems are covered in cork.
- Bark transpiration is extremely low, but because of its bigger area, it is generally measured at a higher rate than lenticular transpiration.
- Bark transpiration happens constantly day and night, same like cuticular and lenticular forms.

Mechanism of Water Loss in Transpiration:

o Water inside the plant's exposed sections need heat energy from a source to condense into vapours. It is the heat energy from the transpiring organ at night and the radiant energy during the day. The temperature of the transpiring organs drops by 2 to 5 °C below the ambient temperature in both situations.

- o Water vapours rarely reach saturation levels in the atmosphere. High DPD, or low water potential, is present in the atmosphere's dry air at 13.4 atm at 99% relative humidity (RH), 140 atm at 90% RH, 680 atm at 60% RH and 2055 atm at 20% RH.
- o A high DPD or low water potential can get past the obstacles that water molecules must overcome to go from the liquid phase to the vapour phase and to exit the transpiring organ.
- o Water vapours are nearly saturated in the transpiring organ's intercellular gaps. Because of the higher DPD of the latter, water vapours are pulled from the sub-stomatal cavities to the outside air when the stomata are open. As a result, the substomatal air's DPD rises and more water vapour is drawn from the intercellular gaps.
- o The wet walls of the mesophyll cells subsequently yield water vapour to the latter. Transpiration through the stomata will persist until they open. The mechanisms underlying lenticular and stomatal transpiration are comparable.
- o Water cannot pass through the cuticle very much. Nonetheless, its molecules imbibition water from the epidermal cells. The atmosphere, which has a high DPD, gradually loses the water that was ingested. Thickness of cuticle reduces imbibition flow.
- o As a result, transpiration cannot pass through a thick cuticle. During the day, cuticle shrinks and thickens, but at night, it stretches and loosens. Cuticular transpiration may therefore occur more frequently at night. The mechanisms underlying cuticular transpiration and bark transpiration are comparable.

Transpiration and Photosynthesis— a compromise (Water Requirement): Water need, transpiration ratio, or transpiration efficiency refers to the volume of water transpired for every unit of dry matter produced by a plant. For succulent or crassulacean plants, it is minimum; for C_4 plants, it is moderate; and for C_3 plants, it is greatest. This ranges from 300 to 600.

It is roughly 50 in CAM plants. Plants with C_3 transpire the most. Rain forests are humid because plants lose a lot of water to evapouration. Condensation and localised rain are the outcomes.

For every gram of dry matter produced by C_3 plants, 600–900 grams of water are wasted. The reason for this is because their mesophyll tissue has a high degree of sponginess because there are fine and wider spaces between the palisade and spongy parenchyma cells, respectively.

C_4 plants have developed a way to limit water loss and enhance CO_2 availability. They have a 300 transpiration ratio. When it comes to fixing carbon, C_4 plants are twice as effective as C_3 plants. For the same quantity of CO_2 fixed, a C_4 plant loses half as much water as a C_3 plant. The stomata of CAM plants are closed during the day.

They keep water in storage as well. When transpiration loss is at its lowest at night, the plants absorb and store carbon dioxide. During the day, photosynthesis uses the absorbed CO_2. The figure denotes the need for rain and irrigation. It is possible to cultivate those crops and plants in places with insufficient irrigation and light rainfall.

Transpiration – A Price Paid for Photosynthesis: The majority of plant physiologists believe that transpiration is a cost of photosynthesis that the plant must bear. Water vapours exit side by side as the stomata open in the light to allow for gaseous exchange (O_2 for respiration and CO_2 for photosynthesis).

Transpiration loses as little water when stomata are closed in the dark. This indicates that transpiration takes place in addition to photosynthesis. Because of this, transpiration is thought to be the cost of photosynthesis that plants bear.

Factors Affecting Transpiration: Water Stress: When transpiration rates rise faster than absorption rates, plants have a water deficit, which causes the leaves to begin to wither. A plant experiences a Y gradient between its guard cells and the mesophyll and epidermal cells that envelop them when an internal water deficit develops.

Because of this gradient, water from the guard cells moves more freely outward and the stomata either completely or partially close as a result of the decreased turgor.

It is also interesting that the mesophyll cells' starch is quickly hydrolysed under the same condition (wilting). Their Ys rise as a result, allowing them to take up water from the guard cells. The increased sugar content in mesophyll cells may also inhibit photosynthesis's use of CO_2, leading to an increase in CO_2 concentration from ongoing respiration.

Increased acidity from a higher CO_2 concentration encourages the conversion of sugar to starch in guard cells. The guard cells' osmotic concentration decreases. Water moves from the guard cells into the parenchyma of the leaf.

The stomata close and the guard cells take on a flaccid form. Therefore, the basic mechanism governing stomata closing due to water deprivation and darkness is the same, with both processes being governed by increasing acidity brought on by a rise in CO_2 concentration.

Carbon Dioxide: The concentration of carbon dioxide affects the stomatal process. When there is an abundance of CO_2, stomata close.

Oxygen: The opening of stomata depends on this gas. Lack of it causes stomatal closure to speed up.

Hormones: Stomatal closure is brought on by ABA, although K+ intake requires cytokinin. Additionally, ABA is only effective in the presence of CO_2.

Light: In general, stomata close in the dark and open in response to light. Most of the time, opening is impacted by blue light that is absorbed by a unique pigment system.

Similarly the phytochrome system may also lead to opening of the stomata. Earlier workers assumed that guard cells when exposed to light and warmth increased their output of osmotically active substances through photosynthesis. The resulting increase in osmotic pressure and turgor resulted in the opening of stomata.

The objection to this assumption is that photosynthesis occurs in the guard cells only at a reduced rate and could not account for the amount of osmotically active substances required to cause opening.

The theory also cannot explain the fact that stomata open in carbon dioxide-free air in the dark as well as in the light when there is no photosynthesis. Light may provide ATP for K^+ transport into guard cells.

Temperature: Stomatal opening rises with temperature up to 25–30°C and falls at even higher temperature when all other parameters are equal.

Most species' stomata close to or fail to open at 0°C. Even in the dark, certain species experience stomatal opening rather than closure at high temperatures of approximately 40°C.

Mineral deficiencies: The stomatal motions slow down when any of the minerals potassium, phosphorus, or nitrogen are lacking. The stomatal aperture is affected differentially by different ions.

Wind: When a leaf is exposed to strong winds, its stomata close. This could be because the guard cells lose water during transpiration, or it could be an indirect consequence of the temperature of the leaves.

Mechanical shock: A variety of shocks, such roughly manipulating leaves, can cause stomatal closure. It has been discovered that when adjacent tissue is injured, stomata may close in light.

Root-shoot ratio: If other factors were in favour of transpiration, the rate of transpiration rose as the root-shoot ratio increased.

Leaf area: There is no correlation between leaf area and water loss; the greater the leaf area, the greater the quantity of water loss. Even though the larger maize plant lost more water overall, the smaller plant lost more water per unit area overall.

Pruning a plant results in a decrease in leaf area, which raises transpiration per unit leaf area of the plant; yet, unpruned plants lose more water overall. This is because a trimmed plant's root system is supplying more water to fewer leaves, which increases the plant's transpiration efficiency.

Leaf structure: Many structural changes, particularly in the leaves, are evident in plants living in arid environment. These include thick cuticles, thick cell walls, well-developed palisades, sunken stomata, hair coverings, etc. These characteristics lessen water loss. Since the stomata are closed in dry conditions, the primary mechanism for water loss is cuticular transpiration.

In xerophytic species, well supplied with water, stomatal transpiration exceeds that of mesophytes. This is due to higher number of stomata per unit area, more extensive venation, greater internal evaporating surfaces exposed to the internal atmosphere in xerophytic leaves as compared to mesophytic leaves.

In addition leaf size and shape, spacing, distribution and structure of stomata, water content of leaves greatly affect the rate of transpiration. In comparison to healthy plants, sick plants transpire at a significantly higher rate.

Availability of soil water: The rate of transpiration lowers if the leaves are not receiving enough water. A protracted state of this kind will cause a water deficit, which will make the plant look wilted.

The supply of water can be inadequate due to the following reasons:

(i) Water in the soil near the wilting coefficient of the roots

(ii) The root system is insufficient to provide the top

(iii) A low soil temperature

(iv) A high concentration of soil solution

Atmospheric pressure: The pressure of other gases has an inverse relationship with the rate of evapouration. The rate of evapouration would double if atmospheric pressure were to drop from one to half an atmosphere.

The lower atmospheric pressure at the summit of mountains causes a rise in transpiration rate.

Advantages of Transpiration: There are advantages of transpiration to the plants and these are discussed below:

1. Creates transpirational pull for transport
2. Remove excess water
3. Transport of minerals:
   - o High rates of mineral absorption are typically caused by high transpiration rates However, deep understanding of diffusion and differentially permeable membranes point towards the independent absorption of minerals.
   - o It is often believed that the transpiration stream is the means by which minerals taken up by the plant from the earth are transported up the plant.
   - o Rather than transpiration rate, the amount of minerals that reach the leaves depends on the rate at which the roots absorb the minerals.
   - o It appears that the rate of transpiration has no effect on the minerals that are available in the leaf. Conversely, the rate of transpiration is essential for the translocation of minerals when they are abundant in the soil.
4. Lowering of leaf temperature:
   - o Water molecules with a relatively high kinetic energy are lost as they evaporate from a leaf's surface, lowering the temperature of that organ. The molecules with the highest kinetic energy evaporate first.
   - o About 600 calories are eliminated via transpiration for every gramme of water that evaporates, according to rough calculations.
   - o Heat can also be expelled by radiation and convection, among other methods. The higher leaf temperature would result in an increased loss of heat by radiation and convection if transpiration stopped.
5. Optimum turgidity:
   - o It's often accepted that a plant's water potential has an ideal level. Both above and below this threshold, a wide range of functions become inefficient or sluggish.
   - o Without transpiration, plants get too turgid and eventually stop growing. In a similar vein, growth stops when the water potential drops significantly.
6. Bringing water to the top of a plant: It's also argued that transpiration's significance stems from the fact that it's necessary for drawing water to the top of the trees.

7. Supplies water for photosynthesis

Disadvantages:

- o It is undeniable that transpiration is detrimental. One of the most vital elements of life and a substrate for photosynthesis, the process that is vital to plant life, is lost. It should be noted that just when transpiration occurs in plants, it does not always imply that the plant body benefits from it.
- o Transpiration is required for wet cell walls and open stomata, both of which are necessary for photosynthesis. Diffusion is used to achieve this. The risk is that withering, which lowers production or perhaps kills a plant, may occur when diffused water levels are higher than those reaching the roots.
- o It is well accepted that wilting may have several deleterious effects which include stomatal closure, dehydration of enzymes, lowering of membrane permeability and reduction of photosynthesis. In brief, transpiration appears to be a ***necessary evil***. Leaves are primarily meant to perform the process of photosynthesis and transpiration is just incidental.
- o Several practices are employed to reduce or at least minimize water losses and these include removal of weeds adjacent to crop plants. Greenhouses are whitewashed to reduce light intensity, temperature and thus transpiration. Periodically the atmosphere is moistened by spraying water to saturate the greenhouse and to reduce transpiration. This reduces the periodicity of watering the crop plants.
- o Transpiration is decreased in vegetatively propagated plants because there is little to no water absorption due to a lack of roots. The cuttings are denudated of leaves or covered with bell jars of plastic covers. In such vegetative propagated portions, transpiration is reduced to avoid wilting of organs or tissues since flaccid cells do not grow.

Anti-Transpirants: During drought conditions the plants show wilting. A single wilted plant reduces yield by 50%. Crop plants cannot be grown in arid and unirrigated environments due to high transpiration and little water supply.

Substances that reduce the rate of transpiration are called ***anti-transpirants***. Even in situations where there is really little water available, anti-transpirants will keep a favourable internal water balance. It will assist foresters in planting trees even in the most arid or desert regions and enable farmers to raise crops profitably in unirrigated locations. According to a number of studies, anti-transpirants enhance numerous vegetable crops' physiological, disease-resistant, quality and yield characteristics in addition to lowering water loss. Anti-transpirants are not advised for high productivity because their major function is to sustain growth and productivity under stressful situations. When expectations are close to zero, these are employed to conserve crops and assist in obtaining a marginal yield.

In an effort to use less water by decreasing transpiration, anti-transpirants are chemicals sprayed on transpiring plant surfaces, or they can be substances that, when applied to plant foliage, can slow down the rate of transpiration. These are physical or chemical substances that, by shutting the stomatal pores, lower the rate of transpiration. In situations of water stress, they are employed in agricultural techniques to regulate water loss through the stomata. Anti-transpirants, then, are substances that, when applied to crop plants, lessen transpiration. A wide range of materials have been tried as anti-transpirants (solarova *et al.*, 1981).

Within the plants some natural anti-transpirants are synthesized and these include ABA and salicylic acid. Within the plant ABA regulates the water economy of plant under water stress conditions. Some studies have demonstrated triazoles as anti-transpirants. CO_2 is also shown to act as an anti-transpirants, high concentrations of CO_2 partially close stomata and increase the rate of photosynthesis.

Types of Anti-transpirants

Stomata closing type

- Herbicides and fungicides with lower concentrations cause stomata to close, acting as anti-transpirants.
- Found to be less transpiring than photosynthesis; yet, it is poisonous to fruits and vegetables.
- A slight increase in CO_2 content from the normal range of 0.03% to 0.05% causes stomata to partially close.
- These reduces the photosynthesis. e.g. Phosphon D, 2,4-D, Atrazine, Simazine, Hydroxysulfonates, Potassium metabisulphite, TIBA, Succinic acid,, Triazine and High CO_2

Film forming type

- This kind prevents the leaf from losing water vapour by creating a thin layer covering on its surface. However, they permit CO_2 to enter the leaf through the lower epidermis.

- Application to leaves induce stomatal closure and thus reduces transpirational losses of water from plants.
- e.g. Tag 9, S- 789 foliate, Hexadecanol, Mobileaf, Waxes, Silicone oils, Folicote, Vapourgard *etc.*

Reflectance type

- The materials are of the reflecting white kind. .
- The reflective layer could last longer than ten days.
- It lowers the leaf temperature and reduces the vapour pressure gradient from leaf to atmosphere.
- Thus decreases the transpiration.
- e.g. Kaolin, Lime water, Calcium bicarbonate *etc.*

Growth retardant type

- Chemicals which reduces or slow down the shoot growth and increases root growth.
- Enable the plants to resist drought conditions through closure of stomata.
- These are useful for improving the water status of the plant.
- e.g. Cycocel, ABA and Ethrel *etc.*

Ideal properties of Anti-transpirants

- o Non toxic
- o Less expensive
- o Specific in function
- o Reducing transpiration without reduction in photosynthesis
- o Stable and long lasting in their effectiveness
- o Anti-transpirants form a supplement, rather than a substitute for, good water management and crop improvement
- o Non-permanent damage to stomatal mechanism
- o Specific effects on guard cells and not to other cells

Effectiveness of anti-transpirants primarily depends on-

- Coverage of lower surface of leaves
- Interaction of anti-transpirant with external environment
- Surface anatomy of leaf
- Rate of formation and growth of new leaves

Benefits of Anti-transpirants

- High water demanding crops growing in areas of water scarcity
- Optimized yield levels under infrequent rainfall situations
- Transport of vegetable seedlings
- Save several fruits from cracking
- Getting normal sized fruits
- Managing drought
- Prevent the adverse effects of water stress on crop growth
- Very useful for farmers with minimum irrigation facilities
- Saving large nurseries when water is scarce in summer months
- Saving of crops with marginal crop productivity under drought

Review of Research work

1. Under conditions of moisture stress, Prakash and Ramachandran (2000) investigated the impact of chemical ameliorants on brinjal's stomatal frequency and water relations. According to their findings, cycocel outperformed potassium chloride in terms of relative water content, whereas potassium chloride excelled in terms of stomatal behaviour and leaf water potential.

Treatment	Vegetative	Flowering	Fruit development	Mean
Control	47.07	58.36	54.41	53.28
Stress	36.74	48.58	45.37	43.56
Stress + Cycocel 1.5%	52.32	67.76	61.74	60.60
Stress + lime wash 2%	39.94	52.11	48.27	46.77
Stress + Potassium chloride 1%	42.64	56.65	51.42	50.23
Mean	43.74	56.69	52.24	
CD (P = 0.05)	4.57	2.46	3.88	

2. According to Wafaa (2002), Kaolin and Nu-film @ 1% were more successful than other film-forming chemicals at reducing the amount of downy mildew lesions, infected area and spore counts that germinated. When administered either before or after inoculation, Kaolin anti-transpirant hindered spore germination and caused the sporangia to collapse and lose their turgidity, according to scanning electron microscopy analysis. Kaolin enhanced vine height and output while providing robust protection against downy mildew in cucumbers grown under protection.

	Disease severity				*P. cubensis* spores counts/cm^2 of infected lesion			
	Pre-inoculation		Post-inoculation		Pre-inoculation		Post-inoculation	
ATs	1%	3%	1%	3%	1%	3%	1%	3%
Kaolin	0.1	0.1	0.8	0.5	97.6	81.0	392.1	298.4
Bio-Film	1.7	1.0	2.3	2.1	625.2	482.1	934.6	702.7
Nu-Film	0.8	0.5	1.6	1.3	343.4	281.3	767.8	624.8
Folicote	3.6	3.3	4.3	4.0	1643.6	1336.4	1989.4	1831.1
Polyacrylamide	2.7	2.1	3.6	3.0	1187.1	981.7	1464.3	1231.6
Control	8.1	8.1	8.1	8.1	4173.6	4173.6	4693.2	4693.2

Plant height and yield of cucumber plants sprayed with film forming anti-transpirants under protected cultivations

	Spring season				Autumn season			
	2000		2001		2000		2001	
ATs	Plant height (cm)	Yield (Kg/m^2)	Plant height (cm)	Yield (Kg/m^2)	Plant height (cm)	Yield (Kg/m^2)	Plant height (cm)	Yield (Kg/m^2)
Kaolin	214.4	10.9	219.4	12.0	198.2	7.14	198.3	7.34
Bio-Film	181.3	8.55	186.4	9.01	171.5	5.56	175.3	5.82
Nu-Film	199.3	8.91	201.3	9.75	186.3	6.88	186.5	6.23
Folicote	168.3	7.80	170.6	8.26	154.6	5.14	158.4	5.27
Polyacrylamide	194.5	9.20	198.4	10.4	181.1	6.55	183.3	6.02
Control	148.9	5.72	150.4	7.21	128.0	3.89	130.0	4.4

3. Karuppaiah *et al.* (2003) studied the effects of several anti-transpirants on brinjal development, photosynthesis and yield characteristics. The anti-transpirants included liquid paraffin, kaolin, salicylic acid and phenyl mercuric acetate. Anti-transpirants had a major impact on the biomass, number of branches, height, leaf area and number of flowers on the plant. The best spray was discovered to be kaolin (7.5%), which increased plant biomass and blossom count. The photosynthetic characteristics of net photosynthetic rate, stomatal conductance, intercellular CO_2 concentration, transpiration and relative water content (RWC) were all impacted by all anti-transpirants. In comparison to salicylic acid (1000 ppm) and liquid paraffin (1.5%), kaolin (7.5%) showed the highest net photosynthetic rate, RWC and least transpiration. Salicylic acid (1000 ppm) and kaolin (7.5%) considerably enhanced fruit output.

Influence of anti-transpirants on Brinjal Photosynthetic characters

Treatments	Net photosynthetic rate (μ mol CO_2 m^{-2} sec^{-1})	Transpiration rate (μ mol H_2O mol^{-1})
T_1 – Control (No spray)	37.32	8.52
T_2 – Control (Water spray)	38.69	8.50
T_3 – PMA 4%	34.82	7.29
T_4 – PMA 5%	32.18	7.99
T_5 – Kaolin 5%	39.36	7.23
T_6 – Kaolin 7.5%	42.51	7.02
T_7 – SA 1000 ppm	40.93	7.42
T_8 – SA 1500 ppm	39.52	7.78
T_9 – liquid paraffin 1%	35.09	8.25
T_{10} – liquid paraffin 1.5%	38.02	8.31
C.D. 5%	0.64	0.09

4. According to Ibrahim and Selim (2010), foliar application of kaolin @ 6% greatly improved the summer squash crops' water usage efficiency, foliage weight, leaf weight, mean fruit weight, total yield per hectare and TSS during the summers of 2008 and 2009.

Treatments	WUE (Kg fruits/ m^3 water)		Foliage weight/plant (g)		Leaves weight/plant (g)	
Kaolin levels	2008	2009	2008	2009	2008	2009
0%	11.31	11.04	1298	1314	1016	1014
3%	12.12	11.88	1370	1402	1038	1077
6%	12.95	12.82	1446	1484	1114	1136
CD at 5%	0.23	0.22	35	37	32	34

5. Francini and associates (2011) A study was done to find out how sensitive bean (*Phaseolus vulgaris* cv. Pinto) plants responded to realistic ozone fumigations when exposed to the commercial antitranspirant Vapour Guard (VG) emulsion (di-1-p-menthene), which is based on pinolene. For four hours, plants treated with the chemical were exposed to up to 150 parts per billion of ozone in the atmosphere; these plants suffered far less damage than untreated controls. The anti-transpirant markedly reduced stomatal conductance and photosynthesis in unozonated plants. The following effects of VG treatment are observed in VG-treated subjects: (1) significant reduction in visible injury;

(2) counteraction of membrane damage; (3) preservation of photosynthetic activity, stomatal conductance and CO_2 storage in the sub-stomatal chamber; (4) stability of the photochemical apparatus is revealed by Fv/Fm and other parameters of chlorophyll fluorescence; and (v) there is no stimulation of antioxidant defence. Unexpectedly, our findings showed that, depending on the mechanism of distribution of VG, there is a significant variation in the protective effects against ozone damage. As previously said, when whole plants (i.e., "both leaves") are treated or left untreated with VG, the plants that did not receive VG treatment exhibited significant changes in the phenomenological, biochemical and eco-physiological parameters that were examined because of ozone toxicity. This is untrue when VG is applied to specific areas (referred to as "half leaves") or "single primary leaves." Physiological metrics, such Amax and Gw, in ozonated and VG-treated single leaves and half leaves are identical to unfumigated controls even if there is obvious damage following ozone fumigation. For the parameters of chlorophyll fluorescence, similar findings were found. A protective effect on the membrane is noted with half-leaves therapy. The behaviour of untreated regions' ozone is altered by the presence of regions (single leaf or half leaf) that have received VG treatment. The observed phenomenon's potential causes are examined. For evaluating ozone damage in the field, the anti-transpirant di-1-p-menthene has shown to be a low-cost, low-technology solution.

6. Moussa (2012) found that applying folicote foliarly at a 15% rate significantly improved the yield, yield-attributing characteristics and water usage efficiency of sweet potatoes grown in low-water locations.
7. Anti-transpirants increase survival of grafted watermelon transplants
    - Dabirian *et al.* (2016) conducted research at Department of Horticulture, Washington State University, Mount Vernon, NWREC on use of commercial anti-transpirant products (film-forming or stomata closing) to determine if they can increase survival of grafted watermelon.
    - Experimental Design was Randomized complete block design with 5 replications, 12 plants per plot, repeated 2 times (29 January and 2 February 2016).
    - Treatments 1) Moisturin: Apply to foliage before grafting, 10% solution (10 parts water:1 part Moisturin) 2) Root-Zone: Apply to soil before grafting, 1.56% solution (2 oz. root-zone in 1 gal water) 3) Moisturin + Root-Zone: Apply Moisturin to foliage (10% solution) and apply Root-Zone to soil (1.56% solution) before grafting 4) Water Control: Apply water to foliage (3 mL per plant) and soil (30 ml per cell) before grafting.

- They concluded 'Root-Zone' and 'Moisturin + Root-Zone' had the greatest survival rate (92% and 89% on average, respectively).
- Stomatal conductance decreased with 'Moisturin + Root-Zone' and 'Root-Zone' applications, indicating less transpiration and was lowest for 'Moisturin + Root-Zone' 1 day after application.

Precautions required while using Anti-transpirants

- Spray after rainfall stops
- Ensure 2 sprays of anti-transpirant at an interval of one week
- Surfactants should be mixed in spray for effective response
- Repeat spray of anti-transpirant if rain follow after one hour of sprays

Major Constraints

- May decrease rate of photosynthesis
- Their effect being varied greatly with environmental factors
- Difficult to cover the entire leaf surface. Some crop plants does not receive proper application of the chemical and thus the effectiveness of the chemical may be reduced
- The lack of knowledge on the type of nozzles and surfactant to be used
- May produce toxicity in leaf
- The cost of certain chemicals forbid their use as anti-transpirants

CONCLUSION

- Anti-transpirants are primarily used to preserve soil water, which lowers the need for frequent irrigation. Applications for this purpose, however, would only be warranted in the event that water costs were high enough and that potential water savings outweighed application losses during irrigation.
- Additional applications for them include extending the range of habitats in which plant types sensitive to water deficit can achieve favourable growth and yield, reducing winter kill, increasing transplanted seedling survival, helping valuable and established plants survive drought conditions and treating plant material for transportation.
- There is evidence to suggest that an anti-transpirant film on foliage could act as a physical barrier against fungal and insect attacks, as well as lessen damage from salt spray and pollution.
- If there is no conflict between the two components, adding an anti-transpirant to a pesticide spray would significantly lower application costs.

- There are still many possible applications for antitranspirants that need to be explored.
- Research must go on to identify the compounds that provide the greatest transpiration decrease with the least amount of photosynthesis reduction, as well as the ideal concentrations and application techniques.

REFERENCES

Dabirian S, Miles C, Scheenstra E and Kreider P (2016) Antitranspirants increase survival of grafted watermelon transplants Department of Horticulture, Washington State University, Mount Vernon, NWREC.

Francini A, Lorenzini G and Nali C (2011) The Anti-transpirant Di-1-p-menthene, a Potential Chemical Protectant of Ozone Damage to plants. *Water Air Soil Pollut.*, 219:459–472.

Ibrahim EA and Selim EM (2010) Effect of irrigation intervals and Anti-transpirant (Kaolin) on summer squash (*Cucurbita pepo* L.) growth, yield, quality and economics. *J. Soil Sci. and Agril. Engineering,* 1(8): 883-894.

Karuppaiah P, Ramesh KS, Shah K and Marimuthu R (2003) Effect of Anti-transpirants on growth, photosynthetic rate and yield characters of brinjal. *Indian J.Pl. Physiol.,* 8(12): 189-192.

Moussa SAM (2012) Effect of folicote anti-transpirant on sweet potato crop and water use efficiency under drip irrigation treatments. *J. Pl. Production.,* 3(12): 2993-3009.

Prakash M and Ramachandran K (2000) Effect of chemical ameliorants on stomatal frequency and water relations in brinjal (*Solanum melongena* L.) under moisture stress conditions. *J. Agron. and Crop Sci.,* 185: 237-239.

Solarova JJ, Paspisilova and Slavik B (1981) Gas exchange regulation by changing epidermal conductance with anti-traspirants. *Photosynthetica,* 15: 365-400

Wafaa Haggag M (2002) Application of epidermal coating Anti-transpirants for controlling cucumber downy mildew in green house. *Pl. Pathol.Bull.,* 11: 69-78.

Sustainable Agriculture and Biodiversity: Innovations and Insights, Pages: 41–84
Edited by: Bibhuti Bhusan Sahoo, Harvinder Kumar Singh, Niranjan Chinara and Amitabh Panda

CHAPTER - 3

STRATEGIC GENETIC GAIN FOR THE DEVELOPMENT OF NUTRACEUTICALS IN CASSAVA (*Manihot esculenta* Crantz)

Sampurna Bhattacharya[1], Arlin Sen[2], Aditya Pratap Singh[3], Sujaya Dewanjee[4], Trupti K Vyas[5], Jayanta Tarafdar[6] and Nilima Karmakar[7*]

[1]*Department of Genetics and Plant Breeding, N M College of Agriculture, Navsari Agricultural University, Navsari-396450, Gujarat*
[2]*Department of Plant Molecular Biology and Biotechnology, N M College of Agriculture, Navsari Agricultural University, Navsari-396450, Gujarat*
[3]*Department of Genetics and Plant Breeding, Bidhan Chandra Krishi Viswavidyalaya, Mohanpur, Nadia-741252, West Bengal*
[4]*RRSS, Bidhan Chandra Krishi Viswavidyalaya, Raghunathpur, Purulia-723133, West Bengal*
[5]*Food Quality Testing Laboratory, N M College of Agriculture, Navsari Agricultural Universty, Navsari-396450, Gujarat*
[6]*Department of Plant Pathology, Bidhan Chandra Krishi Viswavidyalaya, Mohanpur, Nadia-741252, West Bengal,*
[7]*Department of Soil Science and Agricultural Chemistry, N M College of Agriculture, Navsari Agricultural University, Navsari-396450, Gujarat*
[*]*Corresponding email: nilimanau13@gmail.com*

INTRODUCTION

Agricultural Importance of the Crop

Cassava (*Manihot esculenta* Crantz) originated in Latin America and was used to be grown by the Indian population at least 4000 years ago. This starchy crop has a long history to be transferred in Africa and later on different parts of Asia by the European traders as food security crop. Cassava was originally a tropical American crop that the Portuguese brought to Africa in the 16th century (Okogbenin *et al.* 2006). This crop can be grown in less fertile soil under low rainfall area compared to other crops like maize, soybean and different vegetables. The starchy roots of cassava are the major source of carbohydrate worldwide with the maximum production in African countries (Alves *et al.* 2002). In tropical Africa, the popularity of cassava is increasing day by day due to its agricultural advantages. The HIV/AIDS affected households are actually shifting from high to low input farming system, which include cassava to feed rapidly increasing population. Though the roots are rich in carbohydrate but it is poor in protein content (Scott *et al.* 2000). The cassava roots are used not only for human consumption but also as animal feed. Cassava flour are used for production of different bakery products. The carbohydrate rich roots are often subjected to fermentation for alcohol production (Kenyon *et al.* 2006)

Nutritional Composition of the Product/parts

Cassava crop is rich in dietary fibre, mineral like magnesium, vitamins like thiamine, riboflavin, nicotinic acid and citrate, while poor in protein content. In African countries almost 80% of produced cassava is used as human food where as in America 30% of produced cassava is used as livestock feed (Westby, 2001). Starch is the main constituent of cassava which contributes up to 80% of dried weight root (Olomo & Ajibola, 2003). Cassava starch contains lesser amount of lipids compared to cereal starch. The starch granules contain certain types of fatty acids like palmitic, oleic, linoleic and linolenic acids. Cassava contains much lesser amount of phosphorous compared to the other starches like potato or wheat and the phosphorous present in the cassava are in the form of phosphate monoester.

Growing Importance in the Face of Chronic Diseases and Malnutrition

During the time of 1990s, a significant change came in the economical level of tropical and subtropical countries in the world due to globalization, where the cassava is cultivated as main food. Hence, the import of temperate grown maize gained a great importance in tropical countries (Pandey & Gardner, 1992). As a result, an agronomic void was created and which was fulfilled by none other than cassava. Not only the tuber part but also the foliage portion of the cassava plant are of great importance as food or feed as those are enriched in minerals, proteins and vitamins which is also called as tropical alfalfa. Different African communities in central or western Africa consume cassava leaves (Lancaster & Brooks, 1983; Diasolua Ngudi*et al.* 2003). Though cassava can be a very good source of energy in animal diets, but it needs the complementary protein source like soybean, as it is poor in protein content (Tewe & Lutaladio, 2004).

Limitations of Conventional Breeding and Rational for Next-Generation Breeding

There are some factors which may be considered as the restrictions on the efficiency on cassava breeding. Inbreeding is not carried out in cassava breeding which reduces the probability of detection of recessive traits. Allelic frequencies cannot be modified efficiently as there is no clear and defined population. Considering the data obtained from diallel studies, it had been found that cassava crop is highly heterozygous. In this context, if an elite clone is selected, it can be propagated vegetatively. If the elite clones are selected as parent material for production of new plants, then dominance effect is carried and because of distorting effect of dominance the breeding value is not perfectly correlated with their performance. On an average 0.6 percent viable seeds are produced per pollination in case of cassava, hence the production of recombinant seed is difficult. Transfer of even single gene controlled desirable trait from one genotype to another and the powerful and successful techniques like back cross is not successful in cassava due to constant heterozygous state during the breeding process.

The advantages of inbreeding in cassava are as follows:

Selection among the double haploids is not affected by the dominance effect genetically. Additive effects among double haploids are doubled. Germplasm exchange by seeds is easier and common than vegetative cutting (Iglesias *et al.* 1994) and virus free planting material can be available without the meristem culture.

DESCRIPTION ON NUTRITIONAL ELEMENTS/COMPOUNDS

Detailed Chemical Composition: Contents and their Variations

Cassava is mainly the source of starch and also considered rich in dietary fibre, magnesium, riboflavin, thiamine, B-complex vitamin (Westby, 2001). The two main molecules that structured starch molecule are amylose and amylopectin. The amount of amylose in a starch impacts its quality. Amylose content has been shown to vary greatly 0–30.3% in starch content. There are loads of rumoured genetic variability in amylose content. Amylopectin with 5.6 percent of branches is that the principal molecular building block of cassava.

Cyanogenic glycosides: According to the aglycone's chemical makeup, the toxic glycosides may be showing neatness classified into four groups: 1) Glycosides made up of 2-hydroxy-2-phenylacetonitrile or its derivatives (aromatic ring replaced with a hydroxyl group or associate alkyl radical group) (amygdalin type). 2) Glycosides made up of open-chain, saturated aglycones (linamarin type). 3) Glycosides having a covalent bond positioned IX,P to the organic compound cluster within the aglycone (acacipetalin type). 4) Glycosides of the gynocardin sort, that have associate unsaturated, alicyclic aglycone.

Chemical Type, Structure, Biochemical Pathways of Production:

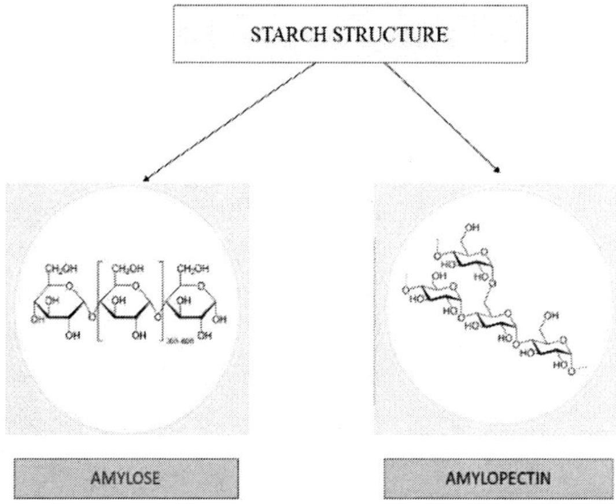

a) **Starch:** $(C_6H_{10}O_5)_n$ is that the formula for starch. Starch could be a carbohydrate comprising aldohexose monomers joined by á- 1,4 linkages.

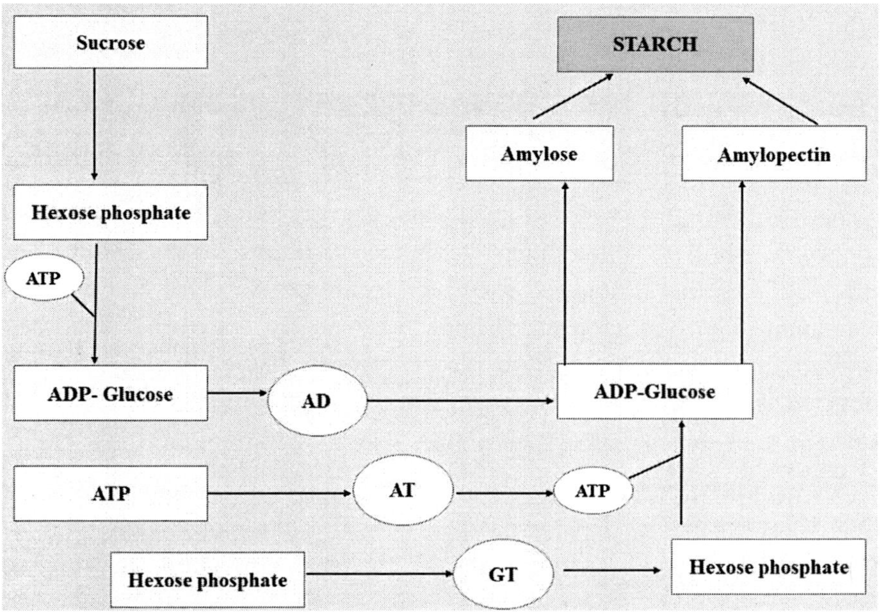

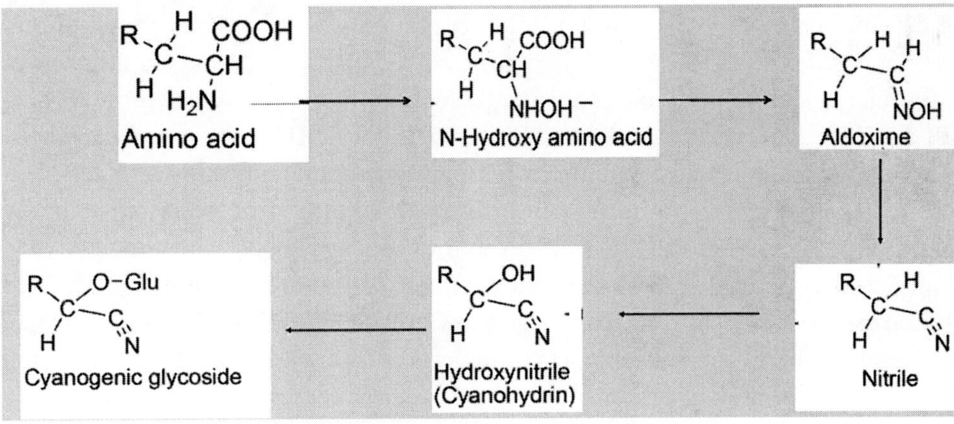

Biosynthesis pathway of starch:

b) Cyanogenic glycoside structure:

The above-mentioned figure includes some abbreviations like: AD: ATP/ADP glucose transporter; AT: Plastidial ATP transporter; GT: glucose -6-phosphate transporter.

Medicinal/physiological Properties and Functions in Relation to Human Health

Cassava contains cyanogenetic organic compound in the majority the elements except seeds. The concentration of cyanogenetic organic compound is principally obsessed with age of the plant, status and genotype variation (McMahon *et al.* 1995). Cassava root contains the organic compound linamarin that is born-again to hepatotoxic chemical compound (HCN) by the activity of linamarinase catalyst. Hence, a hard product "gari" is extremely well-liked in geographical region to get rid of the HCN (Kenyon *et al.* 2006). The cyanide toxicity over cassava consumption results in headache, vomiting, abdomen pains, symptom etc. and generally even causes death (Mlingi*et al.*, 1992). Children are typically more susceptible to complete poisoning than adults since the amount of cyanide is proportional to weight. Consuming cyanide from cassava worsens the illness and hypothyroidism that are brought on by iodine deficiency in areas where it occurs (Delange*et al.* 1994). Ihedioha & Chineme (2003), in their work steered that customers of mandioca were at risk when the fermentation time of cassava mash is reduced to just twenty-four hours. Consuming cassava, which contains residual cyanogen, is linked to a number of health issues (Osuntokun, 1981). Following consumption of cassava, the majority of the cyanide consumed is converted back into salt in a process aided by the catalyst rhodanese, which depletes a portion of the supply of essential amino acids containing S as well as cysteine and cysteine(Osuntokun, 1981; Cardoso *et al.* 2004). These amino acids are crucial to the diet since the body can only get them from the food it consumes. As was discovered in an extensive study of children in the DRC, a deficiency of those S-containing amino acids would restrict supermolecule synthesis and result in aerobatics in growing children(Cardoso *et al.* 2004). According to a study conducted in Nampula Province in Mozambique, a kid should consume between 20 and 140 grammes of fresh cassava flour per day in areas that aren't affected by konzo, compared to 700 to 900 grams(Cardoso *et al.* 2004). Instead of just disregarding the negative elements of this crop, it's critical that the introduction of cassava into new areas coincides with efforts to teach the locals how to properly process the crop to get rid of cyanogen(Madamombe, 2006). It is possible that the rapid rate of population growth in these tropical African nations is a significant factor in the increased cassava production, which emphasises the need for effective cyanide disease protections.

Methods of Nutraceutical Improvement: Agronomic and Post-Harvest Techniques

Introduction: *Manihot utillisima* Pohl and *Manihot esculenta* Crantz are the two most prevalent botanical varieties of cassava. The classification of the tubers as "sweet" or "bitter" cassava depends on their cyanoglucoside content; the sweet variety has a low (140ppm) cyanoglucoside content. Based on its dry weight, the bitter cassava contains more than 140 ppm of cyanoglucosides (linamarine and luteostraline). The fact that the distribution of cyanoglucosides is not uniform. Despite being widely distributed throughout the sweet, low-cyanide cassava tuber, they are mostly concentrated in the peel.

Processing of cassava: The undercooked leaves and raw roots of cassava are unappealing and contain small levels of cyanide, which is poisonous to both humans and animals. The roots are difficult and expensive to transport to metropolitan markets because of their size and high moisture content (70%). Cassava must therefore be processed to lengthen the goods' shelf lives, make shipping and marketing easier, lower the cyanide concentration, and enhance palatability. Cassava processing stabilises seasonal changes in the crop's supply and reduces postharvest losses. Numerous scientists has provided description of the processing techniques used to turn raw cassava into meals in Africa.

Cassava processing unit processes include cleaning, peeling, soaking in water or holding in air for varying lengths of time to allow root fermentation, drying, milling, roasting (garification), steaming, crushing, sifting, and combining in hot or cold water are all steps in the preparation process. A variety of various cassava products that appeal to a wide range of consumers are produced by specific combinations of these unit operations (Bokanga, 1995).

Processing and Nutrition: Even the most toxic cultivars of cassava—those with the highest concentrations of cyanogenic glucoside—often benefit nutritionally from processing. In major part, cassava detoxification takes place during the processing of the roots, when the structural integrity of the cells is typically lost. The hydrolytic enzyme linamarase interacts with the cyanogenic glucosides to begin the production of HCN. According to Akingbala *et al.* 2005, the amount of cyanoglycosides in cassava decreased by almost 98% after fermentation and by about 95% after grating during the production of gari. The HCN produced is likely to entirely evaporate as heating is typically used in the food processing (Bokanga, 1995). Boiling whole bitter cassava roots, however, can make them hazardous because the heat denatures the hydrolytic enzymes without hydrolysing the cyanoglucosides.

Nutritional value of a product and fermentation: When making cassava products, lactic acid fermentation is typically used to further process the cyanogenic glucosides, which results in the glucosides being hydrolysed to sugar and volatile HCN. It is safe to assume that lactic acid fermentation, in addition to having a detoxifying effect on cassava, also benefits cassava products like fufu and fermented flours by providing them with the same advantages it does for milk, nuts, and other proteinous foods, such as improved protein quality, increased protein content, and increased vitamin content. Additionally, it has been also noted for the process of turning cassava into "gari" (Akingbala*et al.* 2005).

3. BRIEF ON GENETIC DIVERSITY ANALYSIS

In one of the earliest global attempts to use molecular markers, genetic diversity, differentiation, and the possibility of heterotic groups were examined in cassava (Fregene *et al.* 2003). The genetic diversity was assessed in 283 cassava landraces from Tanzania (163), Nigeria (29) and the Neotropics (Brazil, Colombia, Peru, Venezuela, Guatemala, Mexico and Argentina) (91) using 67 marker loci. Using 30 SSR markers, a broader diversity investigation of almost 2,300 accessions from the CIAT, Empresa Brasileira de Pesquisa Agropecuária, and International Institute of Tropical Agriculture gene banks was carried out (Hurtado *et al.* 2008). A mean of 7.38 alleles each locus, or 192 total alleles, were found. Every locus was polymorphic in every country (Kawuki*et al.* 2013). In this study, the use of 26 simple sequence repeat (SSR) markers discovered a total of 154 alleles, of which 24% were novel alleles found solely in landraces. Landraces have slightly higher heterozygosity and genetic diversity than elite accessions. Certain of the landraces and elite accessions clustered together, indicating that they shared same alleles (Turyagyenda *et al.* 2012). Nine genes with high-quality sequences had a total of 26 SNPs, with an approximate proportion of one SNP each 121 nucleotides. Using AFLP markers, the Makushi Amerindians of Guyana examined the genetic diversity of 31 kinds of cassava (*M. esculenta* Crantz). As a guide, they used a representative of 38 different kinds from an ex-situ core harvest. There were also wild cassava accessions present. Although the variations' clonality was anticipated given cassava's vegetative reproduction, 21 varieties showed intravarietal polymorphism (Elias *et al.* 2000).

BRIEF ACCOUNT OF MOLECULAR MAPPING OF HR GENES AND QTLs

Association Mapping

In cassava breeding, the main selection criteria are dry matter content and micronutrient density, especially provitamin A carotenoids. In cassava, utilizing 672 clones genotyped at 72,279 single nucleotide polymorphism (SNP) sites, a genome-wide association mapping was carried out (Rabbi *et al.* 2017). A few thousand SSR markers for cassava have been created by multiple organisations using enhanced genomic DNA libraries and expressed sequence tags (ESTs) (Chavarriaga-Aguirre *et al.* 1998; Mba*et al.* 2001; Raji *et al.* 2009; Sraphet *et al.* 2011; Tangphatsornruang *et al.* 2008). The Cassava Genome Database at the University of Maryland offers 384 potential SNPs obtained from genes and 371 potential SNPs produced from the physical map of the cassava. To explain the genetic structure of resistance to provitamin A and dry matter content, several genome-wide association studies (GWAS) have been carried out (Esuma*et al.* 2016; Rabbi *et al.* 2017; Ikeogu*et al.* 2019).

4.2. Glimpses of Molecular Mapping

Welsch *et al.* 2010, found that white and yellow cassava storage roots may be distinguished by an amino-acid alteration caused by an SNP mutation in the *PSY2* gene using the candidate gene strategy. With approximately 1000 polymorphic clones and 99.8% repeatability in the first cassava DArT array created (Xia *et al.* 2005), it provided an affordable high-throughput marker screening technique. 26 SNPs were found by Kawuki*et al.* 2009 after they examined the sequence diversity of nine genes. The Granule-bound starch synthase I (GBSSI) gene's stop codon is produced as a result of a natural deletion, which results in the waxy starch phenotype. Additionally, hybrid mutants with the waxy morphology can be created by silencing the GBSSI gene (Ceballos *et al.* 2007; Zhao *et al.* 2011; Aiemnaka*et al.* 2012). There have been several published linkage maps of between 100 and 510 markers, including simple sequence repeats (SSRs), SSRs produced from expressed sequence tags (ESTs), Randomly amplified polymorphic DNA (RAPDs), and Amplified fragment length polymorphism (AFLPs)(Rabbi *et al.* 2012).The GWAS analysis of colour-chart based variation in root carotenoid content identified a significant locus on chromosome 1 with three markers near 24.1, 24.6, and 30.5 Mb areas, with the top markers being S1 24159583, S1 24636113, and S1 30543962, respectively(Rabbi *et al.* 2022).

Details on QTLS

An important objective of cassava improvement projects around the world is biofortification breeding to boost provitamin A carotenoids in store roots.A biparental QTL mapping analysis using two clones from the Genetic Gain collection (TMS-I961089A and TMEB117) also found a single QTL peak whose confidence interval covered the same *PSY2* gene(Rabbi *et al.* 2014). In a biparental population genotyped using SSR markers, BalyejusaKizito *et al.* 2007 identified a QTL for dry matter composition that also correlates to this area on chromosome 1.2,954 putative SNPs have been found in the resource from ESTs produced from the cassava transcriptome. From these 1,536 SNPs were translated to Illumina's GoldenGate assay, which uses BeadArray technology, and genotyped on a variety of cassava accessions (Ferguson *et al.* 2012a).

5. MARKER-ASSISTED BREEDING FOR HR TRAITS

Cassava (*Manihot esculenta*) is mainly a root vegetable, belongs to the family Euphorbiaceae. Roots and leaves of cassava are used for edible purpose. It is a hardy tuber crop that is a rich and affordable source of carbohydrate. Marker-assisted breeding takes the help of DNA markers associated or linked with desirable traits to select an individual for inclusion in a breeding program early in its development. This new generation breeding technology helps the breeders to select the desirable traits more efficiently in less time than that of those conventional approaches. It is much simpler than the phenotypic screening.

Cassava is a very long duration crop, nearly of 12-18 months. So, naturally it has a long breeding cycle too. Clonal propagation in cassava is not fruitful due to low multiplication rate and with this low seed setting, difficult cross-ability are the barriers for making successful breeding programmes in cassava (Jennings &

Iglesias, 2001; Ceballos *et al.* 2012).The starch that has accumulated in the tuberous storage roots of cassava, which is used by industries to manufacture ethanol and starch-based goods, is the major nutrient in cassava (Tonukari, 2004). According to Fregene *et al.* 2006, several wild relatives of cassava can be served as high protein source, higher dry matter content, and having delayed post-harvest loss characteristics to the modern cultivars. These characters maybe transferred to modern cultivars on the basis of marker assisted selection.The formal introduction of the CR 41-10 variety in Nigeria in 2010 served as the first and only instance of marker-assisted selection (MAS) in cassava (Ceballos *et al.* 2012).

Marker assisted selection can be possible if the marker is directly related or closely related with the gene of interest. In classical form of MAS, single genes or Quantitative Trait Loci (QTLs) are selected but now a days with the help of Next Generation Sequencing (NGS) methods and through SNP genotyping, GBS approaches MAS can be done for more rapid genetic gain in qualitative characters easily (Heffner *et al.* 2009).

A few thousand SSR markers for cassava have been established by several group of scientists using optimized genomic DNA libraries and expressed sequence tags (ESTs)(Chavarriaga-Aguirre *et al.* 1998; Mba*et al.* 2001; Raji *et al.* 2009; Sraphet*et al.* 2011; Tangphatsornruang *et al.* 2008). In six years, varieties could be assessed for release under the cassava molecular breeding programme currently in use.By drastically reducing population sizes at the seedling stage, MAS makes breeding more feasible and enables breeders to deals with more populations.MAS costs are being reduced in the early stages of the breeding plan at CIAT through initiatives to reduce high segregation in seedling nurseries. The markers can be used to locate specific regions of the genome associated with the emergence of heterosis and to compute the genetic differences between inbred lines in order to perform successful crossings(Ceballos *et al.* 2004).

In wild relatives and two accessions of Cassava, GBSSI gene is present. The glycosyl transferase region of this gene is sequenced and four SNPs were introduced, those are different from the wild relatives of Cassava. These were utilised to create allele-specific molecular markers useable in MAS(Blair *et al.* 2007).In order to restore waxy starch in the long term, genotypes that carry the recessive mutant gene may be chosen using such allele-specific markers.For the diagnosis of waxy cassava, two single-nucleotide-amplified polymorphism (SNAP) markers were constructed based on GBSSI sequences (Aiemnaka *et al.* 2012).

Germplasm Characterization

Germplasm characterization simply means the recording of distinctly identifiable and highly heritable characteristics. It is crucial to characterise germplasm in order to inform users of the features of accessions and ensure that the germplasm collection is used to its full potential.

In wild accessions of Cassava many useful genes are present those can be used in future days for quality improvement. Mainly high quantity of protein in *M. esculenta* sub spp. *flabellifolia*, *M. peruviana* and *M. tristis*(CIAT, 2004), less amylose corn starch (3–5%) or waxy starch in *M. crassisepala* and *M. chloristicta*, and late PPD character in an interspecific hybrid between cassava and *M. walkerae*(Bertram, 1994).

A Genome Wide Association Study (GWAS) was done by Esuma*et al.* 2016, in cassava using a panel of partial S_1 and S_2 generation inbreds produced from eight clones. Using this constrained set of parents, they identify a single genomic region on chromosome 1 that is linked or associated with the variation in total carotenoid content.Recently Bredeson *et al.* 2016 reported that,a significant M. *glaziovii* introgression, which frequently occurs in the Genetic Gain collection, is present in this chromosome 1 region.

The CR 41-10 variety of cassava is the first successful example of MAS that is the selection by a set of molecular markers linked to the resistance of Cassava Mosaic disease in Colombia (where CMD is not present) and evaluated in Africa for general adaptation and agronomic performance.

Okogbenin*et al.* 2012 stated that to assess the cassava genetic diversity analysis, several genetic markers can be used. Beeching*et al.* 1993, reported Restriction fragment length polymorphism (RFLPs), Silva *et al.* 2015 suggested Random amplified polymorphic DNA (RAPDs), Fregene*et al.* 2000 reported Amplified fragment length polymorphism (AFLP), Asare*et al.* 2011 suggested Simple sequence repeats (SSRs), Kawuki*et al.* 2009 reported Single nucleotide polymorphism (SNP) markers for cassava genetic diversity analysis.

Though availability of all these marker technologies for germplasm characterization and further breeding opportunities, SNPs are termed as the best one due to their relatively low operational cost than other markers, high abundance and widespread nature in the whole genome, locus specificity and codominance nature, high through-put analysis and low genotyping error rate (Rafalski, 2002;Schlötterer, 2004; Chagnéet *al.* 2007).

5.2 Marker-assisted Gene Introgression

The introduction of one or more advantageous genes from one line (the donor) to another (the recipient) with as little disruption to the recipient's genetic make-up as possible is known as marker-assisted gene introgression.The use of markers in introgression backcross programmes allowed for the simultaneous selection of the ideal genomic background while also introgressing an allele.

In tropical regions, vitamin A deficiencies are higher than the other parts of the globe. The Harvest Plus project seeks to improve the beta-carotene content in cassava so that vitamin A content can also increase through some conventional and genetic transformation methods (Fregene *et al.* 2006). There is a huge source of beta carotene in wild relatives of cassava that can be explored. The cost for improved beta carotene breeding can be lowered if the gene sources and the number of gene involved are known. The broad segregation pattern of the Thai cassava variety MTAI8 (AM320) is an important resource of beta carotene content in cassava (Fregene*et al.* 2006). The use of molecular markers to introgress a single target region of the genome can prevent the need for two to four backcross generations(Frisch *et al.* 1999).The introduction of a naturally occurring mutant that in its wild relatives produces granule-bound starch required a very targeted approach. A new molecular tool to hasten the introduction of advantageous alleles from wild relatives into cassava is the method of sequencing the glycosyl transferase region of the GBSSI gene.

By using molecular marker selection, Hospital *et al.* 1992 was successful in cutting the number of backcross generations from two to one.In a later modelling work, Frisch *et al.* found that the number of backcross generations was decreased from two to four by the use of molecular markers for the introgression of a specific target allele (1999).They came to the conclusion that generation BC_3 might be able to reach the same level of recurrent parent genome as BC_7 achieves without the use of molecular markers through marker-assisted selection.However, the recurrent progenitor in these investigations must be homozygous, which is a requirement that has not yet been met for cassava.

The genotypes developed by CIAT served as donor parent of *CMD2* gene in several crosses for the Cassava Mosaic disease resistance and they have also the characteristics of high dry matter content and yield (Parkes *et al.* 2015).

Introgression of useful traits in Cassava from its wild relatives is complicated through conventional breeding schemes mainly due to long breeding cycle and linkage drag problem. But the use of molecular markers aids to the introgression of a single target region (Frisch *et al.* 1999). To introgress traits for high protein content, waxy starch, delayed PPD, and resistance to whiteflies and hornworms, CIAT currently adapts the ABC-QTL technique. Maintaining optimum levels of -carotene, cyanogenic potential, and amount of dry matter in roots are some more promising MAS targets for upgrading cassava.

Gene Pyramiding

When many genes are packed into a single genotype to incorporate desirable qualities using recombinant DNA technology or traditional breeding, this process is known as gene pyramiding. In case of marker assisted breeding, gene pyramiding is much easier and less time taking than the conventional breeding.

MAS is a great approach for pyramiding genes of relatively huge effect, such as disease resistance genes (Jia *et al.* 2002; Komori *et al.* 2003; Murai *et al.* 2001). Mostly the agronomic traits are quantitative in nature, so they are very much complex in nature. For dealing with them high cost as well as high specificity is needed due to the complex epistasis and environmental influences, this is one of the major reasons of shifting the breeding schemes from conventional to MAB where phenotypic screening is not required.

In case of Cassava breeding, in wild relatives and accessions of Cassava there is a huge source of beta carotene that can be evaluated and use for gene pyramiding with several improved varieties of cassava. High protein source genes, higher dry matter content, and having delayed post-harvest loss characteristics can be included with the genes for proper â- carotene content, cyanogenic potential and dry matter content in roots of cassava.

Introgression of high protein from wild relatives of cassava can be possible through ABC-QTL scheme, that is currently running at CIAT. With this character, introgression of resistance to green mites and cassava mosaic disease, delayed post-harvest physiological losses, increased dry matter content and protein content is also possible. But there is still a long way to go in case of gene pyramiding for reaching the utmost success in cassava.

Limitations and Prospects of MAS and MABCB

Beside their usefulness, these techniques have some limitations, like they are more expensive than the conventional techniques, sometimes recombination between the marker and the gene leads to false positive results etc. But it has several advantages over the conventional techniques like speedy recovery of the recipient parent genome, minimizing the linkage drag, more accuracy in selection etc.

Molecular breeding in Cassava has several prospects over the conventional breeding techniques like Gene pyramiding for resistance genes, higher genetic gain in quantitative traits through Marker Assisted Recurrent Selection (MARS) and genomic selection, less time requirement, easily manageable breeding population etc. (Ferguson *et al.* 2012b). MAB technique increases precision of selection and lowers the need for phenotyping. The reduction of the period needed for the introgression of characteristics from wild relatives is another way that MAS is used in cassava breeding.

The main limitation in case of the use of molecular markers in cassava genetic enhancement is the limited availability of field data and unreliable phenotypic information. Other restrictions are generally associated with the use of biparental mapping populations and inadequate software(Ceballos *et al.* 2012). But now a days, some promising alternatives are there for solving the above mentioned problems, like applying the genomic selection directly on the actual breeding populations (Heffner *et al.* 2009). With addition of this approach, genetic improvement of cassava can be possible by using its homozygous progenitors in the Marker Assisted Breeding programmes.

Although MAS is an indirect selection approach for characteristics, many breeding programmes still find it difficult to execute (Chukwu *et al.* 2019; Cobb *et al.* 2019). Although numerous molecular markers closely associated to significant cassava features have been discovered, there are few instances of their successful application for MAS in cassava breeding. The implementation of MAS in cassava breeding programmes in SSA has been hampered by the time and money required to develop reliable markers. The International Institute of Tropical Agriculture (IITA) and collaborators have created and transformed molecular markers into low-plex-high-throughput marker assays for a number of important gene characteristics, such as provitamin A, CGM, DMC. These indicators' dependability and usefulness are currently being assessed in various genetic contexts and habitats.

MAP-BASED CLONING OF HR GENES/QTLs

A long growth cycle coupled with heterozygous genetic background is the main hindrance in case of root quality improvement in cassava. These factors restrict the flowing of elite genes in cassava. As a result, cassava production fails to fulfil the demand in the regions where people consume it as staple food. The improvement in modern technologies in molecular breeding like microsatellite markers and recombinant technology enabled to remove the aforesaid obstacles. In CIAT, studies on many markers associated complex traits as well as genetic map had been developed recently. The expressed sequence tags mapping (ESTs) and candidate gene approach had been adopted to complex trait mapping in cassava in CIAT. Bacterial artificial chromosome (BAC) in cassava had already been constructed to hasten the positional cloning of genes which were known by phenotypic markers and molecular genetic map so far.

The genome of cassava was of 760 million base pair, hence Bacterial artificial library (BAC) had a 5× coverage of the genome which helped to find out a desirable gene with 95% probability. The BAC library in cassava had been constructed for root quality related genes as well as cloning of genes related to disease and pest resistance. The *Xa21* gene in cassava is a homologue of rice (Song *et al.* 1995), which had been cloned by a degenerative primer. This gene was found to be linked with resistance to cassava bacterial blight (Jorge *et al.*2000). The screening of BAC library with cassava homologue resulted 6 positive BAC clones among which 50% clones were found to be unique by the help of restriction enzyme analysis. Present experiments are going on transformation of bacterial blight susceptible cassava by the help of these unique BAC clones. The BAC cassava library is presently publicly available in CIAT (Fregene*et al.* 2001).

The bacterial artificial chromosome (BAC) system is widely preferable for the construction of large segment of gene cloning (Shizuya*et al.* 1992; Tomkins *et al.* 2004). In addition to positional cloning, BAC libraries are very much significant for, high throughput sequencing, end sequencing etc. Different constructs of BAC like TMS30001 and MECU72are cloned for cassava mosaic disease and white fly resistance respectively. Working with the candidate genes initiated by resistance gene analogues and PCR products were isolated by using degenerate PCR primers are further used to clone in BAC. These are subjected under fingerprinting and organized into contigs. The genetic map is raised by using the contigs as anchors and end sequencing of BAC clones are performed by using the large population of cassava. The unique homologues are identified and used in some genetic complementation experiments based on the contigs' proximity to the resistance QTL genes. A large population of progeny mapping and a huge segregant progeny analysis are two concepts for creating a fine genetic map of cassava (Michelmore *et al.*, 1991). The BAC clones, which have already been fingerprinted and organised into contigs, are being isolated using this fine-map. Sequenced BAC end products are shown as allele-specific markers in fine maps. The co-segregating BAC clones with resistance genes are completely shot-gun sequenced two to three in number, from where the search of sequence is being performed for the resistance genes for genetic complementation experiments (Tomkins *et al.* 2004).

7. GENOMICS-AIDED BREEDING FOR HR TRAITS

Due to its ability to thrive in arid areas, cassava (*Manihot esculenta*) has become one of the most significant starchy root and tuber crops in the tropics. The identification of markers and potential genes for use in cassava breeding programmes can hasten genetic advancement in this vegetatively propagated crop. Using a panel of 5130 clones created at the International Institute of Tropical Agriculture—Nigeria, they conducted a genome-wide association study (GWAS). Through genotyping-by-sequencing, the community was genotyped at more than 100,000 SNP markers (GBS). The genomic areas underpinning the genetic variation for 14 traits—quality (dry matter content and carotenoid content), biotic stress (severity of cassava mosaic disease and cassava green mite), and plant agronomy—were examined (harvest index and plant type). They also had a number of highly heritable leaf, stem, and root agromorphological characteristics. 41 significant connections were found in total. Although several of the identified loci were similar to those that had already been published, they also provide new connection indications for the traits. For each trait-locus combination, they offer a list of advantageous alleles at the most significant SNP, along with candidate genes found among the GWAS results. These resources serve as a basis for the creation of candidate genes for functional validation as well as markers that could be employed in cassava breeding efforts.

In the tropics, cassava, a starchy root crop, is grown for both fresh consumption and industrial processing. The key selection criteria in cassava breeding are dry matter content and micronutrient density, particularly provitamin A carotenoids. In African germplasm, these qualities are adversely associated. Utilising 672 clones genotyped at 72,279 single nucleotide polymorphism (SNP) sites, a genome-wide association mapping was carried out. Indirect methods were utilised to gauge carotenoid concentration fluctuation in the form of root yellowness. On chromosome 1, locations 24.1 and 30.5 Mbp, two important loci for root yellowness have been found. The 24.1 Mbp crest for carotenoids colocalized with a specific locus for dry matter accumulation, which was found. Regarding root yellowness and dry matter content, haplotypes at these loci predicted 70 and 37% of the phenotypic variability, respectively. The major dry matter locus for the white- and yellow-root subpopulations was identified in separate analyses, and evidence of megabasescale linkage disequilibrium (LD) around the major loci for the two traits suggests that physical linkage, rather than pleiotropy, is more likely to be responsible for the negative correlation between the target traits. In addition, potential genes for starch and carotenoid biosynthesis (UDP-glucose pyrophosphorylase and sucrose synthase) were found close to the locus at 24.1

Mbp. These discoveries clarify the genetic structure of the dry matter and carotenoids in cassava and offer a chance to speed up breeding for these features (Rabbi *et al.* 2017).

Cassava is grown because of its ability to withstand droughts and because its store roots are highly carbohydrate-rich. The uneven shape and lack of homogeneity of storage roots place limitations on collection and postharvest processing. Image analysis of storage root images are collected in the field to phenotype the Genetic gain and offspring (C_1) stocks from the International Institute of Tropical Agriculture (IITA) breeding programme. It had already been discovered QTL on chromosomes 1 and 12 for the majority of shape- and size-related phenotypes in the genome-wide association study (GWAS). In a prior investigation, it was established that resistance to cassava mosaic disease (CMD) is correlated with the QTL on chromosome 12. The standard deviation (SD) of each root measurements per clone is reported since root homogeneity is crucial for breeding. It was further discovered that the additional significant QTL for perimeter, fraction, and aspect ratio on chromosomes 6, 9, and 16 using SD measurements. Root size and shape image-extracted traits had higher predictive accuracy than yield trait, for the most part. The purpose of this study was to investigate the viability of the picture phenotyping methodology as well as GWAS and genomic forecast for the size and shape attributes that were retrieved from the images. The results of the technology outlined are encouraging and present a chance to use high-throughput techniques in cassava (Yonis *et al.* 2020).

A lot of work has lately been put into enhancing the starch characteristics of cassava (*Manihot esculenta* Crantz), an essential crop for producing starch. Due to its neutral flavour and pasting qualities, waxy cassava starch provides several benefits beyond starch from other sources. Although the waxy phenotype of starch is thought to be monogenic and recessive in its genetic inheritance, it is plausible that other genomic areas are involved. This study sought to assess the effectiveness of genome-wide association studies (GWAS) in detecting waxy starch-related variations and in identifying alternative genes or genomic areas. In order to accomplish this, an array of 20,956 single nucleotide polymorphisms was used to assess a collection of 382 samples, both waxy and non-waxy (SNPs). The 2% iodine assay for waxy starch was used to determine whether there was of waxy starch in all genotypes. The retrospective CAse-control Retrospective Association Testing (CARAT) methodology was used to conduct the GWAS. The model was expanded to include research on kinship and population structure. At chromosome 2 and in nearby genomic areas, ten SNPs were discovered to be important. The important SNPs are located close to the genes for Beta-amylase and Granule-bound starch synthase I (*GBSSI*). Among the important SNPs, additive and epistatic effects were also discovered; these must be further researched in segregant communities of waxy sources. The areas and genes found in this study may inspire additional research and shed light on biological mechanisms relating to the waxy phenotype (do Carmo *et al.* 2020).

The development of novel genotypes with improved agronomic features, such as higher root output and resistance to biotic and abiotic stresses, has been greatly advanced by cassava breeders. Because these fresh and improved cassava (*Manihot esculenta* Crantz) cultivars in crop production in Nigeria have experienced little to no improvement in their culinary qualities, there is a scarcity of genetic details relating the texture of boiled cassava, particularly with regard to its mealiness, the primary sensory quality attribute of boiled cassava roots.The current study aimed to identify genomic regions and polymorphisms linked to natural variation in root mealiness and other texture-related properties of cooked cassava roots, such as fibre, adhesiveness (ADH), flavour, smell, colour, and hardness. A genome-wide association (GWAS) analysis was performed using phenotypic data from a board of 142 procurements obtained from the National Root Crops Research Institute (NRCRI), Umudike, Nigeria, and an acquisition of 59,792 elevated quality single nucleotide polymorphisms (SNPs) dispersed throughout the cassava genome. Genome-wide association mapping was used to find 80 SNPs on chromosomes 1, 4, 5, 6, 10, 13, 17, and 18 that were substantially linked with root mealiness, fibre, adhesiveness, flavour, fragrance, colour, and hardness. Additionally, it was discovered that the pertinent candidate genes in *M. esculenta* that are close to peak SNPs associated with these features. A review of the cassava reference sequence v6.1 revealed that Manes is close to the SNPs on chromosome 13. 13G026900, a gene known to play a significant impact in the texture of cooked potatoes, is identified as being in charge of cell adhesion, the mealiness or crunchiness of vegetables and fruits, and cell adhesion. This work offers the first explanations of the underlying genetic causes of the texture of boiled cassava roots. The markers and candidate genes discovered in this novel work may be useful genomic resources for use in genomic selection (GS) and marker-assisted selection (MAS), which would accelerate the genetic progress of root mealiness and other cooking attributes in cassava breeding programmes in West Africa, especially in Nigeria, where the intake of boiled and pounded cassava is low(Uchendu *et al.* 2021).

8. RECENT CONCEPTS AND STRATEGIES DEVELOPED

8.1. Gene Editing

The world's population is predicted to nearly triple to nine billion people by the year 2050. The demand for food may be higher than the supply due to scarce resources and global expansion. As a result, the staple crop cassava has been mentioned as one that has the capability to lessen food shortages, poverty, and famine, especially in developing African nations. It is believed that cassava evolved in South America and then moved to Asia and Africa. More than 800 million people worldwide now eat it regularly as a result of this development. *Manihot esculenta* leaves are eaten in some regions of Africa as a source of protein, vitamins, and minerals, but its starchy-dense roots are the main reason people eat it. This crop's drawbacks include its vulnerability to pests and diseases, poor starch quality, and post-harvest quality degradation. Cassava-related problems can be solved using biotechnological techniques like gene editing.

Genome editing tools: Developments in technology for genome editing have showed promise in crops, making the development of new types easier. Because they are quick, efficient, and easy to apply technically, zinc finger nucleases (ZFNs), transcriptional activator-like effector nucleases (TALENs), and much more subsequently CRISPR are significant gene-specific genome editing methods. Due to its simplicity, adaptability, and effectiveness, the genetic modification method using CRISPR/CRISPR-associated proteins has shown more promise in tackling agricultural concerns than its predecessors. Recent genome editing experiments have employed the CRISPR/Cas9 technology to boost the yield of the drought-tolerant crop cassava by introducing disease resistance, fast blooming, herbicide tolerance, and lower cyanide concentrations in the leaves and roots.

How CRISPR-CAS 9 is exploited in Cassava?

The fundamentals published by (Odipio*et al.* 2017) state that the tropical mainstay cassava (*Manihotesculenta*) was modified using CRISPR/Cas9 methodology in order to target the Phytoene desaturase gene (*MePDS*). The gRNA constructs that targeted two sequences in *MePDS* exon 13 were used to target the gene in two cultivars. Both designs caused obvious albino phenotypes in cotyledon-stage somatic embryos reproducing on selection medium after Agrobacterium-mediated transport of CRISPR/Cas9 components into cassava cells, and the plants propagated therefrom, 38 plant lines (19 from each cultivar), out of a sum of 58 (cv. 60444) and 25 (cv. TME 204), were examined for mutagenesis. Between 90% and 100% of plant lines in cv. TME 204 exhibited the albino phenotype, which is a high frequency. Full albinos lacking of green tissue and chimaeras with a combination of white and green tissues made up the observed albino phenotypes. Sequence analysis showed that the targeted *MePDS* region had alterations in 38/38 (100%) of the plant lines tested, with additions, deletions, and substitutions being noted. A putative mono-allelic homozygous line (1/19) was discovered in the cv. 60444, while putative bi-allelic homozygous lines (1/19) and 4 (4/19) were discovered in the cvs. 60444 and TME204, respectively. It was discovered that the leftover plant lines, which mainly consisted of chimaeras, were putatively heterozygous. It was found that, the small (1bp) deletions and/or substitutions of nucleotides in the targeted Phytoene desaturase (*MePDS*) area 50 and/or 30 upstream and/or downstream, respectively. The data presented shows that cassava genome editing using CRISPR/Cas9 is quite effective and manageable, leading to multi-allelic alterations in both cultivars under study. It only takes 6 to 8 weeks for the *MePDS* modification described here to produce physically discernible altered events, and sequencing is not necessary to verify editing at the destination.

Various molecular strategies were adopted to improve cassava, some of which are listed below :

(a) dsRNA hairpins are used to control whiteflies. Hairpin dsRNA could be employed for disease control in the cassava mosaic. The V-ATPase-A mRNA is targeted by the hairpin dsRNA, which stops it from being translated. The whitefly dies as a result of additional transport systems being activated when V-ATPase-A is expressed.

(b) Silencing by small RNA (sRNA) to develop virus resistance in cassava. It is possible to develop resistance to many cassava viruses using sRNA. Resistance to these viruses was produced via the production of miRNAs in cassava that were targeted to the 21 conserved sequence of CBSV.

(c) Creation of the siRNA and siRNA libraries. The expression of several siRNAs likes a reaction when a Xanthomonas cassava infection first begins. Through the use of X. axonopodis-treated cassava and bioinformatics methods to identify the siRNAs that over-express in this situation, a library of these siRNAs was produced.

(d) TILLING is used to identify nucleotide polymorphism.

Nanotechnology

Producing food, goods for industry, and animal feed all involve cassava. In 2017, Thailand produced 30.84 million tonnes of cassava on 1.34 million hectares of land, a 1.15-fold increase over the preceding ten years. Although Thailand's acreage and output only make up 5.45% and 11.04% of the global totals, respectively, the amount and price of the country's cassava crop on the global export market ranges between 58.5-81.2% and 44.4-56.7%. Cassava is also Thailand's main export crop. Cassava has a place in locations where there is a lack of water for farming since it can withstand drought and nutrient-poor soil. However, biotic stress—which includes things like plant diseases, insects, and weeds—or abiotic stress—which includes unfavourable environmental conditions—may affect the growth and development of cassava, which could result in a decline in production.There are currently 28 known varieties of bacterial, viral, or fungal cassava infections. One of the most significant cassava diseases is leaf spot disease. Up to 30% of the cassava crop's output can be lost due to the diseases. However, until a recent epidemic in Brazil, the significant issue of cassava leaf spot disease was frequently ignored. The pathogen *Alternaria* sp. was identified as the cause of the cassava leaf spot disease in 2019. The efficiency of fungicides containing copper, benomyl, thiophanate, and carbendazim in controlling infections varies. As a result, using fungicides to lessen cassava leaf spot damage is a common practise. Thailand does not, however, have this cultivar that is resistant to the leaf spot disease. Therefore, finding a novel way to improve cassava's resilience is essential. Elicitation is one of the most efficient methods for boosting plant immunity since it stimulates the secondary metabolites. The plant has an inherent defence mechanism against harmful environmental elements, such as abiotic and biotic stressors, which varies based on the cultivar and the harmful elements. An elicitor can be used to artificially activate the plant's immune system. Elicitors are biotic or abiotic substances that stimulate innate immunity and plant defences in response to pathogens and stressful environmental conditions. The elicitors could be composts, charcoal, chitosan (CS), plant or microbial extracts, or chemicals.

In recent years, nanotechnology has been applied to a number of agricultural fields, including nano fertilizers, nanobiotechnology, nanomaterials, nano sensors, nano pesticides, nano elicitors, and nano herbicides, in order to increase plants' tolerance to biotic and abiotic stress, improve crop yield and quality, and in particular, to develop sustainable agriculture.To protect plants from bacteria, fungus, viruses, and insects, nanoparticles (NPs) are utilised as transporters of active substances (insecticides, fungicides, herbicides, and RNA-interference) or as protectants (silver, gold, copper, titanium dioxide). Pesticide-based nanoparticles have a longer shelf life, a more precise absorption at the target site, are more soluble, and have less soil leaching.

HOW NANOPARTICLES WORKED IN CASSAVA?

In the instance by Hoang *et al.* 2012, the ionic gelation method was employed to produce three formulations of salicylic acid loaded with chitosan nanoparticles (N1, N2, and N3) and three compositions of silver loaded with CS-NP (N4, N5, and N6).When exposed to CS-NP compositions up to 800 ppm, cassava leaves did not turn poisonous, according to the results of the leaf disc assay method. CS-NP formulations at 25–800 ppm was primarily used as elicitors. The results showed that compared to the other formulations, the CS-NP-loaded SA (N3) and CS-NP-loaded Ag (N6) were more effective at reducing the disease severity and the disease index of leaf spot.N3 at 400 ppm and N6 at 200, 400, and 800 ppm may lower disease incidence by 68.9-73.6% and 37.0-37.7%, respectively, when a fungal pathogen was introduced at 2 and 21 days after spraying elicitors with concentrations of 104 and 105 conidia per mL. It outperformed flutriafol (Flu), pyraclostropin (Pyr), and commercial zinc oxide NP (ZON), which all performed at 55.7 and 14.2%, 59.6 and 14.7%, and 54.7 and 35.4%, respectively.The N3 and N6 treatments encouraged cassava growth in the means of shoot height, number of leaves, number of shoots, largest leaf area, root length, and root weight, much like the positive control group. In order to treat cassava leaf spot disease, this raises the potential of utilising SA or Ag that has been coated with CS-NP at modest dosages (200–400 ppm) as elicitors.The ionic gelation approach has a drawback in that it is challenging to make NPs with homogeneous sizes, which leads to unstable efficiency. The control of plant diseases was also affected differently by the size of the NP in different reports. This work concentrated on reverse research methods, commencing with elicitor creation and continuing via toxicity tests, formulation effectiveness testing, concentration effectiveness monitoring, and the characterization of successful elicitors, in order to address this issue. Instead of concentrating on in vitro conditions, this study motif first investigates the efficacy under in vivo (net-house) scenarios after developing elicitors and doing toxicity assessments. Finally, instead of initially concentrating on all six formulations, only the very successful compositions of CS-NP-loaded SA (N3) and CS-NP-loaded Ag (N6) were characterised. The NP formulations employed in this study were also concentrated for use as elicitors at small doses. Therefore, despite the fact that fungal infections are frequently concentrated in in vitro trials, the effective quantities of NP formulation may not necessarily prevent them. As a result, the research process takes less time and money. Before it is generally endorsed, a field experiment (200-400 ppm) is still required. Another field where this area of research and the CS-NPs-based ionic gelation technique show promise is the management of plant or agricultural disease, particularly in locations where access to conventional procedures is constrained.

9. BRIEF ON GENETIC ENGINEERING FOR HR TRAITS

9.1 Target Traits and Alien Genes

Cassava is a crop with limited amount of micronutrient like iron, zinc as well as vitamin A, which is a staple food for 250 million Africans and critically responsible for health improvement in Sub-Saharan Africa (SSA). These essential micronutrients play a significant role during pregnancy in women as well as metabolic activities in human body (FAO, 2010). Hence, there is prime need of micronutrient biofortification in cassava, especially for the African population.

Review on Achievements of Transgenics

Genetically modified (GM) biofortified cassava may be the actual way to solve this problem. A rapid improvement in cassava breeding, had been found with the advantage in the field of genetic engineering in different countries. As a result, the programmes like Harvest Plus and BioCassava Plus, had achieved a remarkable achievement through molecular breeding (CGIAR, 2008; Sayre *et al.* 2011). The Biocassava plus (BC+) team stressed more on genetically modified (GM) cassava to uplist the nutritional deficiency in iron, zinc, vitamin A, and protein. This project focused on introducing 10 times more provitamin A in GM cassava compared to non-GM cassava.

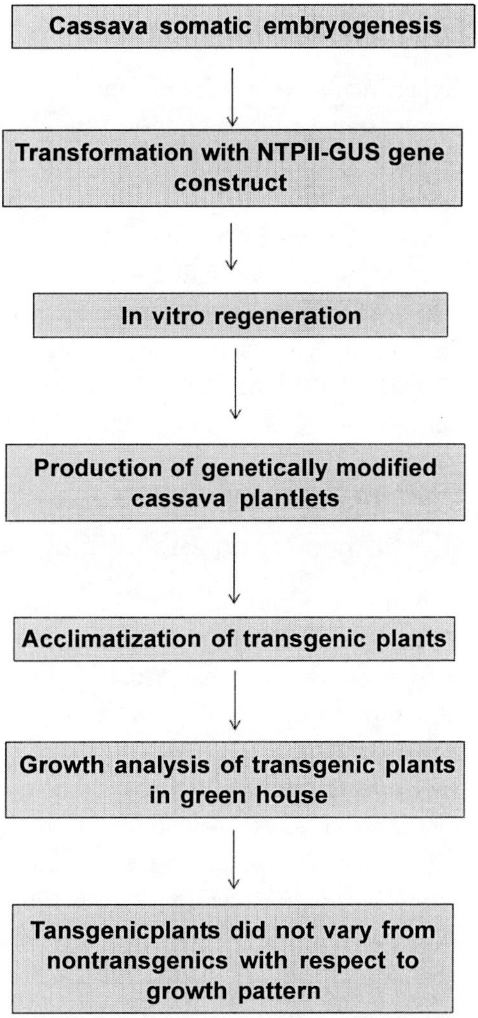

In addition to mineral and vitamin biofortification, scientists now-a-day is also interested to set their focus for improvement of enhanced production. Cassava is able to fix high rate of CO_2 and convert it to sucrose. Being a starchy root crop, it is of utmost importance to uplift the yield potential in the field, which can be achieved by the improved source to sink relationship and sink strength for carbohydrate. The transgenic plants with increased tuber ADP-glucose pyrophosphorylase (AGPase), is a rate limiting enzyme in starch biosynthesis pathway were produced, by expressing modified bacterial *glgC* gene under Class I patatin promoter. The GM cassavas lead to increase starch production. To assure maximum AGPaseactivity, *glgC* gene was modified by site-directed mutagenesis (G336D) to downregulate allosteric feedback by fructose-1,6-bisphosphate. There were 70% higher AGPase activity in transgenics expressed *glgC* gene and increased root biomass production upto 2.6-fold, established that targeted alteration of enzymes involved in source-sink relation can bringabout high carbohydrateaccumulation in sink (Ihemere *et al.* 2006).

Organelle Transformation

Experiments were under taken to compare morphology and quantitative growth among transgenic and non-transgenic plant. The transgenic plants were modified by NPT II marker and GUS reporter genes. The external morphology did not differ between transgenic and non-transgenic plants though the transgenic plants possessed arched leaf lobe in contrast with obovate in non-transgenic plants. As the GM of cassava with NPT II and GUS genes did not alter in growth patterns can be cultivated under same agronomic package and practices (Jelili, 2018). Greater micronutrient concentration offers a relatively cost-effective way to increase daily sufficiency of micronutrient intake and to address micronutrient deficiencies. Genetically modified (GM) technology has a significant potential to alter agriculture.

Biosynthesis and Biotransformation

The functional properties of starch are majorly dependent on the relative amounts of amylose, amylopectin, size of the starch granule as well as covalent modifications. The paste clarity and swelling in starch is dependent on total phosphate content though it does not influence the functional properties. Phosphorylation is a naturally occurring phenomena to the C3 or C6 carbon of amylopectin glucosyl residues. The degree of phosphorylation has been modified by the starch phosphorylating enzyme like GLUCAN WATER DIKINASE (GWD). The transgenic plants were generated either the wild type potato GWD or a redox-insensitive version of the same were overexpressed. In addition to that, RNAi mediated gene silencing was done to endogenousphosphoglucan phosphatise genes like *MeSEX4* and *MeLSF* responsible for leaf starch metabolism in *Arabidopsis thaliana.* Transgenics overexpressed with potato *GWD* gene uplifted the level of phosphorylation in both C6 and C3 carbons. The endogenous gene silencing leads to storage tissue specific increasing ofthe ratio of C3:C6 phosphorylation (Wang *et al.* 2018).

Metabolic Engineering Pathways and Gene Discovery

The industrial usage of starch is well known fact giving rise of many more of the products like dextrin, adhesive bio degradable plastics etc. The products like emulsion stabilizers, consistency modifiers are obtained from modified starch. The modified starch (designer starch) is prepared by varying the two constituents of starch namely amylose and amylopectin and altering its properties. The chemical method of starch modification is expensive as well less eco-friendly. The production of waxy cassava by CIAT is an example of designer starch, which was raised by using the antisense RNA technology. The waxy cassava starch has altered rheological properties compared to other cereal derived waxy starch (Radhika *et al.* 2014; Taylor *et al.* 2004).

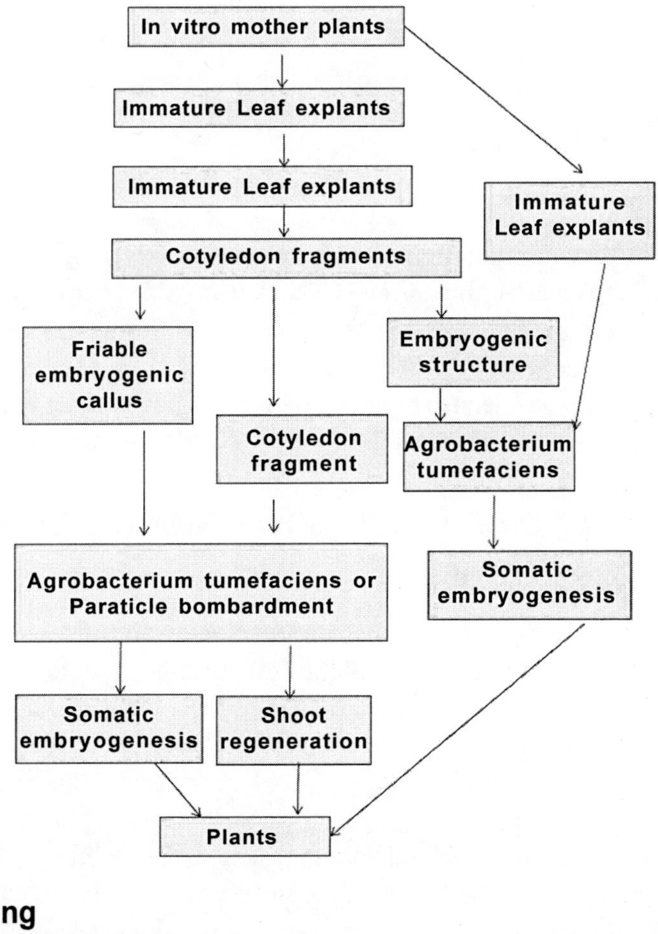

Gene Stacking

According to scientists, BC+ project provided nutrient dense cassava germplasm 60444 apart from â-carotene through GM technology. Along with the increasing of yield, the importance had also been given to raise the disease and pest resistant cassava variety, especially in cassava mosaic disease (CMD). However, in BC+ project a parallel importance had been given to raise disease resistant as well as more nutrient enriched cassava genotypes in Phase II as some improved genotypes yields are getting limited due to CMD infestation.

Gene Silencing

In non-model species where T-DNA insertion lines are not accessible and/or genetic alteration is time-consuming and ineffective, validating genetic function in plants can be difficult. Therefore, when dealing with these restrictions, reverse genetics technologies like virus-induced gene silencing (VIGS) are especially helpful (Lacomme, 2015). Though reliable genetic transformation protocols are available for cassava but due to their lengthy nature and recalcitrant cassava seeds, reverse genetic tools can be marked as more suitable to study the function and nature of cassava genes (Lentz *et al.* 2018).

Prospects of cisgenics

Cis genesis is genetic alteration used to introduce advantageous alleles into a receptive plant from crossable species. The source genes employed in traditional breeding are the identical ones that are transmitted during cis genesis. Linkage drag can be avoided, and available gene variants can be used more effectively. This strategy greatly accelerates breeding by fusing conventional breeding methods with contemporary biotechnology (Hou *et al.* 2014).

BRIEF ACCOUNT ON ROLE OF BIOINFORMATICS AS A TOOL

Gene and Genome Databases

To fulfil the objectives of nutritional improvements and disease resistance and disease resistance in cassava, a prediction of genome size cover of 69% along with 96% of protein coding genes had already been done by 454 based genome sequencing. The prediction of 30,666 genes as well as 3,485 alternate splice forms had also been established by 1.4 M expressed sequence tags (ESTs). Microsatellite (SSR), EST and single nucleotide polymorphisms (SNPs) based maps for cassava genome was already present earlier while genome sequence contributed a high-density linkage map from a cross between two contrast genotypes for resistance against cassava brown streak disease with the objective to successful marker-assisted breeding programs for disease-resistance and nutritional enrichment (Prochnik *et al.* 2012).

Comparative Genome Databases

It is a well-known fact that homeodomain-leucine zipper (HD-Zip) plays a significant role in plant signal transduction under stress condition. An expression study with HD-Zip genes (*MeHDZ01-57*) established that, the pattern of expression of *MeHDZ* genes may be constitutive or no expression or tissue specific in case of both cultivated and wild species. The differential expression of these above mentioned three genotypes had been confirmed under drought and polyethylene glycol treatments by the help of transcriptomic analysis. Moreover, comprehensive responses of twelve different *MeHDZ* genes under several stimulates like cold, salt, and ABA treatments were recorded through quantitative RT-PCR analysis. These findings may prove to be very much significant in future for the characterization of *MeHDZ* gene in cassava for their contribution in abiotic stress signalling (Ding *et al.* 2017).

UDP-glycosyltransferases (UGTs) are widely known for conjugating the sugar molecules with small lyophilic compounds. It is also responsible for plant growth and has significant role in stress response. However, the role of UGTs are yet not clear in cassava. In a recent study 121 *MeUGT* genes had been grouped into 14 groups. UGT family proteins have unique features and tandem duplications are frequent in *MeUGT*, which act as an important driving force for the expanse of these genes. Analysis of Cis-Acting elements revealed at least 14 types of cis-elements are involved in different biotic as well as abiotic stress. Silencing of*MeUGT-14* or *MeUGT-41* genes in cassava promoted the infestation of *Xanthomonas axonopodis* was indicated by transcriptome data for function prediction. These findings firmly established the priority of improvement of *MeUGT*s for better resistance in cassava (Wu *et al.* 2021).

Gene Expression Databases

An approach had been taken for tissue based molecular identification in cassava through RNA sequencing. Characterization of cassava transcriptome revealed similarity between organized embryogenic structure and friable embryogenic callus tissue. Experimental data established that a lacking of genetic expression in the root tissues differentiated cassava roots from other plants and raised a sink (Wilson *et al.* 2017).

Cassava is a highly stress tolerant crop. The transcription factor like *GRAS* regulates the growth, development and stress tolerance in plants. Experiments revealed 77 *GRAS* gene family members from cassava genome which were later categorised under 14 subfamilies. Though the gene structure as well as 'protein motif composition within the same subfamily was proved to be conserved but, segmental duplication was responsible for *GRAS* gene expression in cassava was established by global expression analysis. In addition to that, it was established by qRT-PCR analysis, that *MeGRAS* genes response to different abiotic stresses may also have other functions (Shan *et al.* 2020).

In cassava a 3' RNA sequencing was performed for 150 accessions the National Crops Resources Research Institute, Uganda, on five different tissue types and 8820 differentially expressed genes (DEG) revealed similar expression patterns between different tissues as well as clustering of DEGs in 18 gene modules. Specifically, 27 genes had been identified for differential expression in plant–pathogen interaction pathway cassava mosaic disease (CMD). This experiment helped to understand molecular expression in complex trait like CMD (Ogbonna *et al.* 2021).

Protein or Metabolome Databases

Molecular mechanisms had been identified for several abiotic stress so far in cassava but the cross talk between the stress specific responses is yet to be known. A total of 6103 and 7462 transcripts were found to be significantly regulated under artificially created cold or drought stress in cassava respectively. It was established that abscisic and jasmonic acid pathways shared between these two responses. A number of 2434 differentially expressed genes (DEGs) were identified during the experiment, in which 1130 up-regulated and 841 down-regulated DEGs were included. All these co-induced or co-supressed genes were clustered in different groups for stress signals perception and transduction, as transcription factors (TFs), or transport facilitation according to the function annotation. In addition to this protein kinases, TF families and ubiquitin proteasome system related genes played a key function of crosstalk between cold and drought stress (Li *et al.*2017).

A study on genetic expression profiling on cassava variety Huanan 124 under root bulking stage was carried out in China. Expressed sequence tags (ESTs) were established by sequencing a total of 9,600 cDNA clones. A total of 2,878 putative unigenes were raised from the assembly of resulting ESTs, among which, 62.6% unigenes matched with known ETSs of cassava and the remaining had no 'hits' against the cassava database in integrative PlantGDBdatabase where as 59.59% unigene match were found by Blastx analysis. Hence, a cDNA microarray was developed considering all the 2,878 unigenes to find out the genetic expression profile in different growth stages. The microarray data analysis confirmed different dimensions of starch metabolism in cassava. The experiment established that, in early root development stage starch synthesis and degradation occurs simultaneously while it becomes restricted with the maturity of root due to the decrease in UDP-glucose dehydrogenase activity. The stress responsive genes are expressed constitutively due to the adaptation of cassava under prolonged exposure of environmental stress factors (Li *et al.* 2010).

Integration of Different Data

The heterozygous nature of cassava genome is a well-known fact. To reveal the in-depth context of heterozygosity is much difficult on account of technological limitations of sequencing, but in present days the problem had been overcome by the use of different types long-read sequencing technologies. The majority of the genes were combined into the first cassava genome reference, version 4.1 (v4.1), which was created with the aid of Roche 454 sequencing(Prochnik*et al.* 2012), which through a light to the repeat sequence of cassava genome. Version 5 (v5) was an improved version over the 57% of v4 assembled sequences in 18 chromosomes with the help of a dense composite genetic map (ICGMC 2015; De Carvalho & Guerra, 2002). Version 6 (v6) assembly was raised recentlyby the Illumina sequencing technology by the improvement over v5 by considering 18% more contig sequence that resulted to inclusion of 45% more sequence in the chromosomal scaffold which is the primary reference in genomic analysis in cassava (Bredeson*et al.* 2016; Kuon *et al.* 2019; Wolfe *et al.* 2019; Kayondo*et al.* 2018; Nzuki*et al.* 2017; Ramu*et al.* 2017; Andrade *et al.* 2019).

BRIEF ACCOUNT ON SOCIAL, POLITICAL AND REGULATORY ISSUES

Concerns and Compliances

Manihot esculenta Crantz, also known as cassava, is considered the most important root and tuber crop in tropical regions (Pujol *et al.* 2002; Meireles da Silva *et al.*, 2003). Cassava's relevance for tropical and subtropical agriculture was little investigated until four decades ago, with few exceptions. Despite this, when the International Institute for Tropical Agriculture (IITA) in Nigeria and the International Center for Tropical Agriculture (CIAT) in Colombia were established in the early 1970s and several National Agriculture Research programs were rapidly consolidated, a new era in cassava was started with the development of successful breeding programs, modernization of cultural practices, and the development of new processing techniques (Cock *et al.* 1985; Jennings & Iglesias, 2001). Many other countries have conducted successful research on cassava, including Brazil, Colombia, China, Cuba, India, Indonesia, Nigeria, Thailand, and Vietnam. Millions of people eat cassava as part of their daily diet.

Patent and IPR Issues

Cassava and manioc plants (*Manihot esculanta*) that develop biological characteristics through chemical mutation are covered by patent application WO2019121603. The Dutch company Keygene asserts that it was the first to use chemicals in mutation breeding on cassava. The business asserts ownership of all treated cells, seedlings, and plants, rather than just the chemical application process.There are many countries in Africa, South America, and Asia where cassava (manioc) is an extremely important food. This could limit breeding activities, for example, in developing varieties adapted to new climatic conditions or pests, if such a broad-based patent is granted.

Even if the patent is only granted in Europe, the restrictions may still apply to African nations because many breeding methods entail international cooperation. It is possible to see the broad-based patent claims as an abuse of the patent system intended to restrict breeding resources and rob breeding rights. A simple, nontechnical approach can transform food plants into 'intellectual property' of companies. By declaring random mutations to be patentable inventions, the Administrative Council introduced legal loopholes in its 2017 decision.

Socioeconomic Aspects

Cassava is commonly known as a "drought, war, and famine crop" throughout the developing world because of the versatile nature of the crop. Despite its status as a staple crop, it could potentially contribute to improved food security in a time of climate change (Liu *et al.* 2008; Nassar & Ortiz, 2007; Gallo & Sayre, 2009). The plant can tolerate periods of drought and is easy to propagate, even on poor soils (Hillocks *et al.* 2002; El-Sharkawy, 2003; Dahniya, 1994). In general, tuberous roots are the major food product of the plant, and they can be recovered from the soil up to three years after maturity. In addition to providing social "insurance", this also protects against periods of stress and unrest. Thus, cassava could significantly increase food production without requiring large quantities of agricultural inputs (such as fertilisers, water, and pesticides).

The cassava plant is also becoming an important source of industrial starch (Balagopalan, 2001) and bio-fuel (de Vries *et al.* 2010). Cassava is an indigenous crop in Brazil. It is difficult to unearth the prehistory of this crop in humid tropical environments, but genetic and paleo ethnobotanical evidence indicates that it originated in the Amazon basin and spread throughout the Neotropics 6,500 years ago (Isendahl, 2011). The roots of cassava are cultivated primarily for their starch content, but the leaves are also consumed for their nutritional value (Alves *et al.* 2002). Cassava grows in arid and erratic weather conditions, as well as in soils with acidity and low mineral content. Due to its hardiness, it is suitable for small-scale and low-input agriculture.

Among Brazil's most widely grown crops, cassava is predominantly used for flour production. The northern region traditionally stands out in its per capita and family-owned *casas de farinha*. There are large-scale cultivation and processing industries in the South that produce cassava starch and flour for food and non-food industries (Chuzel, 2001). Historically, cassava has been associated with colonization, particularly in the West. The cassava was credited with facilitating colonization by Jo aoBrigido dos Santos, a 19th century Brazilian historian. Known as the 'Brazilian bread', it was the country's primary food source, particularly for African slaves and indigenous people (Picanço, 2018).

Especially during long voyages, like those undertaken by slave ships, cassava flour was particularly useful. There was also a political and economic significance to this crop: people's ability to produce flour was used as a criterion in 1823 for voting or getting approved as part of the first constitutional project in the independent country, commonly known as the Constituiçao da Mandioca (Cassava Constitution). Besides being used as a currency, cassava was also traded for slaves (Aguiar, 1982). As the cassava crop became part of the resistance against colonialism among runaway quilombo slave communities, its role in colonial projects was reversed. Cassava flour was a typical element of the quilombola economy. Through grinding, sieving, and oven baking, they transformed cassava into flour and other derivatives.

Because of low agricultural productivity, most Sub Saharan African countries have difficulty ensuring a sustainable food supply. Low investment in the agricultural sector is one of the contributing factors. Over the years, agricultural budgets have decreased. The Sub Saharan African countries have a high level of rural poverty, food insecurity, and unemployment. To assist in transforming rural economies in the region, policy makers and donor agencies have designed and implemented agricultural interventions. Nigeria's presidential cassava initiative (PCI) implemented between 2001 and 2007 was one of these interventions.

Problems of production

1. A lack of quality high-yield hybrid cassava varieties. Comparatively to the local popular cultivars, the quality of the cassava hybrids available is poor.

The quality of the cassava product becomes crucial since most of it goes directly to consumers.

2. In addition to frequent attacks by diseases, it faces problems such as tuber rot, cassava mosaic virus, etc.
3. It is difficult to devise varieties that grow well for short periods of time, are drought resistant, shade tolerant, disease resistant, and can be trained to adapt to both highland and lowland environments.
4. Farmers' limited resources limit their ability to adopt proper technology.
5. Tubers of cassava have a poor shelf life.
6. Lack of knowledge about price behaviour, market demand, and marketing channels.
7. Inadequate marketing system.
8. There is a lack of awareness about the diversification of products.
9. Its improvement is limited because the government has no well-defined policies.
10. Cassava exports aren't being organized.

Complexity and Cassava

By 1967, the General Agreement on Tariffs and Trade (GATT) had reduced import duties for Europe's use of cassava (*Manihot esculenta*) and had recognised Brazil and Indonesia as the crop's top exporters. Thailand replaced Indonesia and Brazil as the major exporters of cassava chips, pellets, and starches to Europe in 1975, becoming the dominant exporter of these products globally. In 1985, Thailand supplied 95% of the cassava imported into Europe.The European Economic Community (EEC) moved immediately to try to control the influx of cassava goods in reaction to Thailand's sudden and dramatic rise as a cassava exporter in order to defend rivals in the European grain business. Through managed quotas, Thai exporters' access to European markets was severely constrained in the middle of the 1990s. As warning signs increased in the late 1980s and early 1990s, a range of economic development organisations and analysts forecast direct ramifications for Thai cassava growers. As a result, the production of cassava in Thailand has only experienced a slight decline, and Thai exports continue to rule the global market.

FUTURE PROSPECTS

Cassava is now used more frequently both as a crop with added value and as a food source. The industrial crop cassava is quickly gaining importance. Our nation has enormous potential for growth. The plant offers a number of benefits, including the greatest dry matter rate of any crop plant and the ability to thrive in a wide range of climate and soil conditions. It is nutrient-rich in addition to offering energy, vitamins, minerals, and other things. This data supports the advantages of raising cassava output in India. The following can help to achieve this:

1. Selection of disease-free varieties with high starch content that produce high yields.
2. Identification of and supply of early-bulking cassava varieties that have a short-duration.
3. A research project that aims to improve productivity by managing water, nutrients, and droughts.
4. Developing value-added cassava products, such as plastic from cassava, fish feed binder with cassava, and cell cultures from cassava.
5. The development of cassava food items, such as chips and wafers, as well as integrated management strategies to tackle the disease problem.
6. The opportunity to convert cassava into glucose and dextrose is larger as the demand for these substances increases in the pharmaceutical and food industries; as a result, these opportunities must be taken advantage of.
7. In order to fully use the export potential, organised activities are essential. This entails performing market analyses to ascertain the export potential and evaluating the demand for exports of goods derived from cassava.
8. To estimate the future demand, a spot assessment on cassava's pre- and post-harvesting characteristics is required. Up until now, cassava has mostly been used for the manufacture of fresh food, starch, and sago. According to its current consumption and growth rate, India may require at least 10 million tonnes of cassava by 2025, while starch and sago manufacturing may reach 0.5 million tonnes.

In our nation as a whole, cassava has enormous potential. In order to boost the usage of cassava, research activities are required to create starch extraction methods and to update outdated technology used to produce cassava products. There is an urgent demand for cassava technologies, as well as for extending them to atypical locations. The cassava crop has a number of qualities that make it a good substitute crop for small-scale farmers in remote areas with limited resources, including its capacity to produce carbohydrates effectively, its tolerance of low soil fertility, its capability to recover from biotic stress, and its ûexibility to multispecies agricultural cropping systems. As a result of all these attributes, cassava cultivation and production of value-added products need to be improved urgently.

REFERENCES

Aguiar PD (1982) Cassava: bread of Brazil.

Aiemnaka P, Wongkaew A, Chanthaworn J, Nagashima SK, Boonma S, Authapun J, and Phumichai C (2012) Molecular characterization of a spontaneous waxy starch mutation in cassava. *Crop science* 52(5):2121-2130.

Akingbala JO, Oyewole OB, Uzo-Peters PI, Karim RO and Baccus-Taylor GS (2005)Evaluating stored cassava quality in gari production. *Journal of Food, Agriculture & Environment* 3(1):75-80.

Alves AA, Hillocks RJ, Thresh JM and Bellotti AC (2002) Cassava: Biology, production and utilization. Eds. RJ Hilocks, JM Tresh and AC Bellotti. CAB international P:67-89.

Andrade LRBD, Sousa MBE, Oliveira EJ, Resende MDVD and Azevedo CF (2019) Cassava yield traits predicted by genomic selection methods. *PLoS One* 14(11):e0224920.

Asare PA, Galyuon IKA, Sarfo JK and Tetteh JP (2011) Morphological and molecular based diversity studies of some cassava (*Manihot esculenta*Crantz) germplasm in Ghana. *African Journal of Biotechnology* 10(63):13900-13908.

Balagopalan C (2001) Cassava utilization in food, feed and industry. In Cassava: Biology, production and utilization (pp. 301-318) Wallingford UK: CABI.

Balyejusa Kizito E, Rönnberg Wästljung AC, Egwang T, Gullberg U, Fregene M and Westerbergh A (2007) Quantitative trait loci controlling cyanogenic glucoside and dry matter content in cassava (*Manihot esculenta* Crantz) roots. Hereditas 144(4):129-136.

Beeching JR, Marmey P, Gavalda MC, Noirot M, Haysom HR, Hughes MA and Charrier A (1993) An assessment of genetic diversity within a collection of cassava (*Manihot esculenta* Crantz) germplasm using molecular markers. *Annals of Botany* 72(6):515-520.

Bertram RB (1994) Application of molecular techniques to genetic resources of cassava (*Manihot esculenta* Crantz-Euphorbiaceae): Interspecific evolutionary relationships and intraspecific characterization.

Blair MW, Fregene MA, Beebe SE and Ceballos H (2007) Marker-assisted selection in common beans and cassava. Marker-assisted selection: Current status and future perspectives in crops, livestock, forestry and fish, 81-115.

Bokanga M (1995) Cassava: opportunities for the food, feed and other industries in Africa. Transformation Alimentaire du Manioc. ORSTOM Editions, Paris, 2(1995):557-569.

Bredeson JV, Lyons JB, Prochnik SE, Wu GA, Ha CM, Edsinger-Gonzales E and Rokhsar DS (2016) Sequencing wild and cultivated cassava and related species reveals extensive interspecific hybridization and genetic diversity. *Nature biotechnology* 34(5):562-570.

Cardoso SM, Santana I, Swerdlow RH and Oliveira CR (2004) Mitochondria dysfunction of Alzheimer's disease cybrids enhances Aâ toxicity. *Journal of neurochemistry* 89(6):1417-1426.

Ceballos H, Iglesias CA, Pérez JC and Dixon AG (2004) Cassava breeding: opportunities and challenges. *Plant molecular biology* 56(4):503-516.

Ceballos H, Kulakow P and Hershey C (2012) Cassava breeding: current status, bottlenecks and the potential of biotechnology tools. Tropical plant biology 5(1):73-87.

Ceballos H, Sánchez T, Morante N, Fregene M, Dufour D, Smith AM and Mestres C (2007) Discovery of an amylose-free starch mutant in cassava (*Manihot esculenta* Crantz) *Journal of Agricultural and Food Chemistry* 55(18):7469-7476.

CGIAR (2008) Science Council. Report of the First External Review of the HarvestPlus Challenge Program. Rome, Italy: Science Council Secretariat.

Chagné D, Batley J, Edwards D and Forster JW (2007) Single nucleotide polymorphism genotyping in plants. In Association mapping in plants (pp. 77-94) Springer, New York, NY.

Chavarriaga-Aguirre PP, Maya MM, Bonierbale MW, Kresovich S, Fregene MA, Tohme J and Kochert G (1998) Microsatellites in cassava (*Manihot esculenta* Crantz): discovery, inheritance and variability. *Theoretical and Applied Genetics* 97(3):493-501.

Chukwu SC, Rafii MY, Ramlee SI, Ismail SI, Oladosu Y, Okporie E and Jalloh M (2019) Marker-assisted selection and gene pyramiding for resistance to bacterial leaf blight disease of rice (*Oryza sativa* L.) *Biotechnology & Biotechnological Equipment* 33(1):440-455.

Chuzel G (2001) The cassava processing industry in Brazil: traditional techniques, technological developments, innovations and new markets.

CIAT (2004) Annual Report. Cali, Colombia.

Cobb JN, Biswas PS and Platten JD (2019) Back to the future: revisiting MAS as a tool for modern plant breeding. *Theoretical and Applied Genetics* 132(3):647-667.

Cock JH, Porto MC and El Sharkawy MA (1985) Water Use Efficiency of Cassava. III. Influence of Air Humidity and Water Stress on Gas Exchange of Field Grown Cassava *Crop Science* 25(2):265-272.

Dahniya MT (1994) An overview of cassava in Africa. *African Crop Science Journal* 2(4):337-343.

De Carvalho R and Guerra M (2002) Cytogenetics of *Manihot esculenta* Crantz (cassava) and eight related species. *Hereditas* 136(2):159-168.

de Vries SC, van de Ven GW, van Ittersum MK and Giller KE (2010) Resource use efficiency and environmental performance of nine major biofuel crops, processed by first-generation conversion techniques. *Biomass and Bioenergy* 34(5):588-601.

Delange F, Ekpechi LO and Rosling H (1994) Cassava cyanogenesis and iodine deficiency disorders. In International Workshop on Cassava Safety 375 (pp. 289-294).

DiasoluaNgudi D, Kuo YH and Lambein F (2003) Amino acid profiles and protein quality of cooked cassava leaves or 'saka saka'. *Journal of the Science of Food and Agriculture* 83(6):529-534.

Ding Z, Fu L, Yan Y, Tie W, Xia Z, Wang W and Zhang J (2017) Genome-wide characterization and expression profiling of HD-Zip gene family related to abiotic stress in cassava. P*LoS One* 12(3):e0173043.

do Carmo CD, Brito AC and de Oliveira EJ (2020) Genome-wide association studies for waxy starch in cassava. *Euphytica* 216(5):1-17.

Elias M, Panaud O and Robert T (2000) Assessment of genetic variability in a traditional cassava (*Manihot esculenta* Crantz) farming system, using AFLP markers. *Heredity* 85(3):219-230.

El-Sharkawy MA (2003) Cassava biology and physiology. *Plant molecular biology* 53(5):621-641.

Esuma W, Herselman L, Labuschagne MT, Ramu P, Lu F, Baguma Y and Kawuki RS (2016) Genome-wide association mapping of provitamin A carotenoid content in cassava. *Euphytica* 212(1):97-110.

FAO: FAOSTAT (2010) http//:faostat.fao.org

Ferguson ME, Hearne SJ, Close TJ, Wanamaker S, Moskal WA, Town CD and De Villiers EP (2012a) Identification, validation and high-throughput genotyping of transcribed gene SNPs in cassava. *Theoretical and Applied Genetics* 124(4):685-695.

Ferguson M, Rabbi I, Kim DJ, Gedil M, Lopez-Lavalle LAB and Okogbenin E (2012b) Molecular markers and their application to cassava breeding: past, present and future. *Tropical Plant Biology* 5(1):95-109.

Fregene MA, Suarez M, Mkumbira J, Kulembeka H, Ndedya E, Kulaya A and Kresovich S (2003) Simple sequence repeat marker diversity in cassava landraces: genetic diversity and differentiation in an asexually propagated crop. *Theoretical and Applied Genetics* 107(6):1083-1093.

Fregene M, Bernal A, Duque M, Dixon A and Tohme J (2000) AFLP analysis of African cassava (*Manihot esculenta* Crantz) germplasm resistant to the cassava mosaic disease (CMD) *Theoretical and Applied Genetics* 100(5):678-685.

Fregene M, Morante N, Sánchez T, Marin J, Ospina C, Barrera E and Ceballos H (2006) Molecular markers for Introgression of Useful Traits from Wild Manihot Relatives of Cassava. Marker-Assisted Selection (MAS) of Disease and Root Quality *Traits Root Crop Journal* 32:1-31.

Fregene M, Okogbenin E, Mba C, Angel F, Suarez MC, Chavarriaga P and Tohme J (2001) Genome mapping in cassava improvement: Challenges, achievements and opportunities. *Euphytica* 120(1):159-165.

Frisch M, Bohn M and Melchinger AE (1999) Comparison of selection strategies for marker assisted backcrossing of a gene. *Crop Science* 39(5):1295-1301.

Gallo M and Sayre R (2009) Removing allergens and reducing toxins from food crops. *Current opinion in biotechnology* 20(2):191-196.

Heffner EL, Sorrells ME and Jannink JL (2009) Genomic selection for crop improvement. *Crop Science* 49(1):1-12.

Hillocks RJ, Thresh JM and Bellotti A (Eds.) (2002) Cassava: biology, production and utilization. Cabi.

Hoang S, Berglund SP, Hahn NT, Bard AJ and Mullins CB (2012) Enhancing visible light photo-oxidation of water with TiO2 nanowire arrays via cotreatment with H2 and NH3: synergistic effects between Ti3+ and N. *Journal of the American Chemical Society* 134(8):3659-3662.

Hospital F, Chevalet C and Mulsant P (1992) Using markers in gene introgression breeding programs. *Genetics* 132(4):1199-1210.

Hou H, Atlihan N and Lu ZX (2014) New biotechnology enhances the application of cisgenesis in plant breeding. *Frontiers in plant science* 5:389.

Hurtado P, Ospina C, Marin J, Buitrago C, Castelblanco W, Correa A, Alfonso P, Barrera E, Gutierrez J, Fregene M, Hearne S, Ferguson M, Alves A, Fortes-Ferreira C, De Vicente C (2008)Assessment of the diversity in global cassava genetic resources based on simple sequence repeat (SSR) markers. In First Scientific Meeting of the Global Cassava Partnership GCP-1.

Iglesias C, Hershey C, Calle F and Bolaños A (1994) Propagating cassava (*Manihot esculenta*) by sexual seed. *Experimental agriculture* 30(3):283-290.

Ihedioha JI and Chineme CN (2003) Clinicopathologic implications of shortened fermentation periods in the production of toasted cassava granules (gari) *Plant Foods for Human Nutrition* 58(3):1-11.

Ihemere U, Arias Garzon D, Lawrence S and Sayre R (2006) Genetic modification of cassava for enhanced starch production. *Plant Biotechnology Journal* 4(4):453-465.

Ikeogu UN, Akdemir D, Wolfe MD, Okeke UG, Chinedozi A, Jannink JL and Egesi CN (2019) Genetic correlation, genome-wide association and genomic prediction of portable NIRS predicted carotenoids in cassava roots. *Frontiers in Plant Science* 10:1570.

International Cassava Genetic Map Consortium (ICGMC) (2015) High-resolution linkage map and chromosome-scale genome assembly for cassava (*Manihot esculenta*Crantz) from 10 populations. G3: Genes, Genomes, *Genetics* 5(1):133-144.

Isendahl C (2011) The domestication and early spread of manioc (*Manihot esculenta*Crantz): a brief synthesis. Latin American Antiquity 22(4):452-468.

Jelili TO (2018) Growth analysis of genetically modified cassava (*Manihot esculenta*Crantz) plants expressing NPT II marker and GUS reporter genes. *Agri Res Tech: Open Access J.*

Jennings DL and Iglesias C (2001) Breeding for crop improvement. In Cassava: biology, production and utilization (pp. 149-166) Wallingford UK: CABI.

Jia Y, Wang Z and Singh P (2002) Development of dominant rice blast Pi ta resistance gene markers. *Crop Science* 42(6):2145-2149.

Jorge V, Fregene MA, Duque MC, Bonierbale MW, Tohme J and Verdier V (2000) Genetic mapping of resistance to bacterial blight disease in cassava (*Manihot esculenta* Crantz) *Theoretical and Applied Genetics* 101(5):865-872.

Kawuki RS, Ferguson M, Labuschagne M, Herselman L and Kim DJ (2009) Identification, characterisation and application of single nucleotide polymorphisms for diversity assessment in cassava (*Manihot esculenta* Crantz) *Molecular Breeding* 23(4):669-684.

Kawuki RS, Herselman L, Labuschagne MT, Nzuki I, Ralimanana I, Bidiaka M and Ferguson ME (2013) Genetic diversity of cassava (*Manihot esculenta* Crantz) landraces and cultivars from southern, eastern and central Africa. *Plant Genetic Resources* 11(2):170-181.

Kayondo SI, Pino Del Carpio D, Lozano R, Ozimati A, Wolfe M, Baguma Y and Jannink JL (2018) Genome-wide association mapping and genomic prediction for CBSD resistance in *Manihot esculenta. Scientific Reports* 8(1):1-11.

Kenyon L, Anandajayasekeram P, Ochieng C, Ave C and UK KME (2006) A synthesis/lesson-learning study of the research carried out on root and tuber crops commissioned through the DFID RNRRS research programmes between 1995 and 2005. DFID, United Kingdom.

Komori T, Yamamoto T, Takemori N, Kashihara M, Matsushima H and Nitta N (2003) Fine genetic mapping of the nuclear gene, Rf-1, that restores the BT-type cytoplasmic male sterility in rice (*Oryza sativa* L.) by PCR-based markers. *Euphytica* 129(2):241-247.

Kuon JE, Qi W, Schläpfer P, Hirsch-Hoffmann M, von Bieberstein PR, Patrignani A and Gruissem W (2019) Haplotype-resolved genomes of geminivirus-resistant and geminivirus-susceptible African cassava cultivars. *BMC biology* 17(1):1-15.

Lacomme C (2015) Strategies for altering plant traits using virus-induced gene silencing technologies. *Plant Gene Silencing* 25-41.

Lancaster PA and Brooks JE (1983) Cassava leaves as human food. *Economic Botany* 37(3):331-348.

Lentz EM, Kuon JE, Alder A, Mangel N, Zainuddin IM, McCallum EJ and Vanderschuren H (2018) Cassava geminivirusagroclones for virus-induced gene silencing in cassava leaves and roots. *Plant Methods* 14(1):1-9.

Li S, Yu X, Cheng Z, Yu X, Ruan M, Li W, and Peng M (2017) Global gene expression analysis reveals crosstalk between response mechanisms to cold and drought stresses in cassava seedlings. *Frontiers in plant science* 8:1259.

Li YZ, Pan YH, Sun CB, Dong HT, Luo XL, Wang ZQ and Chen B (2010) An ordered EST catalogue and gene expression profiles of cassava (*Manihot esculenta*) at key growth stages. *Plant Molecular Biology* 74(6):573-590.

Liu J, Fritz S, Van Wesenbeeck CFA, Fuchs M, You L, Obersteiner M and Yang H (2008) A spatially explicit assessment of current and future hotspots of hunger in Sub-Saharan Africa in the context of global change. *Global and Planetary Change* 64(3-4):222-235.

Madamombe I (2006) Is cassava Africa's new staple food?. Africa Renewal 20(2):13-13.

Mba REC, Stephenson P, Edwards K, Melzer S, Nkumbira J, Gullberg U and Fregene M(2001) Simple sequence repeat (SSR) markers survey of the cassava (*Manihot esculenta*Crantz) genome: towards an SSR-based molecular genetic map of cassava. *Theoretical and Applied Genetics* 102(1):21-31.

McMahon JM, White WL and Sayre RT (1995) Cyanogenesis in cassava (*Manihot esculenta* Crantz) *Journal of Experimental botany* 46(7):731-741.

Meireles da Silva R, Bandel G and Sodero Martins P (2003) Mating system in an experimental garden composed of cassava (*Manihot esculenta*Crantz) ethnovarieties. *Euphytica* 134(2):127-135.

Michelmore RW, Paran I and Kesseli R (1991) Identification of markers linked to disease-resistance genes by bulked segregant analysis: a rapid method to detect markers in specific genomic regions by using segregating populations. *Proceedings of the national academy of sciences* 88(21):9828-9832.

Mlingi N, Poulter NH and Rosling H (1992) An outbreak of acute intoxications from consumption of insufficiently processed cassava in Tanzania. *Nutrition Research* 12(6):677-687.

Murai H, Hashimoto Z, Sharma PN, Shimizu T, Murata K, Takumi S and Nakamura C (2001) Construction of a high-resolution linkage map of a rice brown planthopper (*Nilaparvatalugens* Stål) resistance gene *bph2*. *Theoretical and Applied Genetics* 103(4):526-532.

Nassar NMA and Ortiz R (2007) Cassava improvement: challenges and impacts. The Journal of Agricultural Science 145(2):163-171.

Nzuki I, Katari MS, Bredeson JV, Masumba E, Kapinga F, Salum K and Ferguson ME (2017) QTL mapping for pest and disease resistance in cassava and coincidence of some QTL with introgression regions derived from *Manihot glaziovii*. *Frontiers in plant science* 8:1168.

Odipio J, Alicai T, Ingelbrecht I, Nusinow DA, Bart R and Taylor NJ (2017) Efficient CRISPR/Cas9 genome editing of phytoene desaturase in cassava. *Frontiers in plant science* 8:1780.

Ogbonna AC, Ramu P, Esuma W, Nandudu L, Morales N, Powell A and Mueller LA (2021) A population based expression atlas provides insights into disease resistance and other physiological traits in cassava (*Manihot esculenta* Crantz) *Scientific reports* 11(1):1-16.

Okogbenin E, Egesi CN, Olasanmi B, Ogundapo O, Kahya S, Hurtado P and Fregene M (2012) Molecular marker analysis and validation of resistance to cassava mosaic disease in elite cassava genotypes in Nigeria. *Crop Science* 52(6):2576-2586.

Okogbenin E, Egesi C and Fregene M (2006) Marker Aided Introgression of CMD Resistance in Latin American Germplasm for the broadening of the Genetic base of cassava in Nigeria. NRCRI, *Umudike Annual Report*, 2-6.

Olomo V and Ajibola O (2003) Processing factors affecting the yield and physicochemical properties of starch from cassava chips and flour. *Starch Stärke* 55(10)4:76-481.

Osuntokun BO (1981) Neurological syndromes, management and prognosis in sickle-cell anaemia. *Tropical Doctor* 11(1):2-7.

Pandey S and Gardner CO (1992) Recurrent selection for population, variety, and hybrid improvement in tropical maize. *Advances in Agronomy* 48:1-87.

Parkes E, Fregene M, Dixon A, Okogbenin E, Boakye-Peprah B and Labuschagne M (2015) Developing Cassava Mosaic Disease resistant cassava varieties in Ghana using a marker assisted selection approach. *Euphytica* (3):549-556.

Picanço M (2018) Na roça, na mesa, navida: umaviagempelastrajetórias da mandioca, no e além do nordesteparaense. EditoraPaka-Tatu.

Prochnik S, Marri PR, Desany B, Rabinowicz PD, Kodira C, Mohiuddin M and Rounsley S (2012) The cassava genome: current progress, future directions. *Tropical plant biology* 5(1):88-94.

Pujol B, Gigot G, Laurent G, Pinheiro-Kluppel M, Elias M, Hossaert-McKey M and McKey D (2002) Germination ecology of cassava (*Manihot esculenta* Crantz, Euphorbiaceae) in traditional agroecosystems: seed and seedling biology of a vegetatively propagated domesticated plant. *Economic Botany* 56(4):366-379.

Rabbi IY, Kayondo SI, Bauchet G, Yusuf M, Aghogho CI, Ogunpaimo K and Kulakow P (2020) Genome-wide association analysis reveals new insights into the genetic architecture of defensive, agro-morphological and quality-related traits in cassava. *Plant Molecular Biology* 1-19.

Rabbi IY, Kulembeka HP, Masumba E, Marri PR and Ferguson M (2012) An EST-derived SNP and SSR genetic linkage map of cassava (*Manihot esculenta* Crantz) *Theoretical and Applied Genetics* 125(2):329-342.

Rabbi IY, Udoh LI, Wolfe M, Parkes EY, Gedil MA, Dixon A and Kulakow P (2017) Genome wide association mapping of correlated traits in cassava: dry matter and total carotenoid content. *The plant genome* 10(3), 09.

Rabbi I, Hamblin M, Gedil M, Kulakow P, Ferguson M, Ikpan AS and Jannink JL (2014) Genetic mapping using genotyping by sequencing in the clonally propagated cassava. *Crop Science* 54(4):1384-1396.

Radhika NK, Sheela MN, Devi AA, Sreekumar J, Kumar TM and Chakrabarti SK (2014) Genetic modification for designer starch from cassava. *Journal of Tropical Agriculture* 52(1):1-6.

Rafalski A (2002) Applications of single nucleotide polymorphisms in crop genetics. *Current opinion in plant biology* 5(2):94-100.

Raji AA, Anderson JV, Kolade OA, Ugwu CD, Dixon AG andIngelbrecht IL (2009) Gene-based microsatellites for cassava (*Manihot esculenta* Crantz): prevalence, polymorphisms, and cross-taxa utility. *BMC Plant Biology* 9(1):1-11.

Ramu P, Esuma W, Kawuki R, Rabbi IY, Egesi C, Bredeson JV and Lu F (2017) Cassava haplotype map highlights fixation of deleterious mutations during clonal propagation. *Nature Genetics* 49(6):959-963.

Sayre R, Beeching JR, Cahoon EB, Egesi C, Fauquet C, Fellman J and Zhang P (2011) The Bio Cassava plus program: biofortification of cassava for sub-Saharan Africa. *Annual review of plant biology* 62(1):251-272.

Schlötterer C (2004) The evolution of molecular markers—just a matter of fashion?. *Nature reviews genetics* 5(1):63-69.

Scott GJ, Rosegrant MW and Ringler C (2000) Global projections for root and tuber crops to the year 2020. *Food polic*y 25(5):561-597.

Shan Z, Luo X, Wu M, Wei L, Fan Z and Zhu Y (2020) Genome-wide identification and expression of GRAS gene family members in cassava. *BMC plant biology* 20(1):1-16.

Shizuya H, Birren B, Kim UJ, Mancino V, Slepak T, Tachiiri Y and Simon M (1992) Cloning and stable maintenance of 300-kilobase-pair fragments of human DNA in *Escherichia coli* using an F-factor-based vector. *Proceedings of the National Academy of Sciences* 89(18):8794-8797.

Silva LI, Vidigal Filho PS, Costa TR, Moiana LD and Gonçalves-Vidigal MC (2015) Molecular characterization of traditional sweet cassava accessions from the periurban region Toledo, Western Paraná., *Southern Brazil. J Glob Biosci* 4:1268-1278.

Song WY, Wang GL, Chen LL, Kim HS, Pi LY, Holsten T and Ronald P (1995) A receptor kinase-like protein encoded by the rice disease resistance gene, *Xa21*. *Science* 270(5243):1804-1806.

Sraphet S, Boonchanawiwat A, Thanyasiriwat T, Boonseng O, Tabata S, Sasamoto S and Triwitayakorn K (2011) SSR and EST-SSR-based genetic linkage map of cassava (*Manihot esculenta* Crantz) *Theoretical and Applied Genetics* 122(6):1161-1170.

Tangphatsornruang S, Sraphet S, Singh R, Okogbenin E, Fregene M and Triwitayakorn K (2008) Development of polymorphic markers from expressed sequence tags of *Manihot esculenta* Crantz. *Molecular Ecology Resources* 8(3):682-685.

Taylor N, Chavarriaga P, Raemakers K, Siritunga D and Zhang P (2004) Development and application of transgenic technologies in cassava. *Plant molecular biology* 56(4):671-688.

Tewe OO and Lutaladio N (2004) Cassava for livestock feed in sub-Saharan Africa.

Tomkins J, Fregene M, Main D, Kim H, Wing R and Tohme J (2004) Bacterial artificial chromosome (BAC) library resource for positional cloning of pest and disease resistance genes in cassava (*Manihot esculenta* Crantz) *Plant molecular biology* 56(4):555-561.

Tonukari NJ (2004) Cassava and the future of starch. Electronic journal of biotechnology 7(1):5-8.

Turyagyenda LF, Kizito EB, Ferguson ME, Baguma Y, Harvey JW, Gibson P and Osiru DSO (2012) Genetic diversity among farmer-preferred cassava landraces in Uganda. *African Crop Science Journal* 20(1).

Uchendu K, Njoku DN, Paterne A, Rabbi IY, Dzidzienyo D, Tongoona P and Egesi C (2021) Genome-wide association study of root mealiness and other texture-associated traits in cassava. *Frontiers in plant science* 12.

Wang W, Hostettler CE, Damberger FF, Kossmann J, Lloyd JR and Zeeman SC (2018) Modification of cassava root starch phosphorylation enhances starch functional properties. *Frontiers in plant science* 9:1562.

Welsch R, Arango J, Bär C, Salazar B, Al-Babili S, Beltrán J and Beyer P (2010) Provitamin A accumulation in cassava (*Manihot esculenta*) roots driven by a single nucleotide polymorphism in a phytoene synthase gene. *The Plant Cell* 22(10):3348-3356.

Westby A (2001) Cassava utilization, storage and small-scale processing. In Cassava: *Biology, production and utilization* (pp. 281-300) Wallingford UK: Cabi.

Wilson MC, Mutka AM, Hummel AW, Berry J, Chauhan RD, Vijayaraghavan A and Bart RS (2017) Gene expression atlas for the food security crop cassava. *New Phytologist* 213(4):1632-1641.

Wolfe MD, Bauchet GJ, Chan AW, Lozano R, Ramu P, Egesi C and Jannink JL (2019) Historical introgressions from a wild relative of modern cassava improved important traits and may be under balancing selection. *Genetics* 213(4):1237-1253.

Wu C, Dai J, Chen Z, Tie W, Yan Y, Yang H and Hu W (2021) Comprehensive analysis and expression profiles of cassava UDP-glycosyltransferases (UGT) family reveal their involvement in development and stress responses in cassava. *Genomics* 113(5):3415-3429.

Xia L, Peng K, Yang S, Wenzl P, Carmen de Vicente M, Fregene M and Kilian A (2005) DArT for high-throughput genotyping of cassava (*Manihot esculenta*) and its wild relatives. *Theoretical and Applied Genetics* 110(6):1092-1098.

Yonis BO, Pino del Carpio D, Wolfe M, Jannink JL, Kulakow P and Rabbi I (2020) Improving root characterisation for genomic prediction in cassava. *Scientific reports* 10(1):1-12.

Zhao SS, Dufour D, Sánchez T, Ceballos H and Zhang P (2011) Development of waxy cassava with different biological and physico chemical characteristics of starches for industrial applications. *Biotechnology and bioengineering* 108(8):1925-1935.

Sustainable Agriculture and Biodiversity: Innovations and Insights, Pages: 85–129
Edited by: Bibhuti Bhusan Sahoo, Harvinder Kumar Singh, Niranjan Chinara and Amitabh Panda

CHAPTER - 4

ADVANCED BREEDING APPROACHES IN VEGETABLE CROPS

Binod Kumar Jena, Amitabh Panda*, Tulasi Majhi and Swaraj Meher

Krishi Vigyan Kendra (OUAT), Bhawanipatna, Kalahandi-766001, Odisha
**Corresponding email: amitabhpandap70@gmail.com*

INTRODUCTION

Vegetable Worldwide Overview

Vegetables are an essential part of the human diet, encompassing various crop species whose stems, leaves, fruits, flowers, roots, or seeds provide vital nutrients. They can be eaten raw or cooked and are typically low in carbohydrates and fats while rich in dietary fiber, vitamins, minerals, and phytochemicals (Singh and Lebeda, 2007; Giovannoni *et al*., 2017; Dias and Ryder, 2011; Dias, 2012a; Dias, 2012b). Due to their rich nutrient profile, vegetables are considered protective foods, contributing to the prevention of diseases such as cardiovascular problems, certain cancers, diabetes, and obesity (Dias, 2012b). The definition of "vegetable" varies across culinary and botanical classifications. Asia stands out as the largest producer of a variety of vegetables, including tomatoes, brinjal, cucumbers, lettuce, peppers (chilli) and spinach (FAOSTAT, 2020). In terms of per capita production, Asia leads in chilli peppers, spinach, brinjal, and cucumbers. A global survey identified 402 cultivated vegetable crops from 69 families and 230 genera, with leafy vegetables making up the majority (53%), followed by vegetable fruits (15%) and root vegetables (17%). Many vegetables have multiple edible parts, and most are marketed fresh due to their perishable nature. India, which ranks second globally in vegetable production after China, has significantly increased its production since the 1970s due to dedicated development efforts (Aditika *et al*., 2017). In India, more than 40% of the population is vegetarian (National Family Health Survey - 2015-16), and thus vegetables are regarded as a staple. Asia, producing and consuming over 70% of the world's vegetables, has seen a considerable rise in per capita consumption over the past two decades. This increase is attributed to higher incomes and greater awareness of nutritional benefits. Vegetable farming is viewed as a reliable income source for farmers, as these short-duration crops allow for multiple harvests within a growing season. Given urbanization and shrinking agricultural land, vegetable cultivation is becoming increasingly viable economically. The rising global awareness of the health benefits of vegetables is reflected in their increasing consumption. As production practices evolve, and with the recognition of vegetables' economic viability, they remain a cornerstone of both diets and agricultural systems worldwide.

Reasons for Vegetable Improvements

The rapidly growing global population, coupled with food insecurity and malnutrition, particularly concerning essential micronutrients and vitamins, poses significant challenges for many developing nations. Micronutrient malnutrition, often termed "hidden hunger," affects individuals who may consume sufficient calories but lack nutritional quality. Deficiencies in iron, zinc, iodine, selenium, and vitamin A are prevalent among malnourished populations. Malnutrition during pregnancy and childhood can lead to severe health consequences, including increased morbidity and mortality, as well as physical and mental impairments. India, in particular, has one of the highest rates of childhood stunting and wasting, with chronic energy-protein malnutrition affecting approximately one-third of all children globally (FAO, 2013; International Institute for Population Sciences, 2016).

Recent data from the National Health and Family Survey (2015-16) indicates that around 27% of women and 23% of men in rural India are malnourished (Kalia *et al.*, 2016; Selvakumar and Kalia, 2018; Singh *et al.*, 2020). Additionally, the Comprehensive National Nutrition Survey (2016-18) reported that 34.7% of children under five are stunted, while 33.4% are underweight (Kalia *et al.*, 2016; Selvakumar and Kalia, 2018; Singh *et al.*, 2020). The economic burden of hidden hunger in adults leads to significant losses, estimated at around 2.4% of India's GDP, equivalent to $15–46 billion. Globally, 2 billion people are malnourished, with about 795 million facing undernutrition. Among children under five, 155 million are stunted, 52 million are wasted, and 17 million are severely wasted. Malnutrition contributes to a loss of 11% of GDP in Asia and Africa, according to the International Food Policy Research Institute (IFPRI, 2013).

As protective foods, vegetables are rich in carbohydrates, fats, proteins, vitamins, amino acids, and minerals, making them the most accessible source of essential nutrients. While carbohydrates, fats, and proteins are necessary for energy, vitamins and minerals, though required in smaller quantities, are vital for physiological processes and metabolic activities. The WHO-FAO highlights the importance of vegetables for a balanced diet. Vitamin A deficiency (VAD) is a critical public health concern in India, with about 25% of the world's blind population residing there (UN-SCN, 2004).

Key Needs for Vegetable Improvement

1. **Increasing World Population**: As the global population continues to rise, so does similarly the demand for food, particularly among vegetarians. Improving vegetable yield is crucial to ensure a sufficient and nutritious food supply for this expanding global population.
2. **Balanced Diet**: Vegetables are essential for a healthy, balanced diet, providing a rich source of vitamins and minerals. Enhancing their nutritional value is vital for addressing dietary deficiencies and promoting overall health.
3. **Climate Change Resilience**: Climate change affects weather patterns and increases pest and disease prevalence, presenting challenges for agriculture. Developing vegetable cultivars that can withstand these changes is key to maintaining stable yields and ensuring food security.
4. **Post-Harvest Attributes**: Enhancing post-harvest qualities such as shelf life and transportability can significantly reduce food waste. Improving these attributes increases the economic sustainability of vegetable production.
5. **Consumer Preferences**: Consumers taste and market demands are ever-evolving. Vegetable improvement aims to develop cultivars with enhanced flavor, color, texture, and nutritional profiles to meet these changing preferences.
6. **Economic Sustainability**: Improved vegetable varieties can yield higher outputs, lower production costs, and increase economic returns for farmers, contributing to the overall sustainability of agriculture or agrarian families.
7. **Resistance to Biotic and Abiotic Challenges**: Developing varieties that are resilient to both biotic (pests and diseases) and abiotic (drought, salinity, extreme temperatures) stressors contribute to a more robust agricultural system.
8. **Genetic Diversity Conservation**: Maintaining genetic diversity is crucial for vegetable breeding programs. A diverse genetic pool is essential for ensuring long-term agricultural sustainability and adapting to unforeseen challenges.

Historically, vegetable production has faced many biotic stresses from pathogens, pests, and weeds, leading to high pesticide usage since the 1950s.

Despite the effectiveness of these synthetic pesticides in reducing crop losses, the challenges posed by pests, pathogens, and weeds persist, resulting in substantial pre-harvest losses. Moreover, the use of pesticides raises concerns about health risks for growers and consumers, as well as threats to environmental sustainability. As consumers increasingly demand high-quality, safe food, the implications of climate and environmental changes on vegetable production become more pressing (Karkute *et al.*, 2017).

Through crossbreeding and modern biotechnological techniques, breeders can enhance vegetable varieties by introducing specific genes to improve yield, quality, and stress tolerance (Meyer and Puruggana, 2013). The goal of genetic enhancement is to produce crops that meet market demands while maintaining agronomic performance. However, the current reliance on a narrow genetic base can make crops more vulnerable to various stresses (Prakash *et al.*, 2017). Traditional breeding methods can be slow, more resource requiring and unpredictable and use of biotechnological approaches allow for more targeted improvements in short time span and less land and labour resources without unintended consequences (Wamiq *et al.*, 2023).

CONVENTIONAL BREEDING IN VEGETABLES

The earliest form of breeding practiced by humanity was the artificial domestication of crop species from wild ancestors. Through domestication, humans inadvertently manipulated the genomes of these crops to develop desired traits, resulting in modern varieties that differ significantly from their wild relatives. For example, teosinte, a small wild ancestor of maize, has given rose to the much larger modern maize through selective breeding. This process, known as selection, has been a fundamental method of crop development, evolving from ancient practices to contemporary techniques involving advanced phenotyping and genotyping. Conventional breeding methods for vegetable improvement typically include controlled pollination, selection, and the development of new varieties over multiple generations. These methods, which have been used for centuries, continue to be combined with modern approaches. Utilizing crop wild relatives (CWR) in pre-breeding efforts poses challenges, as these relatives often have traits like disease and pest resistance, drought and heat tolerance and untapped novel traits that are valuable but not easily integrated into cultivated varieties. Pre-breeding aims to extract these desirable traits from CWR and introduce them into breeding lines that can cross with elite varieties. For example, two quantitative trait loci (QTLs) associated with glucoraphanin content have been identified in *Brassica villosa*, which are being incorporated into cultivated *Brassica* crops. Similarly, wild tomato germplasm such as *S. pimpinellifolium* is a rich source of β-carotene, while other species provide novel traits like purple pigmentation. When genetic diversity is lacking within a crop population, it can be artificially generated through hybridization, where two distant genotypes are crossed to produce diverse progenies. Thus, hybridization creates a broad array of genetic variation, which can be refined through selection to achieve homozygosity and develop improved varieties. Hybridization not only occurs artificially but also naturally, contributing to genetic variation through spontaneous mutations, which can lead to heritable changes.

In crop breeding, the progeny from hybridization are referred to as hybrids, and the increased vigor or superior performance of progeny surpassing the parents for a target trait in the desired direction resulting from these crosses is known as heterosis. Heterosis is a sought-after outcome in breeding, as it confers advantages over the parent plants, enhancing both technological and economic aspects of vegetable production. Leveraging hybrid vigor is crucial for addressing challenges such as climate change and improving food and nutrition access. Hybrids can significantly boost quality and productivity, positively impacting livelihoods by increasing production and enhancing the nutritional value of crops (Singh *et al.*, 2023). The advancements in scientific understanding and hybrid seed production have transformed the agriculture sector. The term "heterosis" was first coined by American botanist George Harrison Shull in 1914 to describe the superior performance of F1 hybrids compared to their inbred parents. The use of hybrid vigor in cucumber was proposed in the early 20th century, and the first commercial F1 hybrid brinjal was developed in Japan in 1925. While initial hybrid production was limited by high costs, the growing body of research demonstrating significant heterosis in various vegetable crops has motivated breeders to explore new hybrids and more efficient seed production methods.

A variety of breeding strategies can enhance vegetable yield, appearance, quality, and nutraceutical content, including recurrent selection, backcrossing, pedigree methods, F1 hybridization, polyploidy, and mutation breeding. For instance, F1 hybrids of muskmelon have shown increased β-carotene levels (Moon *et al.*, 2002). In India, efforts to produce biotic stress-resistant varieties in vegetables have been recognized as sustainable agricultural practices. Various hybrids have been created through methods documented by researchers such as Meena and Meena (2014) and Singh and Chaubey (2013). Historically, hybrid seed production relied on hand pollination and emasculation, but these methods yield limited seeds, making them economically unfeasible. Consequently, developing techniques that maximize outcrossing and minimize self-pollination in hybrid seed production is essential for reducing costs and improving the efficiency of crop breeding activity (Kumar and Singh, 2004).

Summary of Crossbreeding Methods

Crossbreeding methods for vegetables, particularly self-fertilizing types like tomato, brinjal, peas, etc include various strategies such as mass selection, pedigree, bulk, single seed descent, doubled haploids (DHs), backcrossing hybridization, and population improvement through recurrent selection. While pedigree remains the dominant method, hybridization and population improvement are also important. Inbred lines, resulting from extensive self-pollination or sib-mating, are nearly homozygous and serve roles in genetic research, allele discovery, and as parents for hybrids and synthetic cultivars. For outcrossing vegetables like cucurbits and onions, managing inbreeding depression and exploiting heterosis are crucial when developing composite, hybrid, and synthetic cultivars. DNA marker-aided breeding aids in both inbred line development and population improvement for these crops. Mutation and genetic recombination facilitate breeding in asexual crops such as potatoes and cassava, while analytical breeding through ploidy manipulation enhances genetic diversity. Selection based on phenotypes in genetically diverse populations is a key aspect of vegetable breeding, allowing adaptation to local environments over growing seasons. Quantitative genetics helps understand the interplay of genetic and environmental factors affecting traits, thereby directing different feasible and potent breeding strategies. Trait heritability, gene interactions, and genotype × environment (G x E) interaction relationships guide the choice of breeding methods for a specific vegetable crop. The use of DNA markers and phenotyping data further aids in analyzing the genetic diversity of target traits and predicting offspring breeding values.

The development of hybrid cultivars, particularly F1 hybrids, is a significant achievement in vegetable breeding, harnessing heterosis for increased yields and quality. In F1 hybrids, recessive alleles from one parent are often suppressed by dominant alleles from another, though the molecular basis of hybrid vigor remains unclear. Genetic diversity among breeding lines can define heterotic groups and inform hybrid yield predictions. There are two leading hypotheses regarding heterosis i) Dominance hypothesis in which both parents have some dominant alleles for a trait at different loci and in crossing all the dominant alleles present in F1 hybrid show increased vigour or yield and ii) Overdominance hypothesis in which both the mating parents are said to contain some favourable alleles and deleterious alleles at different loci and after hybridization in F1 the deleterious allele of one parent is suppressed by the favorable allele gained by the offspring from another parent. So both are about the dose effect of a trait for hybrid vigour describing the inheritance of quantitative traits carotene, anthocyanin, or antioxidant contents.

Self-fertilizing vegetables, like tomatoes, have successfully produced F1 hybrids, with their commercial viability dependent on the benefits of heterosis and effective pollination. Strategies like cytoplasmic male sterility, genetic male sterility, and cytoplasmic genetic male sterility methods have been used at different times for hybrid seed production in vegetable crops but cytoplasmic genetic male sterility is being extensively used for commercial hybrid seed production. A novel approach, Hybrid-Enabled Line Profiling (HELP), integrates existing and new breeding methodologies to focus on the most promising crosses. This strategy enhances efficiency and addresses yield stagnation in major self-fertilizing crops, promising significant improvements in vegetable production (Ginkel and Ortiz, 2018).

HETEROSIS BREEDING IN VEGETABLES

Genetic Mechanisms of Heterosis

To produce hybrid seeds, various strategies have been developed, particularly focusing on emasculation and pollination techniques. Among these, leveraging genetic mechanisms such as self-incompatibility (SI), gynoecism, and male sterility has emerged as most promising and cost-effective approaches. These mechanisms are highly reliable by producing pure hybrid seeds and also economically viable, making them increasingly popular among breeders and hybrid seed-producing organizations. By utilizing these genetic mechanisms, breeders can achieve improved yields and desirable traits while ensuring seed purity. This strategy also reduces the reliance on labor-intensive practices like emasculation and manual pollination, streamlining the hybrid seed production process.

Self-Incompatibility (SI)

Self-incompatibility (SI) is a genetically controlled mechanism found in certain plant species that prevents self-pollination. This process involves the rejection of pollen from the same plant by the pistil, which encourages outcrossing and helps maintain genetic diversity. Initially described by Koelreuter in 1764, SI is observed in many vegetable crops, including cabbage, cauliflower, and tomato. Darwin (1977) also contributed to the early discussions on SI, while significant advancements in understanding its genetic basis were made by Dodds *et al.* in 1997.The SI mechanism means that viable pollen grains cannot fertilize ovules that carry the same S-alleles (S-self-incompatible), thus inhibiting self-fertilization. This phenomenon is particularly pronounced in cole crops like *Brassica oleracea*, which display a strong self-incompatibility system.

Genetics of Self-Incompatibility

The discovery of self-incompatibility (SI) has led to identifying numerous S-locus alleles. For instance, *Brassica rapa* has around 30 S-locus alleles, while *Brassica oleracea* has more than 50. In the broader Brassicaceae family, there are 30 to 40 S-alleles documented, and over 100 S-haplotypes have been recorded across *Brassica oleracea* and *Brassica rapa*. Classical genetic analysis has categorized Brassica S-alleles into two groups based on their phenotypic effects: high-activity alleles (producing 0 to 10 pollen tubes per self-fertilized stigma) and low-activity alleles (producing 10 to 30 pollen tubes). Key components involved in SI reactions include the S-locus cysteine-rich (SCR) protein and the pollen coat protein S-cystine-rich/'S' Protein 11 and the stigma-specific S-receptor kinase (SRK), which have been reported to regulate the self-incompatibility response in Brassicaceae family (Sakamoto and Nishio, 2001; Kachroo *et al*., 2002; Ivanov and Gaude, 2009). Additionally, mechanisms such as the activity-regulated cytoskeleton-associated protein1 (ARC1) and U-box proteins in the Brassica pistil, as well as other distinct determinative genes, also play roles in SI. Molecular interactions, described as "lock and key" mechanisms, further contribute to this process. Understanding these regulatory systems provides valuable insights for breeders and opens new avenues for molecular research in plant breeding.

Detection and Maintenance of Self-Incompatibility (SI)

Several methods have been developed for detecting self-incompatibility (SI) in plants, utilizing both traditional and modern approaches:

1. **Conventional Breeding Methods**:
    - **Test Crosses**: Involve crossing with known inbred lines to assess SI.
    - **Diallel and Reciprocal Crosses**: Evaluate the compatibility of different genotypes.
2. **Pollen Grain and Stigma Analysis**:
    - **Pollen Grain Staining and Stigma Darkening**: Visual techniques to assess pollen viability and compatibility.
    - **Seed Setting/Fertility Index**: Measures the success rate of fertilization in various crosses.
3. **Molecular Techniques**:
    - **Molecular Markers**: Techniques like PCR-RFLP, PCR-CAPS, and RAPD are used to identify and characterize S-alleles.
    - **SCAR Markers**: Help in improving self-incompatible lines and accelerate marker-assisted selection.

4. **Advanced Techniques**:
   - o **Fluorescence Microscopy**: Increasingly popular for assessing pollen and stigma interactions at a cellular level.
   - o **Isoelectric Focusing**: Used for analyzing the proteins involved in SI.

These methods help in detecting SI and to assist in the maintenance of self-incompatibility by allowing breeders to identify and select for desired traits effectively. The integration of molecular techniques, particularly with the advent of advanced markers, has significantly enhanced the ability to characterize S-alleles and streamline breeding programs focused on SI crops.

Commercial Exploitation of Self-Incompatibility (SI): Pearson (Pearson, 1933) was the first to propose the use of SI for hybrid seed production in *Brassica*. Succeeding studies on SI for use in hybrid seed production in different species of *Brassica* group were cabbage (Odland and Noll, 1950; Adamson, 1965), broccoli (Nieuwhof, 1965), and kale (Thompson, 1964). There was extensive research for evaluation of SI lines and hybrids in cabbage (Gusakova and Gorodilvo, 1996), cole crops (Ruffio-Chable and Gaude, 2000) and cauliflower (Hoser-Krauze,1979; Ram, 1975; Hoser-Krauze 1989; Hoser-Krauze 1982; Gangopadhyay *et al.*, 1997; Sharma *et al.*, 2001; Singh *et al.*, 2002; Sharma *et al.*, 2002; Selvakumar *et al.*, 2011). For hybrid seed production in *Brassica* single, double, three-way, and triple crosses were preferred. It was observed that double, three-way, and triple crosses were more cost-effective than a single-cross hybrid seed production method. The associated problem with hybrid seed production in SI system is lack of uniformity in seed production and labor-intensive pollination and inbreeding. The strategies to address these issues related to hybrid production in SI such as:

- **Use of Isogenic Lines:** Ensures uniform hybrid production (Gerstel, 1950, Hinata, 1975).
- **Controlled Pollination Techniques:** Techniques like double pollination or the application of 3-5% CO_2 (Hinata, 1975) and NaCl (Kucera, 1990; Carafa and Carratu, 1997) have been explored to enhance fertilization success.
- **Asexual Propagation:** Meristem culture can help in producing uniform plants.
- Use of parental lines: Lines with synchronized flowering, similar morphologies, and pollination via stored pollen may be used for better success.
- **Selection of inbreds:** Inbreds selection having high levels of SI and minimum morphological variations is of significant importance. Type-IV S-allelic interaction (co-dominance between the pollen and stigmas) is required for three- and four-way crosses of SI to maintain higher genetic purity in hybrid seeds (Wallace, 1979).

Exploitation of Gynoecism for Hybrid Seed Production: It is the development of femaleness in crop plants means the plant will prefer pistilate flowers than having stamens which will favour cross-pollination and hybrid seed production. This characteristic is common in the Cucurbitaceae family. In cucumber (Kaloo, 1988, Kumar and Singh, 2004; Robinson *et al.*, 1976) had demonstrated the inheritance of gynoecism. The interaction between the mutated gene (recessive for gynoecious expression) and the dominant intensifier for the female sex gene (In-F) was found to control the inheritance of pistillate or femaleness (Robinson *et al.*, 1976). Similarly, in watermelon, the interaction between two genes, a and g, governs stable gynoecism (Robinson and Decker-Walters, 1997). Monogenic dominant gene inheritance of gynoecism is found in cucumber (Pati *et al.*, 2015) and it is under the control of a single recessive gene (gy-1) in bitter gourd (Ram *et al.*, 2006; Behera *et al.*, 2006; Mishra *et al.*, 2015). The following are the important commercial implications of gynoecism, particularly in hybrid breeding:

1. **Hybrid Breeding Techniques**: The production of male flowers in gynoecious cucumber lines has been enhanced through the application of growth regulators. Robinson and Robinson (1999) emphasized this method for commercial-scale hybridization.
2. **Growth Regulators**: Initial studies used silver nitrate ($AgNO_3$) for male flower induction in cucumber, later replaced by silver thiosulfate (Ag_2S2O_3) due to its lower phytotoxicity. Optimal concentrations are recommended—silver nitrate at 50–100 ppm or silver thiosulfate at 25–50 ppm during the 2–3 leaf stage.
3. **Environmental Considerations**: The effectiveness of gynoecious lines is significantly influenced by environmental conditions. Ideal temperatures above 30°C and photoperiods of up to 12 hours have been identified. However, the impact of photoperiod diminishes at higher temperatures, underscoring the need for careful management of growing conditions.
4. **Gene-Environment (G x E) Interactions**: The expression of gynoecism can vary with environmental factors, highlighting the importance of studying gene-environment interactions in the development of gynoecious varieties.

Male Sterility

After understanding hybrid vigour or heterosis in crop plants, most of the breeders and seed-producing agencies are engaged in hybrid seed production in vegetables and field crops due to the high production and quality of hybrids of the concerned crop than open-pollinated varieties. Thus, now most of the commercial varieties of leading vegetables are F1 hybrids. Incorporating male sterility reduces the labour required for hybrid seed production and ensures high varietal purity. Male sterility is the inability of a flower to produce fertile pollen grains which will enable cross-fertilization with fertile pollen grains of other plants and hybrid seed production with high purity. In tomato hybrid seed production is commercially exploited after it described the first mutant since 1915 (Crane, 1915). Male sterility is a crucial trait in hybrid seed production, facilitating cross-pollination and enhancing genetic diversity. By leveraging genetic insights and employing morphological markers, breeders can enhance hybrid production in various vegetable crops, thus contributing to agricultural sustainability and productivity. The genetic bases of hybrid seed production using an array of male sterility systems have been studied and these are summarized as follows:

1. **Genic Male Sterility (GMS)**: This system relies on specific genetic factors present in the nucleus (nuclear fertility restorer *Rf* gene) to produce male sterility. It has been utilized in crops like tomato, chilli, onion, and carrot but is limited in its broader application only.
2. **Cytoplasmic Male Sterility (CMS)**: Particularly significant in the Brassicaceae family, CMS is widely used in cabbage and cauliflower. This system allows for the efficient production of hybrid seeds by eliminating the need for male-parent plants.
3. **Cytoplasmic Genetic Male Sterility (CGMS)**: Employed in crops such as hot pepper, onion, and carrot, CGMS combines both cytoplasmic and genetic factors to control male sterility. This system of hybrid seed production is called as A, B, R system of hybrid seed production wherein A line is the male sterile line, B is the maintainer line and R is the fertility restorer line. Here there are two types of continuous production systems every year. In the first production system, A X B is done where A line is maintained year after year where from plants of A-line, A-line seeds and from B line, B line seeds are produced and this plot is known as maintainer plot. The second production system is A X R and it is a hybrid seed production plot where hybrid seeds are harvested from the A line and restorer seeds are harvested from the R line. The ratio of A:R is around 6:2 to 8:3 depending on the synchronization of flowering of A and R lines and the pollen load of R line.
4. **Temperature-sensitive Genetic Male Sterility (TGMS)** and **Photoperiod-sensitive Genetic Male Sterility (PGMS)**: These systems have been documented in various vegetables but are not yet widely exploited.

The genetic and molecular bases of male sterility have been exploited in the following several vegetable crops:

- **Tomato:** Studies have explored GMS and CMS mechanisms (Jeong *et al.*, 2014; Atanassova and Georgieve, 1986). Different genes controlling male sterility including male-sterile10 (ms-10), ms-15, ms-32, positional sterile-2 (ps-2), exserted stigma (ex) and 7B-1 have been studied, and several ideas and systems for their efficacious application have been developed and tested (Crane, 1915; Atanassova and Georgiev, 2002; Jeong *et al.*, 2014; Zhang *et al.*, 2016; Pucci *et al.*, 2017; Cao *et al.*, 2019; Liu *et al.*, 2019). However, until present, male-sterile lines have not been used on a large scale in tomato hybrid seed production.

- **Brinjal**: Investigations into genetic male sterility have revealed significant insights (Pathak and Jaworski, 1989).
- **Hot and Sweet Peppers**: Extensive research has characterized male sterility systems (Despande *et al.*, 1983; Singh *et al.*, 2014).
- **Okra and Muskmelon**: Studies have focused on the genetics of male sterility in these crops (Pitchaimuthu *et al.*, 2012; Lal *et al.*, 2007).
- **Watermelon** (Zhang *et al.*, 1996; Stern *et al.*, 1986).
- **Radish** (Kim *et al.*, 2007; Ogura, 1968; Tanak *et al.*, 2012; Park *et al.*, 2013).
- **Onion** (Singh and Khar, 2021; Sato, 1998; Khar *et al.*, 2022).
- **Garlic** (Etoh, 1985).
- **Carrot** (Kitagawa *et al.*, 1994; Shemesh Mayer *et al.*, 2013).
- **Cucumber** (Han *et al.*, 2018).
- **Bean** (Macknzie and Chase, 1990; Johns *et al.*, 1992; Janska *et al.*, 1998).
- **Broccoli** (Ruffio-Chable *et al.*, 1987).
- **Cauliflower** (Nieuwhof, 1961; Ruffio-Chable *et al.*, 1993; Chamola *et al.*, 2013).
- **Cabbage** (Swarup and Gill, 1964).

The use of male sterility systems of hybrid seed production faces several challenges, including:

- **Identification of Male Sterile Plants**: The lack of efficient identification systems in GMS restricts its utilization. Morphological markers, such as the potato leaf marker in tomato and anthocyanin-less stems in chilli, have been proposed to aid in recognizing male sterile plants.
- **Environmental Influence**: The performance of male sterile lines can be affected by environmental factors, necessitating careful management during breeding.

Novel Male Sterility–Fertility Restoration System: Singh *et al.* (2015) introduced a novel male sterility and fertility restoration system designed to enhance hybrid seed production, particularly in crops where seeds are of high economic significance. This innovative approach aims to address the limitations of existing systems, such as Barnase–Barstar, chemical methods, and both cytoplasmic (CMS) and genic male sterility (GMS) mechanisms of hybrid seed production. In this novel male sterility and fertility restorer system there has been functional complementation of a TATA-box mutant TGTA promoter and a TATA-binding protein mutant3 (TBPm3) in the sophisticated expression-cassette-based system along with modifications for regulatory control as below:

1. **TATA-box Mutant TGTA Promoter**: This promoter drives the expression of genes essential for the system's function.
2. **TATA-Binding Protein Mutant (TBPm3)**: Modifications to this protein facilitate regulatory control over male fertility.
3. **HFR1NT131 Fragment**: The inclusion of this long hypocotyl fragment with TBPm3 (forming HFR1NT131-TBPm3) enables the expression of COP1 (Constitutive Photomorphogenesis 1) specifically in the male parent.
4. **Interaction Dynamics**: The system operates through the interaction of COP1 with HFR1(hypocotyl in the Far-Red1), leading to the degradation of BECLIN1, a factor crucial for tapetal development. This degradation results in normal tapetal function, restoring fertility in male plants. It is to note that presence of a protein BECLIN1 in male plants degrades the pollen formation and makes the plant completely (100%) sterile.

Advantages

- **Biological Safety**: The system is designed to be biologically safe, reducing environmental concerns associated with chemical methods.
- **Production of Pure Hybrids**: It allows for the production of genetically pure hybrids, which is essential for marketability and crop performance.

Marker-Assisted Selection (MAS) for C-GMS Line Development: At present commercial growers need genetic pure hybrid seeds for high-value production of vegetables. Therefore, it is necessary to have genetic confirmation of male sterility and fertility restoration of parents of hybrid seed production which in turn would enable pure production of hybrids. Now genetic or DNA markers are employed to distinguish the male sterility and fertility restoration factors of various crop plants. There is not much progress regarding MAS for C-GMS line development in vegetable crops and a few are enumerated below:

1. **CMS-Specific Markers**:
   - o Two CMS-specific SCAR (Sequence Characterized Amplified Region) markers have been developed to differentiate between N-cytoplasm and S-cytoplasm.
   - o AFLP (Amplified Fragment Length Polymorphism) markers have also been identified that are linked to the fertility restorer gene Rf, enhancing selection efficiency (Kim *et al.*, 2005).
2. **CAPS and SCAR Markers in Chilli Pepper**:
   - o A CAPS marker (E-AGC/M-GCA122) associated with the Pr locus, which contributes to the partial restoration of fertility in CMS, has been reported in chilli peppers (Lee *et al.*, 2009).
   - o The SCAR marker CRF3S1S demonstrated 100% efficiency in differentiating fertility restorers from maintainers, enabling rapid detection of restorers in germplasm (Kim *et al.*, 2005).
3. **Onion Cytoplasm Preferences**:
   - o In onion, S-type cytoplasm is preferred due to its environmental stability, governed by the Ms gene (Havey and Bark, 1994).
   - o Various PCR-based markers have been employed for identifying the cytoplasm and Ms locus, including RFLP, CAPS, SCAR, and SNPs (Single Nucleotide Polymorphisms). Recent advancements include new primers that facilitate the identification of both cytoplasm and the Ms locus (Kim *et al.*, 2015; Gokce *et al.*, 2002). Despite the development of effective markers, limitations remain in the identification of the Ms locus. Approximately 95% of Indian onion cultivars were found to possess a homozygous recessive Ms locus (Khar *et al.*, 2022).
4. **New PCR Markers**:
   - o New PCR markers specific to the Ms locus, such as AcSKP1 and AcPMS1, have been reported, aiming to improve the accuracy of selecting for fertility restoration in onion (Huo *et al.*, 2015; Kim *et al.*, 2015).

MUTATION BREEDING FOR VEGETABLE IMPROVEMENTS

Resembling to breeding of field crops vegetable breeding is categorized into three sub-types such as recombination breeding, mutation breeding, and transgenic breeding. The fundamental principle is to create new mutated alleles in crop plants. Hugo de Vries coined the term "mutation" in 1900 and referred it to heritable phenotypic alterations. Mutation is a sudden heritable change in DNA sequence that leads to a change in the characteristics of an organism. It is the phenomenon of a sudden intermittent change in the hereditary character. Mutation breeding is the use of induced mutations for crop enhancement. In vegetables, breeding mutagenesis has considerable importance. Mutagenesis is all about treating a biological material here vegetable crop with a mutagen to induce mutations. The entire operation of the mutation induction & isolation of mutants for crop improvement is called 'Mutation breeding'. In addition to changing linkage groups, mutations can cause both qualitative and quantitative variations quite quickly by changing both alleles at recognized locations and newly unidentified loci (Konzak *et al.*, 1977).

In recent years, the availability of genomic sequences from many plant species and the development of a wide array of molecular-genetic technologies have enhanced our ability to detect or engineer such variations at specific genetic loci (reverse genetics), greatly expanding our capacity for both probing gene function and genetic engineering. McCallum and coworkers in 2000 introduced a new reverse genetic strategy that combines the high density of point mutations provided by traditional chemical mutagenesis with rapid mutational screening to discover induced lesions. TILLING (Targeting Induced Local Lesions IN Genomes) combines chemical mutagenesis with a sensitive mutation detection instrument. The general applicability of TILLING makes it appropriate for the genetic modification of vegetable crops. Utilizing viable morphological mutations and induced variability for chlorophyll in the M_2 generation proved to be the most reliable method for utilizing beneficial mutations for effective crop improvement (Kumar *et al.*, 2007). Mutation breeding has been successful in the following crops:

1. **Cauliflower Mutants**: The spontaneous mutation resulting in "orange curd" in cauliflower due to Or mutant was first found in Ontario, Canada by Bradforsh in 1970. The cultivars containing this mutant Or gene were found to be rich in β-carotene (Li *et al.*, 2001). This trait was later introgressed into Indian cauliflower varieties, enhancing their nutritional profile. Kalia *et al.*, ((2016); Selvakumar *et al.*, (2018), and Singh *et al.*, (2020) developed Pusa Kesari Vit.A-1 and promising introgression lines in cauliflower rich in β-carotene content (8-20 ppm) in Indian cauliflower.
2. **Cabbage colouration:** Zhang *et al.* (2015); Cermak *et al.*, (2015); Liu *et al.*, (2017) and Tian *et al.*, (2017) found SCAR markers linked to "Or" gene-inducing β-carotene accumulation in Chinese cabbage. Zou *et al.* (2016) performed fine mapping of the Or gene and identified the BrPro1 molecular marker in the promoter region of Bra031539 (predicted to encode CRTISO, a carotenoid isomerase specifically required for carotenoid biosynthesis) that can be used for early identification of orange head materials.

3. **Tomato Variants**: The discovery of high pigment (hp) variants in tomatoes (*Solanum lycopersicum*), such as hp-1,hp-1w, hp-2, hp-2j, and hp-2 dg., led to increased carotenoid content. These mutations were mapped to the SlDDB1 gene, showcasing the potential for enhancing nutritional quality in tomato fruits (Jones *et al.*, 2012).
4. **Anthocyanin Accumulation**: In cauliflower and other crops, mutations in the purple (Pr) gene resulted in significant anthocyanin accumulation, leading to vivid purple colors and enhanced antioxidant properties.

Several notable quality mutants in vegetables with significant agricultural value have been identified:

- **"Or" Gene in Cauliflower**: This gene is linked to orange curd color and increased β-carotene levels. The introgression of the "Or" gene into Indian cauliflower varieties has resulted in cultivars rich in β-carotene (8-20 ppm) (Selvakumar *et al.*, 2018; Singh *et al.*, 2020).
- **"Pr" Gene for Purple Curd**: The dominant "Pr" gene, which encodes a MYB transcription factor, is responsible for anthocyanin accumulation in curds. This gene has been used to develop purple cauliflower varieties (Chiu *et al.*, 2010).
- **Sweet Potato Mutants**: Orange-fleshed sweet potato mutants have been recognized for their high β-carotene content (30-100 ppm), providing a valuable source of nutrition (van Jaarsveld *et al.*, 2005).

Mutation breeding offers several benefits including i) rapid generation of genetic diversity, ii) potential to create novel phenotypes with desirable traits, and iii) enhanced tracking of agricultural genetic variation and biodiversity. As breeding techniques evolve, integrating mutation breeding with genomic technologies can further enhance the development of multi-colored and nutritionally rich vegetable varieties. Mutation breeding remains a vital strategy in vegetable crop improvement, capable of yielding varieties that meet the nutritional and aesthetic preferences of consumers while maintaining agricultural biodiversity.

MOLECULAR BREEDING OF VEGETABLES

Conventional breeding methods, are undoubtedly the foundation of genetics and breeding knowledge and often involve lengthy and resource-intensive processes. Recent advancements in genetics and genomics have introduced innovative tools and techniques that enhance the efficiency and effectiveness of plant breeding programs.

Molecular breeding refers to the use of molecular biology techniques for the breeding of different crops or organisms. Molecular breeding is a very sophisticated approach to plant breeding that increases the accuracy and effectiveness of breeding programs. This methodology has been widely applied to enhance various traits in vegetables, including yield, quality, disease resistance, and stress tolerance. The foremost requirement of molecular breeding is high-density genome analysis of different accessions of a crop species or organism, high accuracy of genotyping of DNA sequences of the whole genome, preparation of accession-specific genetic maps, thorough analysis of all the genetic maps to prepare a reliable complete genome map, performing of a genetic database of the genome in bioinformatics that could be accessible to researchers and learners. The next step is the identification of profuse stable DNA segments all over the genome that may or may not be linked to an identified phenotype of the species and these are called molecular markers or DNA markers. These markers have a specific position in the genome of the species and are inherited from generation to generation if no mutation occurs for that DNA fragment. Concerning these standard-identified markers the different phenotypic traits of new accessions of the species studied about evolution, genetic diversity, mutation analysis, identification of new genes/QTLs and markers, identification of new varieties, genome-wide association study, etc. Marker-aided selection, according to Foolad and Sharma (2005), is the process of choosing a trait-based more on genetics than phenotype. Using markers could speed up selection and increase the productivity of breeding cycles (Holland, 2014).

The male sterility and fertility restorer genes in chillies were found using the RAPD technique (Kumar *et al.*, 2002). In the Indian long-day onion population, a DNA marker for male sterility and hybrid production has been produced and found for the first time. Using marker-assisted selection, we can use the detected marker for hybrid production and male sterility transfer to other vegetable lines or genotypes in the future (Saini *et al.*, 2015). Through marker-assisted selection, the Ty2 and Ty3 genes were inserted into the plant, producing tomato hybrids that are immune to the tomato leaf curl virus (Prasanna *et al.*, 2015).

It was possible to identify seven SSR DNA segments that are practical in marker-assisted breeding to create cucumber germplasm whose fruits have high β-carotene content. According to Song and colleagues (2010), the carotenoid content in common cucumbers is low, ranging from 22 to 48 ßg/100 g. On the other hand, the Xishuangbanna gourd has a high carotenoid content, with 700 ßg/100 g measured on a flesh weight basis. Therefore, the Xishuangbanna gourd offers a special advantage as a germplasm that can be utilized to increase the nutritional content of cucumbers. The development of superior vegetable varieties with desirable features has advanced substantially.

Key Advances in Genomic Tools

1. **Molecular Markers**: Tools such as microsatellites (SSR) and single nucleotide polymorphisms (SNPs) have become integral to vegetable breeding. These markers allow breeders to select traits based on genetic information rather than solely phenotypic expression, significantly accelerating the breeding process.
2. **Marker-Assisted Selection (MAS)**: Techniques like MAS, marker-assisted backcrossing (MABC), and marker-assisted recurrent selection (MARS) have gained traction in precision breeding. These methods leverage genetic markers to improve selection accuracy and efficiency, particularly in complex traits (Collard and Mackill, 2008).
3. **Genomic Selection (GS)**: GS utilizes genotyping and genomic data to predict an individual's breeding value (GEBV -genomic estimated breeding value). This approach is especially beneficial for traits with low heritability, as it captures small genetic effects across the genome, allowing for more informed selection (Heffner *et al.*, 2009).
4. **High-Throughput Sequencing**: New-generation sequencing technologies are expanding access to genetic markers and enabling detailed genetic analysis of vegetable crops. This facilitates rapid identification of genes and quantitative trait loci (QTLs) associated with desirable traits (Ortiz, 2015).
5. **Genome-Wide Genetic Diversity Studies:** One of the main challenges in agricultural genetics is to access and use the tremendous genetic variation present in germplasm collections and wild relatives. A significant part of this variation remains untapped because of the difficulties in effectively identifying genetic differences in large collections. Some traits, with high heritability and simple characterization, are easy to select. However, desirable allelic variants and genetic combinations for complex traits are difficult to identify. Recent advances in genotyping are enabling genome-wide diversity studies capable of better capturing the spectrum of variability in natural and breeding populations.

BIOINFORMATICS

Next-generation sequencing (NGS) technologies are facilitating sequencing projects, but have brought new challenges, as millions of short DNA reads have to be analyzed and assembled (Metzker, 2010). Also, genetic maps, genotypes, or expression information at a genomic scale have to be processed to obtain the relevant biological information. Therefore, it is necessary to develop new bioinformatics tools (algorithms and software), that allow the analyses of huge amounts of genome-wide data, but it is also necessary to change the approaches used to understand complex biological traits (Pop and Salzberg, 2008). The field of sequence analysis has a long tradition and has enabled the assembly of many genome sequences obtained by Sanger sequencing. Nowadays, the huge amount of sequence reads generated by NGS and the low quality of individual reads requires new software tools and algorithms that allow efficient dealing with these data (Li and Harner, 2010).

Most of the mid-to high-throughput genotyping platforms described above are being used for studies on diversity and population structure in the corresponding crops. By using representative diversity panels, polymorphism rates for individual SNP markers, minor allele frequencies (MAFs), etc. are estimated, facilitating the selection of those SNPs with biological interest and highly polymorphic in the different groups. For example, the Infinium arrays developed in some of these crops are being used to create haplotype maps for vast germplasm collections, such as the 18,000 accessions of the USDA soybean germplasm collection. GWAS work has been successfully applied in tomato (2n = 24), soybean (2n = 40), brinjal (2n = 24), Chilli (2n = 24), Okra (2n = 130), Cucumber (2n = 14), cauliflower (2n = 18), Cabbage (2n = 18) and cowpea (2n = 22), etc.

Innovative Breeding Approaches

1. **Linkage Mapping and Genome-Wide Association Studies (GWAS)**: Techniques like linkage mapping and GWAS are crucial for detecting and mapping genes associated with specific traits. Nested association mapping (NAM) and multi-parent advanced generation inter-cross (MAGIC) populations enhance genetic analysis (Tian *et al.*, 2011; Cavanagh *et al.*, 2008).
2. **Genomic Prediction**: Utilizing pedigree, phenotypic, and SNP data, genomic prediction allows for early screening and selection, reducing reliance on expensive phenotyping. This method enhances the precision of field trials and can adapt breeding strategies based on diverse genetic backgrounds (Abera Desta and Ortiz, 2014).
3. **High-Throughput Phenotyping**: Precision phenotyping technologies are increasingly adopted in breeding programs. By measuring data points across various wavelengths, these tools enable rapid and accurate assessment of phenotypic traits and facilitate efficient selection processes.

Applications in Specific Crops

1. In onions, a DNA marker has been developed to facilitate male sterility and hybrid production and allow the transfer of these traits to other genotypes (Saini *et al.*, 2015).
2. In tomatoes, the insertion of the Ty2 and Ty3 genes through MAS has led to the development of hybrids resistant to the tomato leaf curl virus (Prasanna *et al.*, 2015).
3. Carotenoid Enhancement

Molecular markers have been used to identify cucumber germplasm with high β-carotene content. Research by Ansari (2015) highlighted seven SSR (Simple Sequence Repeat) markers that are effective in breeding programs.

The Xishuangbanna gourd, with a carotenoid content of 700 μg/100 g, serves as a valuable germplasm source for increasing the nutritional quality of cucumbers, which typically contain lower levels of carotenoids (22-48 μg/100 g) (Obel *et al.*, 2022)

Molecular breeding is transforming vegetable improvement by enhancing trait selection and accelerating the breeding process. Technology offers better opportunities to develop more nutritious and resilient vegetable varieties. However, addressing regulatory and public acceptance issues will be essential to fully realize the potential of these advanced breeding methods. The integration of genomics into vegetable breeding is poised to reshape the field significantly. Here are some anticipated trends:

1. **Phenomics and E-Typing**: Shifting from traditional phenotyping to high-throughput phenomics will enhance genetic gains. Utilizing managed environments, such as greenhouses, can accelerate the breeding cycle, allowing for multiple crop generations within a year.
2. **Genetic Diversity Utilization**: Genomic prediction models can be employed to evaluate germplasm from genebanks, helping identify useful genetic variations for introgression into breeding populations.
3. **Precision Breeding**: As genomic tools become more refined, breeding programs will increasingly focus on precision strategies that combine genetic and phenotypic data for optimized trait selection.

Genomics-led breeding represents a transformative approach to vegetable crop improvement, enhancing efficiency and precision. By applying molecular markers, advanced genomic techniques, and high-throughput phenotyping, breeders can develop new varieties that meet the far-reaching general and specific demands of modern agriculture. Successes of some genomic breeding works have been presented below:

Diversity Analysis in Tomato

Diversity analysis in tomatoes harnesses genomics, phenomics, and breeding informatics to precisely screen for target traits, thereby accelerating the identification of desired characteristics and their associated genes. Research integrating both genome and phenome data has significantly advanced the understanding of tomato evolution and domestication, positioning tomato as a key species for studying quantitative trait variation (Ortiz, 2015).

Key Techniques and Findings

1. **DNA Marker-Aided Breeding**: This method has been pivotal in introducing and pyramiding host plant resistance in tomato varieties. By using next-generation sequencing, researchers can identify single nucleotide polymorphisms (SNPs) that facilitate high-throughput genotyping across different tomato species, cultivars, and segregating offspring. This aids in verifying hybrid purity, comparing genetic linkage maps, and elucidating phylogenetic relationships.
2. **Phenotypic Variability and Genetic Diversity**: Understanding the phenotypic variability and genetic diversity in tomatoes is crucial for effectively utilizing genetic resources. This knowledge informs breeding strategies aimed at sustainability and resilience, ensuring that cultivated varieties can withstand environmental challenges.
3. **Pangenomics and Novel Allele Discovery**: Genomic resequencing of various landraces and cultivars, coupled with pangenomic analyses, has led to the discovery of novel alleles. Bioinformatics tools allow researchers to identify genes that may have been lost during domestication. For instance, a comprehensive tomato pan-genome, which includes sequences from 725 phylogenetically and geographically diverse accessions, identified a rare allele in the TomLoxC promoter. This allele was selected during domestication but plays a role in apocarotenoid production, significantly contributing to the desirable flavor profile of tomatoes (Gao *et al.*, 2019).
4. **Exotic Germplasm Utilization**: The exploration of exotic germplasm has become increasingly important for tomato breeding. By identifying and integrating genes from these diverse sources, breeders can develop new lines with improved traits, enhancing flavor, disease resistance, and adaptability.

The integration of genomics and phenomics into tomato breeding offers the following advantages:

- **Enhanced Trait Selection**: Precise screening techniques allow for the efficient identification of genetic variations associated with desirable traits, streamlining the breeding process.

- **Increased Genetic Resilience**: Exploitation of the genetic diversity in landraces and wild relatives, breeders can develop varieties that are more resilient to biotic and abiotic stresses.
- **Flavor and Quality Improvement**: The genetic basis of flavor and other quality traits enables the development of tomato varieties that meet consumer preferences and nutritional needs.

Diversity analysis in tomatoes through advanced genomic and phenomic approaches has opened up new avenues for the improvement of this most important crop. By unraveling genetic diversity and understanding the underlying genetic mechanisms, researchers can develop new strategies to enhance tomato quality, sustainability, and resilience to meet global demands.

Genomic Prediction for Selection in Potato

Potato (*Solanum tuberosum*) serves as a model for genetic enhancement in polysomic polyploid vegetables, particularly tuber crops. Its unique vegetative propagation allows each tuber to be genetically identical to its parent, enabling the fixation of favorable traits in the F_1 hybrid generation. This process, combined with market-driven quality, agronomic performance, and host plant resistance, defines the challenges and opportunities in potato breeding (Eriksson *et al.*, 2016).

Challenges in Potato Breeding

Potato breeding is a lengthy phenotypic process that requires about 10 years to develop a new cultivar, 5 to 10 years to select suitable parents for breeding, and a success rate of 1 in 10,000 seedlings per year for registered cultivars from standard crosses, and 1 in 100,000 when involving wild species. These problems in potato breeding by phenotypic evaluation could successfully be overcome by a current marker-assisted selection of traits or homozygosity of testing progenies or lines efficiently with the use of fewer resources and time.

Marker-Aided Breeding in Potato: Marker-aided breeding has revolutionized potato selection by significantly reducing the time of selection and increasing the selection efficiency of potato tuber. Marker-aided potato breeding has enabled to study:

- **Host plant resistance**: Successfully targeting major genes for resistance to pests like cyst nematodes and viruses such as Potato virus Y.
- **Tuber quality traits**: Enhancing characteristics like size, shape, and nutritional content.

Genomic Tools and Techniques for Potato Breeding

1. **SNP-Based Genetic Maps**: Dense genetic maps generated from single nucleotide polymorphisms (SNPs) help identify quantitative trait loci (QTL) and their associated genetics, providing a clearer understanding of trait inheritance.
2. **Potato Genome Sequence**: The availability of the potato genome sequence offers comprehensive tools for genome-wide assays and gene discovery. This facilitates the development of marker haplotypes that span QTL regions, essential for targeted introgression breeding.
3. **Genomic Estimated Breeding Value (GEBV)**: GEBV represents a holistic approach to breeding, allowing for the estimation of breeding values based on genomic data. This method enhances the selection of elite clones and improves genetic gains over time.

The integration of genomic prediction in potato breeding offers several key benefits:

- **Increased Efficiency**: By using genomic tools, breeders can accelerate the identification of superior traits, thereby shortening the breeding cycle.
- **Enhanced Genetic Gains**: The use of GEBV and marker-assisted selection enables more informed decisions, leading to greater improvements in desirable traits.
- **Better Adaptation to Market Needs**: With precise genomic data, breeders can tailor varieties to meet evolving market demands, improving both quality and resistance to environmental stresses.

Genomic predictions in potatoes present a transformative approach in potato breeding, enhancing the efficiency of selecting desirable traits. By strong adherence to advanced genetic tools, breeders can genetically handle the complexities of polyploid genetics, ensuring the development of high-quality new potato cultivars aligned with consumer preferences. This genomic-led strategy will play a crucial role in meeting the future demands for potato production globally.

Genomic Estimated Breeding Value (GEBV) in Potato Breeding

Genomic Estimated Breeding Value (GEBV) is a powerful tool for enhancing the selection process in potato breeding, particularly for quantitative traits. By integrating GEBV with marker-aided selection, breeders can significantly improve the efficiency of identifying elite germplasm, thereby shortening the breeding cycle and early development of a quality potato variety. The benefits of GEBV are:

1. **Accelerated Cycle Length**: Traditional potato breeding can be time-consuming, but with GEBV, breeders can quickly evaluate the potential of several thousand clones from a few hundred seedlings each year. This allows for more rapid progress, as promising hybrids can be advanced to field trials sooner (Bradshaw, 2017).
2. **Resource Optimization**: Early testing by GEBV will enable substantial genetic gains while conserving time and resources, enabling breeders to focus on the most promising candidates.

Prediction Accuracy Improvements

- **Initial Models**: Early GEBV models for predicting yield in potatoes showed a prediction accuracy ranging from 20% to 40% (Habyarimana *et al.*, 2017; Slater *et al.*, 2016).
- **Enhanced Models**: The introduction of models that account for both additive and dominance effects improved prediction accuracy significantly, reaching 50% to 80% (Endelman *et al.*, 2018; Stich and Van Inghelandt, 2018). This advancement highlights the importance of incorporating various genetic factors to refine predictions. GEBV has harnessed great progress in potato breeding and still some challenges to meet yet:
    1. **Unreliable Predictions Across Populations**: While genomic prediction appears feasible, the reliability of these predictions remains an issue (Sverrisdóttir *et al.*, 2018). The high allelic diversity in potatoes complicates the creation of robust training sets.
    2. **Need for Larger Training Sets**: To enhance prediction reliability, there is a necessity to expand the training sets used for developing GEBV models. A broader genetic representation can help capture the complex genetic architecture of potatoes, ultimately improving the accuracy of breeding value estimates.

GEBV represents a significant advancement in potato breeding, offering a more efficient pathway to identify and select desirable traits. By leveraging improved predictive models and addressing the challenges associated with allelic diversity, breeders can enhance the overall effectiveness of potato breeding programs. This genomic approach holds great promise for developing new potato varieties that meet market demands and withstand environmental challenges.

Molecular Markers, Genomic Selection and Transcriptome Analysis in Vegetables

Molecular markers, genome sequencing, and transcriptome analysis are pivotal tools in modern plant breeding, facilitating the discovery and characterization of traits critical for vegetable improvement. These genomic approaches allow researchers to explore genetic variation, enhance breeding efficiency, and understand complex traits at a molecular level.

Genomics by Sequencing (GBS)

Genomics by sequencing (GBS) is an accessible method for discovering single-nucleotide polymorphisms (SNPs) and mapping genetic traits. It plays a crucial role in:

- **Population Studies and Germplasm Characterization**: GBS aids in understanding genetic diversity within and across populations, essential for effective breeding programs.
- **Trait Mapping**: Identifying quantitative trait loci (QTL) helps locate genetic regions linked to specific traits, enabling more targeted breeding efforts.

Molecular Markers

Molecular markers are specific and DNA fragments (DNA sequences) present at a specific position of the genome show polymorphism and could be identified in the genome by using molecular techniques. There are different types of molecular markers being used in genomics and transcriptome analysis. Molecular markers, such as RAPD (Random Amplified Polymorphic DNA), RFLP (Restriction Fragment Length Polymorphism), SSR (Simple Sequence Repeat), CAPS (Cleaved Amplified Polymorphic Sequence), and SNPs (Single Nucleotide Polymorphisms), are widely utilized for:

- **Tagging Important Traits**: These markers facilitate the selection and cloning of genes associated with desirable traits, including disease resistance and nutritional quality.
- **Linkage Studies**: Markers can be used to study the genetic linkage between traits and specific genes, thereby enhancing breeding strategies.

For example, in Brassica crops, expressed sequence tags (ESTs) enable high-throughput transcriptome analysis, contributing to the identification of markers linked to quality traits.

Transcriptome Analysis

Transcriptome sequencing is a cost-effective alternative for understanding gene expression and its relation to phenotypic traits. Notable achievements in transcriptome include:

- **Enzymatic Browning in Sponge Gourd**: In the cultivar 'Fusi-3', transcriptomic changes associated with browning were analyzed, identifying 11 genes related to enzymatic browning, including those in the PPO, PAL, POD, CAT, and SOD families.
- **Starch and Sugar Synthesis in Pea**: In pea, three Mutant alleles at three loci r (referred to ADP-glucose pyro-phosphorylase), rb (starch branching enzyme) and bsg (phosphor-glucomutase) affect starch and sugar synthesis.
- **Fruit Ripening in Watermelon**: Transcriptomic studies revealed 795 novel genes and numerous differentially expressed genes (DEGs) linked to carotenoid biosynthesis, plant hormone pathways, and sugar metabolism.

Applications in Breeding

Molecular markers have been instrumental in bio-fortifying vegetable crops:

- **Cauliflower Bio-Fortification**: Researchers at IARI have successfully implemented marker-assisted selection (MAS) to incorporate the carotene-enhancing 'Or' gene into Indian cauliflower varieties (Kalia *et al.*, 2016; Selvakumar *et al.*, 2018; Singh *et al.*, 2020).
- **Linking Quality Traits**: SCAR markers associated with the 'Or' gene and ISSR markers linked to glucosinolate content in Brassica have been identified, facilitating the breeding of nutritionally enhanced vegetables.

The integration of molecular markers, genome sequencing, and transcriptome analysis in vegetable breeding enhances the understanding of genetic variation and trait inheritance. These techniques not only streamline the breeding process but also pave the way for developing high-quality, nutritionally rich vegetable varieties. Continued advancements in genomic technologies promise to further transform vegetable breeding, addressing global food security and nutritional needs.

Detection of QTLs in Vegetables

Quantitative Trait Loci (QTL) mapping is a critical tool in understanding the genetic basis of complex traits in vegetable crops. By developing linkage maps, researchers can identify the physical positions of genes within chromosomes and identify both major and minor gene effects on various traits. The mapping of polygenes (QTLs) started with Sax's (1923) findings on the linkage between seed coat color and seed size in common beans (*Phaseolus vulgaris*). This initiated the association of quantitative trait phenotypes with genetic markers, though early efforts faced challenges due to the vast number of polygenes and their scattered genomic locations. Over time, clusters of minor genes associated with quantitative traits were identified, leading to the establishment of QTLs. The following are different QTL mapping techniques:

1. Linkage Mapping: Traditional biparental mapping populations (such as F_2, Backcross, DH, and RIL) offer insights into QTL location but often suffer from low mapping resolution due to limited recombination events.
2. Multi-parent Advanced Generation Intercross (MAGIC): This method provides a more saturated genetic map, enhancing resolution and allowing for better identification of QTLs.
3. Association Mapping: Utilizing linkage disequilibrium in natural populations or germplasm collections, association mapping offers several advantages:
   - Higher Mapping Resolution: More precise localization of QTLs.
   - Less Time-Consuming: Quicker identification of genetic associations.
   - Greater Allelic Diversity: Increased potential for discovering beneficial alleles.

Examples of QTL Detection in Vegetable Crops

- **Tomato:** QTLs for late blight resistance and yield have been identified using various mapping populations. Specific QTLs associated with glandular trichomes were detected on chromosome 1 using SNP markers (Bennewitz *et al.*, 2018).
- **Cucumber:** QTLs for traits such as fruit peduncle length (Song *et al.*, 2016) and resistance to cucumber mosaic virus (Shi *et al.*, 2018) have been mapped to chromosome 6 using SSR markers.

Quality Traits in Tomato

Tomato is highly valued for its nutritional benefits, providing significant amounts of vitamins A and C. Several QTLs linked to important quality traits have been mapped, including:

- **Fruit Size**: QTLs like fw1.1 (17%), fw1.2 (13%), fw2.1 (12%; associated with locule number), fw2.2 (23%), fw3.1 (12%), and fw11.3 (37%; associated with fasciation) contribute to variations in fruit size.
- **Fruit Shape**: Genes such as pr (pyriform), o (ovate), bk (beaked), n (nipple-tip), f (fasciated), and lc (locule number) have been identified and mapped, showing significant effects on fruit morphology.
- **Fruit Color**: The health benefits of lycopene, a major carotenoid responsible for the red color of tomatoes, have led to the identification of genes (hp-1, hp-2, dg, Ogc) linked to high lycopene content.

Table 1: Disease Resistance Genes and QTLs in Tomato

Disease (Pathogen)	Gene/QTL	Chromosome	MAS Assay	Citation
Fusarium Wilt (*Fusarium oxysporum*)	I	11	CAPS	Bohn and Tucker, 1939; Stall and Walter, 1965
	I-2	11	SCAR	Simons *et al.*, 1998; Arens *et al.*, 2010
	I-3	7	CAPS/SCAR	Catanzariti *et al.*, 2015; Zhang and Dilip, 2021
Verticillium Wilt (*Verticillium albo-atrum*)	Ve1	9	ARMS-PCR	Kawchuk *et al.*, 2001; Arens et al., 2010
	Ve2	9	ARMS-PCR	Kawchuk *et al.*, 2001; Arens *et al.*, 2010
Leaf Mold (*Cladosporium fulvum*)	Cf-2	6	SSR	Dixon *et al.*, 1996; Grushetskaya *et al.*, 2007
	Cf-4	1	SNP/InDel	Thomas *et al.*, 1997; Kim *et al.*, 2017a
	Cf-5	6	SSR	Dixon *et al.*, 1998; Grushetskaya *et al.*, 2007
	Cf-9	1	SNP/InDel	Jones *et al.*, 1994; Kim *et al.*, 2017a
Late Blight (*Phytophthora infestans*)	Ph-1	7	Unknown	Clayberg *et al.*, 1965
	Ph-2	10	CAPS	Moreau *et al.*, 1998; Zhi *et al.*, 2021
	Ph-3	9	CAPS	Robbins *et al.*, 2010; Zhang *et al.*, 2014a
	Ph-5	10	Unknown	Merk *et al.*, 2012
Bacterial Speck (*Pseudomonas syringae*)	Pto	5	CAPS/SCAR	Ronald *et al.*, 1992; Yang and Francis, 2005; Zhang and Dilip, 2021
Bacterial Spot (*Xanthomonas* Race T1-T4)	Rx1	1	Unknown	Wang *et al.*, 1994; Yu *et al.*, 1995; Yang and Francis, 2005
	Rx2	1	Unknown	Wang *et al.*, 1994; Yu *et al.*, 1995; Yang and Francis, 2005

Implications for Marker-Assisted Selection (MAS)

The identification of QTLs opens up opportunities for developing linked molecular markers, which can significantly enhance the efficiency of marker-assisted selection. This facilitates the breeding of improved vegetable varieties with desirable traits, ultimately contributing to agricultural sustainability and productivity.

QTL mapping is an invaluable approach in present-day vegetable breeding, allowing for the identification of genetic loci associated with complex traits. The evolution of mapping techniques, from traditional linkage mapping to advanced association mapping, has enhanced our ability to dissect the genetic basis of traits. By leveraging this information, breeders can more effectively utilize marker-assisted selection to develop high-quality vegetable crops.

Marker Assisted Selection (MAS)

Marker Assisted Selection (MAS) is an innovative breeding technique that employs molecular markers to facilitate the selection of target traits in crops. This method offers significant advantages over traditional phenotypic selection, particularly in the context of vegetable breeding.

Advantages of MAS

1. **Early Trait Detection**: MAS allows for the identification of target traits at the seedling stage, enabling the selection of plants for hybridization within the same growing season.
2. **Effectiveness Across Heritability**: This method is effective for both high and low heritable traits, making it versatile for various breeding goals.
3. **Increased Selection Efficiency**: Compared to phenotypic selection, MAS enhances selection efficiency, allowing breeders to make more informed choices.
4. **Accelerated Breeding Cycles**: By streamlining the selection process, MAS reduces the time required to develop new varieties.
5. **Homozygote and Heterozygote Distinction**: Markers such as SSR can distinguish between homozygous and heterozygous alleles, minimizing the need for selfing during the backcross process.

Approaches to MAS

MAS is typically implemented through three main strategies:

1. **Marker-Assisted Backcrossing (MABC)**:
   - o Used to introgress large-effect QTLs and oligogenes.

- o Facilitates recovery of the recurrent parent's genome while minimizing linkage drag by eliminating unnecessary donor parent genomic regions flanking the target gene.

2. **Marker-Assisted Recurrent Selection (MARS)**:
   - o Enhances population improvement by increasing the frequency of beneficial QTLs that contribute significantly to the trait expression.
   - o Focuses on enriching the breeding population with desired genetic variants.
3. **Genomic Selection (GS)**:
   - o Utilizes genomic information to enrich the population with QTLs, regardless of their immediate effect on trait expression.
   - o Increases the representation of minor QTLs, enhancing genetic diversity and potential for trait improvement.

Applications of MAS in Vegetable Breeding

MAS has numerous applications in vegetable breeding programs. For example:

- **Tomato Leaf Curl Virus Resistance**:
  - o Six resistance genes (Ty1, Ty2, Ty3, Ty4, Ty5, Ty6) have been identified.
  - o The Ty3 gene was successfully introgressed into the genetic backgrounds of popular tomato varieties like Pusa Ruby and Pusa Rohini using linked molecular markers (P6-25 and SCAR-1).
- **Watermelon Powdery Mildew Resistance**:
  - o Three microsatellite markers (MCPI_11, BVWS02441, CYSTSIN) linked to a powdery mildew resistance gene were identified by Gama *et al.* (2015).
  - o These markers can facilitate indirect selection for the resistance gene in new breeding populations.

Marker assisted selection (MAS) represents a significant advancement in vegetable breeding, enabling faster and more accurate selection of desirable traits within a limited period using fewer resources. By employing strategies like MABC, MARS, and GS, breeders can enhance the efficiency of their programs and develop improved varieties that meet consumer demands and withstand environmental challenges. The successful application of MAS in crops like tomatoes and watermelons exemplifies its potential to enhance crop resilience and productivity in future prospective.

Table 2: Genes and QTLs for Marker-Assisted Selection in Chilli

Trait	Locus/QTL	Chromosome	MAS Assay	Citation
CMV Resistance	Cmr1	2	CAPS, HRM	Kang *et al.*, 2010; Kim *et al.*, 2017b
	cmr2	8	KASP	Choi *et al.*, 2018
TMV Resistance	L4	11	HRM	Yang *et al.*, 2009
PVY Resistance	pvr1/pvr2	4	SCAR	Yeam *et al.*, 2005; Kim *et al.*, 2017b
	pvr4/pvr7	10	SCAR, RAPD	Arnedo-Andrés *et al.*, 2002; Venkatesh *et al.*, 2018
	pvr6	3	SCAR	Hwang *et al.*, 2009
TSWV Resistance	Tsw	10	CAPS	Moury *et al.*, 2000
Bacterial Wilt Resistance	Bw1	1	SSR	Mimura *et al.*, 2012
Bacterial Spot Resistance	Bs2	9	AFLP, SCAR	Kim *et al.*, 2017b
	Bs3	2	SCAR	Pierre *et al.*, 2000
Anthracnose Resistance	Anthracnose resistant locus	12	SSR	Suwor *et al.*, 2015
Powdery Mildew Resistance	PMR1	4	HRM, SCAR	Jo *et al.*, 2017
RKN Resistance	Me1	9	CAPS, HRM	Uncu *et al.*, 2015; Wang *et al.*, 2018b

Future Prospects of Advanced Breeding Approaches In Vegetables

The landscape of vegetable genetic enhancement is evolving rapidly, transitioning from traditional breeding methods to more advanced techniques. This progression moves through several stages: i) selection with unknown loci, ii) selection by controlled crosses, iii) DNA marker-assisted breeding, particularly applied in crops like tomato, potato, and cassava, and iv) ideotype-based selection and transformation, significantly enhancing efficiency through genetic engineering in species such as tomato, eggplant, and squash, along with genomic selection, particularly in cassava and potato. A structured approach to sustainable genetic gains begins with defining breeding objectives in collaboration with end-users, leading to the development of a designed product portfolio. This portfolio guides subsequent actions, including the identification of useful traits in breeding populations or gene banks. Effective management of genetic variation for these traits should integrate both genetic insights and omics technologies, facilitating the transformation of genes into usable forms (e.g., lines, clones, populations) for targeted crossbreeding.

Genetic engineering, including transgenic breeding and genome editing, plays a crucial role when desirable traits are not present in existing gene banks or breeding populations, provided that biosafety regulations and intellectual property rights are properly managed. Modern tools for assessing functional diversity in gene pools enable innovative vegetable breeding, especially in the face of climate change. Plant genetic resources remain vital for mining allelic variations linked to target traits. Future crop improvement will increasingly depend on leveraging genetic recombination to combine diversity within crop populations. Unlocking functional diversity through advanced omics, high-throughput phenotyping, and e-typing for key agronomic traits—such as crop phenology, plant architecture, yield, climate resilience, host plant resistance, and input-use efficiency—will enhance the genetic modification of vegetable crops in the 21st century. Additionally, speed breeding techniques will further accelerate the development of improved varieties, ensuring that vegetable crops remain resilient and productive in a rapidly changing environment. As we move forward, integrating these advanced breeding approaches will be crucial for developing sustainable, high-quality vegetable varieties that meet the demands of both producers and consumers, while addressing global challenges such as food security and climate resilience.

REFERENCES

Abera Desta, Z and Ortiz, R (2014) Genomic selection: Genome-wide prediction in plant improvement. *Trends Plant Sci.* 19: 592–601.

Adamson RM (1965) Self and Cross-Incompatibility in Early Round-Headed Cabbage. *Can J. Plant Sci.* 45, 493–497.

Aditika K, Priyanka HS and Sharma A (2017) Vegetable Improvement in India; Recent Past, Present and Future: A Review *Int. J. Curr. Microbiol. App. Sci.* 6(8), 3246-3255.

Ansari AM (2015) Molecular markers in vegetable improvement. Horticultural Biotechnology Research, 1, pp-5-10, https://updatepublishing.com/journal/index.php/hbr/article/view/2846.

Arens P, Mansilla C, Deinum, D, Cavellini L, Moretti A, Rolland S, vander Schoot H, Calvache D, Ponz F, Collonnier C, Mathis R, Smilde D, Caranta C, and Vosman B (2010). Development and evaluation of robust molecular markers linked to disease resistance in tomato for distinctness, uniformity and stability testing. *Theor. Appl. Genet.* 120:655–664.

Atanassova B and Georgiev H (2002) Using genic male sterility in improving hybrid seed production in tomato (*Lycopersicon esculentum* Mill.). *Acta Hortic.* 579, 185–188.

Atanassova B and Georgiev, H (1986) Investigation of Tomato Male Sterile Lines in Relation to Hybrid Seed Production. *Acta Hortic.* 190, 553–557.

Behera T, Dey S and Sirohi P (2006) DBGy-201 and DBGy-202: Two Gynoecious Lines in Bitter Gourd (Momordica charantia L.) Isolated from Indigenous Source. *Indian J. Genet. Plant Breed*, 66, 61–62.

Bennewitz S, Bergau N and Tissier A (2018) QTL mapping of shape of type VI glandular trichomes in tomato. *Front Plant Sci.* 9:1421.

Bohn GW and Tucker, CM (1939) Immunity to fusar-ium wilt in the tomato. *Science* 89: 603–604. doi:10.1126/science.89.2322.603

Bradshaw JE (2017) Review and analysis of limitations in ways to improve conventional potato breeding. *Potato Res.* 60: 171–193.

Cao X, Liu XY and Wang XT (2019) B-class MADS-box TM6 is a candidate gene for tomato male sterile-1526. *Theor. Appl. Genet.* 132, 2125–2135.

Carafa MA and Carratu G (1997) Stigma Treatment with Saline Solutions: A New Method to Overcome Self-Incompatibility in Brassica oleracea L. *J. Hortic. Sci.*, 72, 531–535.

Catanzariti AM, Lim GT and Jones DA (2015) The tomato I-3 gene: a novel gene for resistance to Fusarium wilt disease. *New Phytol.* 207: 106–118. doi:10.1111/nph.13348

Cavanagh C, Morell M, Mackay I, and Powell W (2008). From mutations to MAGIC: resources for gene discovery, validation and delivery in crop plants. *Curr. Opin. Plant Biol.*11: 215–221. doi:10.1016/j.pbi.2008.01.002

Cermak T, Baltes NJ, Cegan R, Zhang Y and Voytas DF (2015) High-frequency, precise modification of the tomato genome. *Genome Biology*, 16:232.

Chamola R, Balyan HS and Bhat SR (2013) Transfer of Cytoplasmic Male Sterility from Alloplasmic Brassica Juncea and B. Napus to Cauliflower (*B. oleracea* Var. Botrytis) through Interspecific Hybridization and Embryo Culture. *Indian J. Genet. Plant Breed.* 73, 203–210.

Chiu L-W, Zhou X, Burke S, Wu X, Prior RL and Li L (2010) The Purple Cauliflower Arises from Activation of a MYB Transcription Factor, *Plant Physiology*, 154(3) 1470–1480, https://doi.org/10.1104/pp.110.164160

Clayberg CD, Butler L, Rick, CM, and Robinson RW (1965) Report of tomato genes of January1965. Clough GH, Hamm PB. Coat protein transgenic resistance to watermelon mosaic and zucchini yellow mosaic virus in squash and cantaloupe. *Plant Dis.* 1995; 79: 107–1109.

Collard BCY, and Mackill DJ (2008) Marker-assisted selection: an approach for precision plant breeding in the twenty-first century. *Philos. Trans. R. Soc. Lond. B Biol. Sci.* 363: 557–572. doi:10.1098/rstb.2007.2170

Crane MB (1915) Heredity of types of inflorescence and fruits in tomato. *J. Genet.* 5, 1–11.

Darwin C (1977) The Different Forms of Flowers on Plants of the Same Species. *D. Applet.*

Deshpande AA, Pathak CS and Singh DP (1983) Types of Male Sterility in Chilli Pepper (*Capsicum* Spp.) *Capsicum Newsl.* 2, 104–105.

Dias JS and Ryder E (2011) World vegetable industry: production, breeding, trends. *Hort. Rev.* 38: 299–356.

Dias JS (2012a) Major classes of phytonutriceuticals in vegetables and health benefits: a review. *J. Nutr. Therap.* 1: 31– 62.

Dias JS (2012b) Vegetable breeding for nutritional quality and health benefits. In: Carbone K, editor. Cultivar: Chemical Properties, Antioxidant Activities and Health Benefits. Hauppauge: *Nova Science Publishers Inc.* 1–81.

Dixon MS, Jones DA, Keddie JS, Thomas CM, Harrison K and Jones JDG (1996) The tomato Cf-2 disease resistance locus comprises two functional genes encoding leucine-rich repeat proteins. *Plant Cell* 84: 451–459. doi:10.1016/S0092-8674(00)81290-8

Endelman JB, Carley CAS, Bethke PC, Combs JJ and Clough ME (2018) Genetic variance partitioning and genome-wide prediction with allele dosage information in autotetraploid potato. *Genetics.* 2018; 209: 77–87.

Eriksson D, Carlson-Nilsson U, Ortíz R and Andreasson E (2016) Overview and breeding strategies of table potato production in Sweden and the Fennoscandian region. *Potato Res.* 59: 279–294.

Etoh T (1985) Studies on the Sterility in Garlic, *Allium sativum* L. *Mem. Fac. Agric.* Kagoshima Univ. 21: 77–132.

FAO (2013) The State of Food and Agriculture, Food and Agriculture Organization, Rome

Foolad MR and Sharma, A (2005) Molecular markers are selection tools for tomato breeding. *Acta Horticulturae* 695(695):225-240, DOI:10.17660/Acta Hortic. 2005.695.25

Gama RNC, Santos CAF, Dias RCS, Souza RRC and Queiróz MA (2015) Microsatellite markers linked to powdery mildew resistance locus in watermelon. *Australian Journal of Crop Science.* 2015; 9:92-97.

Gangopadhyay KK, Gill HS and Sharma SR (1997) Heterosis and Combining Ability Studies in Early Group of Indian Cauliflower Involving Self-Incompatible Lines. *Veg. Sci.* 24, 26–28.

Gao L, Gonda I, Sun H, Ma Q and Bao K (2019) The tomato pan-genome uncovers new genes and a rare allele regulating fruit flavor. *Nat. Genetics.* 51: 1044- 1051.

Gerstel, DU (1950) Self-Incompatibility Studies in Guayule; Inheritance. *Genetics* 35: 482–506.

Ginkel M and Ortiz R (2018) Cross the best with the best, and select the best: HELP in breeding selfing crops. *Crop Science.* 58: 17-30.

Giovannoni J, Nguyen C, Ampofo B, Zhong S and Fei Z (217) The Epigenome and Transcriptional Dynamics of Fruit Ripening. *Ann. Rev. Plant Biol.* 68, 61–84.

Gökçe AF, McCallum J, Sato Y and Havey MJ (2002) Molecular Tagging of the Ms Locus in Onion. *J. Am. Soc. Hortic. Sci.* 127: 576–582.

Grushetskaya ZE, Lemesh VA, Poliksenova VD and Khotyleva LV (2007) Mapping of the Cf-6 tomato leaf mould resistance locus using SSR markers. *Russ. J. Genet.* 43: 1266–1270. doi:10.1134/S1022795407110099

Gusakova GA and Gorodilov NA (1996) Selection of White Head Cabbage [Brassica oleracea Capitata] on the Basis of Self Incompatability. Russian, *Food and Agriculture Organization*: Rome, Italy.

Habyarimana E, Parisi B and Mandolino G (2017) Genomic prediction for yields, processing and nutritional qualitytraits in cultivated potato (*Solanum tuberosum* L.). *Plant Breed.* 136: 245–252.

Han Y, Zhao F, Gao S, Wang X, Wei A, Chen Z, Liu N, Tong X, Fu X, Wen C, Zhang Z and Du S (2018) Fine mapping of a male sterility gene ms-3 in a novel cucumber (*Cucumis sativus* L.) mutant. *Theoretical and Applied Genetics*, 131: 449-460.

Havey MJ and Bark OH (1994) Molecular Confirmation That Sterile Cytoplasm Has Been Introduced into Open-Pollinated Grano Onion Cultivars. *J. Am. Soc. Hortic. Sci.* 1994, 119, 90–93.

Heffner EL, Sorrells ME and Jannink JL (2009) Genomic selection for crop improvement. *Crop Sci.* 49: 1–12. doi:10.2135/cropsci2008.08.0512

Hinata K (1975) Effect of CO_2 Gas for Overcoming Self-Incompatibility of Different S Genotypes in *Brassica oleracea* L. *Incompat. Newsl.* 5: 60–62.

Holland JB (2014) Implementation of Molecular Markers for Quantitative Traits in International Service for the Acquisition of Agri-Biotech Application. Global Status of Commercialized Biotech/GM Crops, 2015.

Hoser-Krauze J (1989) Comparison of Suitability of Sources of Cytoplasmic Male Sterility (CMS) and Self-Incompatibility for Hybrid Breeding of Cauliflower. Biul. *Warzyniczy Suppl.* 1–29.

Hoser-Krauze, J. Inheritance of Self-Incompatibility and the Use of It in the Production of F1 Hybrids of Cauliflower (Brassica oleracea Var Botrytis L SubvarCauliflora Dc). Genet. Pol. 1979, 20, 341–367.

Hoser-Krauze J, Gabryl J and Antosik J (1982) Influence of the Cytoplasm of Indian Self-Incompatible Lines on the Earliness and Quality of F1 Cauliflower Curds. *Eucarpia Crucif. Newsl.* 7, 12.

Huo YM, Liu BJ, Yang YY, Miao J, Gao LM, Kong SP, Wang ZB, Kitano H and Wu X (2015) AcSKP1, a multiplex PCR based co-dominant marker in complete linkage disequilibrium with the male-fertility restoration (Ms) locus and its application in open pollinated populations of onion. *Euphytica* 204:711–722. https://doi.org/10.1007/s10681-015-1374-7

Ivanov R and Gaude T (2009) Brassica Self-Incompatibility: A Glimpse below the Surface. *Plant Signal. Behav.* 4: 996–998.

Janska H, Sarria R, Woloszynska M, Arrieta-Montiel M and Mackenzie SA (1998) Stoichiometric Shifts in the Common Bean Mitochondrial Genome Leading to Male Sterility and Spontaneous Reversion to Fertility. *Plant Cell.* 10: 1163–1180.

Jeong HJ, Kang JH, Zhao MA, Kwon JK, Choi HS, Bae JH, Lee HA, Joung, YH, Choi D and Kang BC (2014) Tomato male sterile 10 (35) is essential for pollen development and meiosis in anthers. *J. Exp. Bot.* 65: 6693–6709.

Jo J, Venkatesh J, Han K, Lee HY, Choi GJ, Lee HJ, Choi D, and Kang BC (2017) Molecular mapping of PMR1, a novel locus conferring resistance to powdery mildew in pepper (*Capsicum annuum*). *Front. Plant Sci.* 8: 2090. doi:10.3389/fpls.2017.02090

Johns C, Lu M, Lyznik A and Mackenzie S (1992) A Mitochondrial DNA Sequence Is Associated with Abnormal Pollen Development in Cytoplasmic Male Sterile Bean Plants. *Plant Cell* 4: 435–449.

Jones M, Prunier FP, Marcel F, Piednoir E (2012) Characterization of alleles of tomato light signalling genes generated by tilling. *Phytochemistry*, 9: 78-86.

Kaboli S and Babazada H (2018) CRISPR Mediated Genome Engineering and its Application in Industry. *Curr. Issues Mol. Biol.* 26: 81–92.

Kachroo A, Nasrallah ME and Nasrallah JB (2002) Self-Incompatibility in the Brassicaceae: Receptor-Ligand Signaling and Cell-to-Cell Communication. *Plant Cell* 2002, 14, 227–238.

Kalia P, Soi S and Muthukumar P (2016) Marker-assisted introgression of the Or gene for enhancing β-carotene content in Indian cauliflower. *Acta Horticulturae* 1203:121-128.

Kaloo G (1998) Vegetable Breeding; CRC Press Inc.: Boca, Raton, FL, USA, p. 1.

Karkute SG, Singh AK, Gupta OP, Singh PM and Singh B (2017) CRISPR/Cas9 Mediated Genome Engineering for Improvement of Horticultural Crops. *Frontiers in Plant Science* 8:1635.

Kawchuk LM, Hachey J, Lynch DR, Kulcsar F, van Rooijen G, Waterer DR, Robertson A, Kokko E, Byers R, Howard RJ, Fischer R, and Prufer D (2001) Tomato Ve disease resistance genes encode cell surface- like receptors. *Proc. Natl. Acad. Sci.* U.S.A. 98: 6511–6515. doi:10.1073/pnas.091114198

Khar A, Zimik, M, Verma P, Singh H, Mangal M, Singh MC and Gupta AJ (2022) Molecular Marker-Based Characterization of Cytoplasm and Restorer of Male Sterility (Ms) Locus in Commercially Grown Onions in India. *Mol. Biol. Rep.* 49: 5535–5545.

Kim B, Hwang IS, Lee HJ, and Oh CS (2017a) Combination of newly developed SNP and InDel markers for genotyping the Cf-9 locus conferring disease resist-ance to leaf mold disease in the tomato. *Mol. Breed.* 37:59.

Kim DH, Kang JG and Kim BD (2007) Isolation and Characterization of the Cytoplasmic Male Sterility-Associated Orf456 Gene of Chili Pepper (*Capsicum annuum* L.). *Plant Mol. Biol.* 63: 519–532.

Kim DH and Kim BD (2005) Development of SCAR Markers for Early Identification of Cytoplasmic Male Sterility Genotype in Chili Pepper (*Capsicum annuum* L.). *Mol. Cells.* 20, 416–422.

Kim H, Yoon JB and Lee J (2017b) Development of flu- idigm SNP type genotyping assays for marker-assisted breeding of chili pepper (*Capsicum annuum* L.). *Hortic. Sci. Technol.* 35: 465–479.

Kim S, Kim CW, Park M and Choi, D (2015) Identification of Candidate Genes Associated with Fertility Restoration of Cytoplasmic Male-Sterility in Onion (*Allium cepa* L.) Using a Combination of Bulked Segregant Analysis and RNA-Seq. *Theor. Appl. Genet.* 128: 2289–2299.

Kitagawa J, Posluszny U, Gerrath JM and Wolyn DJ (1994) Developmental and Morphological Analyses of Homeotic Cytoplasmic Male Sterile and Fertile Carrot Flowers. *Sex. Plant Reprod.* 7: 41–50.

Koelreuter J (1764) Descriptions Piscium Rariorum e Museo Petripolitano Exceptorum Continuatio. Novi Comment. Acad. Sci. Imp. Petropolitanae 9: 420–470.

Konzak CF, Nilan RA and Kleinhofs A (1977) Artificial mutagenesis as an aid in overcoming genetic vulnerability of crop plants. In: Genetic diversity in plants. Plenum, New York, 163-177p.

Kucera V (1990) Overcoming Self Incompatibility in Brassica oleracea with a Sodium Chloride Solution. Sborník ÚVTIZ Zahrad. 17: 13–16.

Kumar S, Singh PK (2004) Mechanisms for Hybrid Development in Vegetables. *Journal of New Seeds* 6(4): 300-407.

Kumar S, Singh V, Kumar S, Rai M and Kalloo G (2002) Rapd protocol for tagging of fertility restorer and male sterility genes in chilli (*Capsicum annuum* L.). *Vegetable Science* 29(2): 101-105.

Lal T, Vashisht VK and Dhillion NPS (2007) Punjab Anmol: A New Hybrid of Muskmelon (*Cucumis melo* L.). *J. Res. Punjab Agric.* 44, 83.

Lee YP, Kim S, Lim H, Ahn Y and Sung SK (2009) Identification of Mitochondrial Genome Rearrangements Unique to Novel Cytoplasmic Male Sterility in Radish (*Raphanus sativus* L.). *Theor. Appl. Genet.* 118: 719–728.

Li L, Paolillo DJ, Parthasarathy MV, Dimuzio EM and Garvin DF (2001) A novel gene mutation that confers abnormal patterns of betacarotene accumulation in cauliflower (*Brassica oleracea* var. botrytis). *Plant Journal* 26, 59-67.

Li H and Harner DS (2010) A survey of sequence alignment algorithms for next-generation sequencing. *Brief. Bioinform.* 2: 473-483.

Liu Z, Liang J, Zheng S, Zhang J, Wu J, Cheng F, Yang W and Wang X (2017) Enriching Glucoraphanin in *Brassica rapa* Through Replacement of BrAOP2.2/ BrAOP2.3 with Non-functional Genes, *Frontiers in Plant Science* 8: 1329.

Liu XY, Yang MX and Liu XL (2019) A putative bHLH transcription factor is a candidate gene for male sterile 32, a locus affecting pollen and tapetum development in tomato. *Hortic. Res.* 6, 88.

Mackenzie SA and Chase CD (1990) Fertility Restoration Is Associated with Loss of a Portion of the Mitochondrial Genome in Cytoplasmic Male-Sterile Common Bean. *Plant Cell* 2: 905–912.

Meena OP and Meena NK (2014) Role of hybrids in vegetable production. *Popular Kheti* 2(3): 7-14.

Merk HL, Ashrafi H and Foolad MR (2012) Selective genotyping to identify late blight resistance genes in an accession of the tomato wild species *Solanum pimpinelli- folium. Euphytical* 87:63–75.doi:10.1007/s10681-012-0729-6

Metzker M (2010) Sequencing technologies-the next generation. *Nat. Rev. Genet.*, 11, 31.

Meyer RS and Purugganan MD (2013) Evolution of crop species: Genetics of domestication and diversification. *Nat. Rev. Genet.* 14: 840–852.

Moon SS, Verma VK and Munshi AD (2002). Gene action of quality traits in muskmelon (*Cucumis melo* L.). *Vegetable Science* 29(2): 134-136.

Moreau P, Thoquet P, Olivier J, Laterrot H and Grimsley N (1998). Genetic mapping of Ph-2, a singlelocus controlling partial resistance to Phytophthora infes- tansin tomato. MPMI 11: 259–269. doi:10.1094/MPMI.1998.11.4.259

Nieuwhof M (1961) Male sterility in some cole crops. *Euphytica*, 10(3): 351-356.

Nieuwhof M (1965) Breeding Hybrid Varieties of Vegetables. *Breed. Hybrid Var. Veg.* 1497: 98–107.

Obel HO, Cheng C, Tian Z, Njogu MK, Li J, Du S, Lou Q, Zhou J, Yu X, Ogweno JO and Chen J (2022) Transcriptomic and Physiological Analyses Reveal Potential Genes Involved in Photoperiod-Regulated β-Carotene Accumulation Mechanisms in the Endocarp of Cucumber (*Cucumis sativus* L.) *Fruit. Int J Mol Sci.* (20):12650. doi: 10.3390/ijms232012650. PMID: 36293506; PMCID: PMC9604348.

Odland ML and Noll CJ (1950) The Utilization of Cross-Compatibility and Self-Incompatibility in the Production of F1 Hybrid Cabbage. *Proceedings. Am. Soc. Hortic. Sci.* 55: 391–402.

Ogura H (1968) Studies on the New Male-Sterility in Japanese Radish, with Special Reference to the Utilization of This Sterility towards the Practical Raising of Hybrid Seeds. *Mem. Fac. Agric. Kagoshima Univ.* 6: 39–78.

Ortiz R (2015) Plant Breeding in the Omics Era. New York: *Springer International Publishing.* 249.

Park JY, Lee YP, Lee J, Choi BS, Kim S and Yang TJ (2013) Complete Mitochondrial Genome Sequence and Identification of a Candidate Gene Responsible for Cytoplasmic Male Sterility in Radish (*Raphanus sativus* L.) Containing DCGMS Cytoplasm. *Theor. Appl. Genet.* 126: 1763–1774.

Pati K, Munshi, AD and Behera, TK (2015) Inheritance of Gynoecism in Cucumber (*Cucumis sativus* L.) Using Genotype GBS-1 AS Gynoecious Parent. *Genetika* 47: 349–356.

Pearson OH (1933) Breeding Plants of the Cabbage Group. *Calif. Agric. Exp. Stn. Bull.* 532, 3–22.

Phatak SC and Jaworski CA (1989) UGA 1-MS male-sterile eggplant germplasm. *Hortscience* 24:1050.

Pierre M, Noël L and Lahaye T (2000) High-resolution genetic mapping of the pepper resistance locus Bs3 governing recognition of the Xanthomonascampestris pv vesicatora AvrBs3 protein. *Theor Appl Genet* 101, 255–263. https://doi.org/10.1007/s001220051477

Pitchaimuthu M, Dutta OP and Swamy KRM (2012) Studies on inheritance of Genetic Male Sterility (GMS) and hybrid seed production in okra [*Abelmoschus esculentus* (L.) Moench.]. *Journal of Horticultural Sciences*, 7(2), 199-202.

Pop M and Salzberg SL (2008) Bioinformatics challenges of new sequencing technology. *Trends Genet.* 24, 142-149.

Prakash K, Kumar R, Prakash NR, Singh S and Singh L (2017) Pre-breeding in the omics era: a review in vegetable crops. *Chem Sci Rev Lett* 6(22): 752-762.

Prasanna HC, Kashyap SP, Krishna R, Sinha DP, Reddy S and Malathi VG (2015) Marker assisted selection of Ty-2 and Ty-3 carrying tomato lines and their implications in breeding tomato leaf curl disease resistant hybrids. *Euphytica* 204, 407-418.

Pucci A, Picarella ME and Mazzucato A (2017) Phenotypic, genetic and molecular characterization of 7B-1, a conditional male-sterile mutant in tomato. *Theor. Appl. Genet.* 130, 2361–2374.

Ram D, Kumar S, Singh M, Rai M and Kalloo G (2006) Inheritance of Gynoecism in Bitter Gourd (*Momordica charantia* L.). *J. Hered.* 97, 294–295. [CrossRef]

Ram HH (1975) Level of Self-Incompatibility in Cauliflower (*Brassica oleracea* Var Botrytis L.). *South Indian Hortic.* 64–66.

Robbins MD, Masud MAT, Panthee DR, Gardner RG, Francis D and Stevens MA (2010) Marker-assisted selection for coupling phase resistance to tomato spotted wilt virus and *Phytophthora infestans* (Late Blight) in tomato. HARVEST.

Robinson RW and Decker-Walters DS (1997) *Cucurbits. Cab Int.* 226, 103–108.

Robinson RW and Robinson RW (1999) Rationale and Methods for Producing Hybrid Cucurbit Seed Rationale and Methods for Producing Hybrid Cucurbit Seed. *J. New Seeds*, 1, 1–47.

Ronald PC, Salmeron JM, Carland FM and Staskawicz BJ (1992) The cloned avirulence gene avrPto induces disease resistance in tomato cultivars containing the *Pto resistance gene. J. Bacteriol.* 174: 1604–1611. doi: 10.1128/jb.174.5.1604-1611.1992

Ruffio-Chable V and Gaude T (2000) S-haplotype polymorphism in *Brassica oleracea*; Proceedings of the International Symposium on Molecular Markers for Characterizing Genotypes and Identifying Cultivars in Horticulture; Montpellier, France. 257–261.

Ruffio-Chable V, Bellis H and Herve Y (1993) A Dominant Gene for Male Sterility in Cauliflower (Brassica oleracea Var. Botrytis): Phenotype Expression, Inheritance, and Use in F1 Hybrid Production. *Euphytica*, 67, 9–17.

Ruffio-Chable V, Bellis H and Herve Y (1987) Male Sterile Broccoli (Brassica Oleracea Var. Italica) Induced by in Vitro Mutagenesis. *Crucif. Newsl.* 12, 50–51.

Saini JK, Saini R and Tewari L (2015) Lignocellulosic agriculture wastes as biomass feedstocks for second generation bioethanol production: concepts and recent developments. *Biotech.* 5, 337-53.

Sakamoto K and Nishio T (2001) Distribution of S Haplotypes in Commercial Cultivars of *Brassica rapa. Plant Breed.* 120, 155–161.

Sato Y (1998) PCR Amplification of CMS-Specific Mitochondrial Nucleotide Sequences to Identify Cytoplasmic Genotypes of Onion (*Allium cepa* L.). *Theor. Appl. Genet.* 96, 367–370.

Sax K (1923) The association of size differences with seed-coat pattern and pigmentation in Phaseolus vulgaris. *Genetics.* 8: 552-60.

Selvakumar R and Kalia P (2018) Nutra-rich Carrot Production in Tropical and Temperate Regions of India. Findings in Agricultural Research and Management (FARM) Journal 2(3):7-13.

Selvakumar P, Sinha SN and Pandita VK (2011) Hybrid Seed Production in Cauliflower (*Brassica oleracea* Var Botrytis Subvar Cauliflora). *Indian J. Agric. Sci.* 77, 649–651.

Sharma SR, Singh R, Vinod SDK and Edison SJ (2001) Self-Incompatibility Level in Different Maturity Groups in Indian Cauliflower; Proceedings of the Diamond Jubily. Symposium–Hundred Years Post Mendelian Genetic Plant Breeding; New Delhi, India. 155–156.

Sharma SR, and Kumar V (2002) Breeding for cytoplasmic male sterility in (*Brassica oleracea* L. var. italica Plenck). *Indian Journal of Genetics and Plant Breeding*, 62(02), 165-166.

Shi L, Yang Y, Xie Q, Miao H, Bo K and Song Z (2018) Inheritance and QTL mapping of cucumber mosaic virus resistance in cucumber. *PLoS ONE*, 13:7.

Shull GH (1952) Beginnings of the Heterosis Concept. *Heterosis* 23, 31–33.

Singh B and Chaubey T (2013) Vegetable research in India: An overview. *Progressive Horticulture* 45, 9-35.

Singh S, Selvakumar R, Mangal M and Kalia P (2020) Breeding and genomic investigations for quality and nutraceutical traits in vegetable crops. *Indian Journal of Horticulture* 77(1):1-40.

Singh B, Singh A, Pal AK and Banerjee MK (2002) Evaluation of Self Incompatibility in Indian Cauliflower *(Brassica oleracea* L. Var. Botrytis). *Veg. Sci.* 29, 142–145.

Singh H and Khar A (2013) Onion (*Allium cepa* L.) Hybrid Breeding In India: Status and Prospects. *Acta Horticulturae*. 2023, 1362, 547–552.

Singh H and Khar A (2021) Perspectives of Onion Hybrid Breeding in India: An Overview. *Indian J. Agric. Sci.* 91, 1426–1432.

Singh P, Cheema DS, Dhaliwal MS and Garg N (2014) Heterosis and Combining Ability for Earliness, Plant Growth, Yield and Fruit Attributes in Hot Pepper (*Capsicum annuum* L.) Involving Genetic and Cytoplasmic-Genetic Male Sterile Lines. *Sci. Hortic.* 168, 175–188.

Singh P, Cheema DS, Dhaliwal MS, Garg N, Jindal SK and Chawla N (2015) Combining Ability and Heterosis for Quality and Processing Traits in Chili Pepper (*Capsicum annuum* L.) Involving Male Sterile Lines. *J. Crop Improv.* 29, 379–404.

Singh RJ and Lebeda A (2007) Landmark research in vegetable crops. In Genetic Resources, Chromosome Engineering, and Crop Improvement. Volume 3 – *Vegetable Crops.* Singh, R. J., Ed. CRC Press: Boca Raton, FL, 1–15.

Slater AT Cogan NOI, Forster JW, Hayes BJ and Daetwyle HD (2016) Improving genetic gain with genomic selection in autotetraploid potato. *The Plant Genome.* 9: 3.

Song H, Chen J, Staub JE and Simon PW (2010) QTL analysis of orange color and carotenoid content and mapping of carotenoid biosynthesis gene in cucumber. *Acta Horticulturae.* 871, 607-614.

Song ZC, Miao H, Zhang S, Wang Y, Zhang SP and Gu XF (2016) Genetic analysis and QTL mapping of fruit peduncle length in cucumber. *PLoS ONE*. 11:167-845.

Stall RE and Walter JM (1965) Selection and inherit- ance of resistance in tomato to isolates of races1and 2 of the Fusarium wilt organism. *Phytopathology* 55: 1213–1215.

Stern DB, Bang AG and Thompson WF (1986) The Watermelon Mitochondrial URF-1 Gene: Evidence for a Complex Structure. *Curr. Genet.* 10, 857–869.

Stich B and Van Inghelandt D (2018) Prospects and potential uses of genomic prediction of key performance traits in tetraploid potato. *Front. Plant Sci.* 9: 159.

Sverrisdóttir E, Sundmark EHR, Johnsen HØ, Kirk HG and Asp T (2018). The value of expanding the training population to improve genomic selection models in tetraploid potato. *Front. Plant Sci.* 9: 1118.

Swarup V, Gill HS (1964) The Use of Marker Gene in Hybrid Seed Production in Cabbage. *Curr. Sci.* 33, 315–316.

Tanaka Y, Tsuda, M, Yasumoto K, Yamagishi H and Terachi T (2012) A Complete Mitochondrial Genome Sequence of Ogura-Type Male-Sterile Cytoplasm and Its Comparative Analysis with That of Normal Cytoplasm in Radish (*Raphanus sativus* L.). BMC Genomics 2012, 13, 352.

Thomas PE, Kaniewski WK and Lawson EC (1997) Reduced field spread of potato virus in potatoes transformed with the potato leafroll virus coat protein gene. *Plant. Dis.* 81: 1447–1453.

Thompson KF (1964) Triple-Cross Hybrid Kale. *Euphytica* 13, 173–177.

Tian F, Bradbury PJ, Brown PJ, Hung H, Sun Q, Flint-Garcia S, Rocheford TR, McMullen MD, Holland JB and Buckler ES (2011). Genome-wide association study of leaf architecture in the maize nested association mapping population. *Nat. Genet.* 43: 159–162. doi:10.1038/ng.746

Tian S, Jiang L, Gao Q, Zhang J, Zong M, Zhang H, Ren Y, Guo S, Gong G and Liu F (2017). Efficient CRISPR/Cas9-based gene knockout in watermelon. *Plant Cell Rep.* 2017, 36, 399–406. [CrossRef]

van Jaarsveld PJ, Faber M, Tanumihardjo SA, Nestel P, Lombard CJ and Benadé AJ (2005) Beta-carotene-rich orange-fleshed sweet potato improves the vitamin A status of primary school children assessed with the modified-relative-dose-response test. *Am J Clin Nutr.* 81(5):1080-7. doi: 10.1093/ajcn/81.5.1080. PMID: 15883432.

Venkatesh J, An J, Kang WH, Jahn M and Kang BC (2018). Fine mapping of the dominant potyvirus resistance gene Pvr7 reveals a relationship with Pvr4 in *Capsicum annuum*. Phytopathology 108: 142–148. doi:10.1094/PHYTO-07-17-0231-R

Wallace DH (1979) Procedures for Identifying S-Allele Genotypes of Brassica. *Theor. Appl. Genet.* 54, 193–201.

Wamiq M, Yadav S, Pant M, Kumar M, Singh S, Kumar D, Kaushik K, Kumar A and Ahamad S (2023). Vegetable improvement in India: A review. *The Pharma Innovation Journal* 12(6), 4783-4787.

Yang W and Francis DM (2005) Marker-assisted selec- tion for combining resistance to bacterial spot and bac- terial speck in tomato. *JASHS* 130: 716–721. doi:10.21273/JASHS.130.5.716

Zhang J, Li H, Zhang M, Wang Q, Li L and Zhang L (2015) Fine mapping and identification of candidate Br-Or gene controlling orange head of Chinese cabbage (*Brassica rapa* L. spp. Pekinesis). *Planta.* 241:1381-1394.

Zhang J and Dilip PR (2021) Development of codominant SCAR markers to detect the Pto, Tm2-2, I3 and Sw5 genes in tomato (*Solanum lycopersicum*). *Plant Breed.*140: 342–348. doi:10.1111/pbr.12902

Zhang LY, Huang ZJ, Wang XX, Gao JC, Guo YM, Du YC and Hu H (2016) Fine mapping and molecular marker development of anthocyanin absent, a seedling morphological marker for the selection of malesterile 10 in tomato. *Mol. Breed.* 36, 107.

Zhang XP, Rhodes BB, Baird WV, Skorupska HT and Bridges WC (1996) Development of Genic Male-Sterile Watermelon Lines with Delayed-Green Seedling Marker. *Hort Sci.* 31, 123–126.

Zhang C, Liu L, Wang X, Vossen J, Li G, Li T, Zheng, Z, Gao J, Guo Y, Visser RGF, Li J, Bai Y, and Du Y (2014a) The Ph-3 gene from Solanum pimpinellifolium encodes CC-NBS-LRR protein conferring resistance to Phytophthora infestans. *Theor. Appl. Genet.* 127: 1353–1364. doi:10.1007/s00122-014-2303-1

Zhi X, Shu J, Zheng Z, Li T, Sun X, Bai J, Cui Y, Wang X, Huang Z, Guo Y, Du Y, Yang Y, Liu L and Li J (2021) Fine mapping of the Ph-2 gene conferring resistance to late blight (*Phytophthora infestans*) in tomato. *Plant Dis.* 105:851–858.doi:10.1094/PDIS-03-19-0679-RE

Zou CL, Zheng Y, Wang P, Zhang X, Wang YH, Liu ZY and Feng H (2016) Fine mapping and characterization of the or gene in Chinese cabbage (*Brassica rapa* L. ssp pekinensis). *Genet. Mol. Res.* 15.

Sustainable Agriculture and Biodiversity: Innovations and Insights, Pages: 130–158
Edited by: Bibhuti Bhusan Sahoo, Harvinder Kumar Singh, Niranjan Chinara and Amitabh Panda

CHAPTER - 5

DIVERSITY OF INDIAN FOREST MUSHROOMS: SOCIOBIOLOGICAL INSIGHTS FROM TRIBAL MYCOPHILY

Pradeep Kumar Badhai, Harvinder Kumar Singh* and Chandra Shekhar Shukla

Mushroom Research Laboratory, Department of Plant Pathology, College of Agriculture, Indira Gandhi Krishi Vishwavidyalaya, Raipur-492012, Chhattisgarh
**Corresponding email: harry.0452@gmail.com*

INTRODUCTION

Mushrooms are epigeal or hypogeal fruiting body of macrofungi (Ascomycota and Basidiomycota). Globally an estimated 1,069 species of mushrooms have been reported to be used for food purposes (Boa, 2004). The fruiting bodies of these macrofungi visible with the naked eyes as well as easily harvested by hand, develop from the underground fungal mycelium (Chang and Miles, 1992). Most ethnic groups on the African continent eat wild mushrooms, and almost 300 species have been documented as being eaten to some extent by humans (Rammeloo and Walleyn, 1993). Some researchers in India reported 27,000 fungal species (Chang and Miles, 2004; Cowan, 2001).Over 2,000 edible wild mushrooms have been studied throughout the world, including 283 edible sources in India (Adhikari, 2000; Purkayastha and Chandra, 1985) out of which some are cultivated. Fungi have held great importance as the primary decomposers of lignocellulolytic substrates and the main stores of carbon in soil and dead organic matter (Averill *et al*., 2014).A diverse mushroom flora is associated with the biodiversity of woody flora. Generally, they bloom at the onset of the rainy season and are an integral part of the forest ecosystem. They perform a vital role in forest ecosystems, the global carbon cycle in particular. Some macrofungi are mycorrhizal with roots of higher plants determining the dynamics of forest ecosystem. In tropical environments, symbiotic termites have a significant influence on plant litter decomposition and carbon cycling (Wood and Thomas, 1989; Yamada *et al.*, 2005). Termitophilic mushrooms play an important role in the breakdown of dead plant materials and have a substantial influence on carbon cycling in tropical ecosystems (Matsumoto and Abe 1979; Wood and Sands 1978). A majority of mushroom pickers in rural areas where mushrooms are abundant pick mushrooms for consumption and additional income (Kone *et al.,* 2013). These fungi may also have therapeutic promise in the control of blood pressure and blood lipids, the immunological response, apoptosis, and infection (Tidke and Rai, 2006; Garda - Lafuente *et al.,* 2010). In addition, indigenous people use mushrooms as an aphrodisiac and tonic, as well as to treat many ailments (Devkota, 2006). They are economically and ecologically significant because of their edibility, medicinal properties, mycorrhizal, and parasitic associations with forest trees. Mushrooms have a long history with humans and have a significant biological and economic contribution. For many indigenous cultures across the world, wild mushrooms are an essential natural source of food and revenue (Semwal *et al.*, 2014; Ao *et al.*, 2016; Borah *et al.*, 2018). The traditional markets of India sell many mushroom species, some of which have been exploited as food or medicine (Tanti *et al.*, 2011). Wild edible mushrooms have spiritual and socioeconomic ties to tribal existence, and knowledge of their uses is passed down from generation to generation (Pradhan *et al.*, 2012). In the past, traditional uses of fungi have been documented in different parts of India, such as Central India (Rai *et al.*, 1993), Madhya Pradesh (Harsh *et al.*, 1996), Assam (Sarma *et al.*, 2010), and Northern Odisha (Tayung and Panda, 2015).

Chhattisgarh is the richness of mushroom flora as it sustains high plant endemicity. It has the largest tropical forest cover as well as diversity and richness in ecological systems within the country. During the monsoon season, the high humidity provides ideal agro-climatic conditions for mushroom growth. Several edible mushrooms growing in their natural habitats are being collected across Chhattisgarh by the local people and tribes of several districts during the rainy season for their consumption or sale. They had indigenous knowledge of mushrooms and their use. It is home to many primitive tribes. They are mostly Abujhamarias, Baigas, Bharias, Bhil, Gond, Maria, Muria, Saharia, Kol, Hill Korwa, and Kamar (Tiwari, 1994). Tribals are well acquainted with the habitat and period of occurrences of the fungi (Harsh *et al.*, 1993). They collect them while gathering other food, fodder, and fuel items for the family from the forests. This work, therefore, aims at investigating the sociobiological aspects of wild edible mushrooms from Bastar, Chhattisgarh India.

STUDY AREA AND COLLECTION OF WILD EDIBLE

Chhattisgarh is divided into three agroclimatic zones *viz*., Chhattisgarh Plains, Northern Hilly regions, and Bastar Plateau. The state lies between latitude 17°462 N to 24°62 N and longitude 80°15'E to 84°51'E. The state determines a Sal tree dominates the State forest cover with a population of about 35–40% in mixed and 90–95% in true Sal forests. However, Bastar is an exception being the single confined area of Chhattisgarh hugely dominated by the teak tree (*Tectona grandis*) (Shukla *et al*., 2017). A preliminary survey of wild macrofungi was carried out during the rainy season of 2018 and 2019 in the field as well as local markets of several districts of Bastar Division of Chhattisgarh state. Ethno-medicinal and gastronomical information concerning mushrooms was collected using a questionnaire (Table 1), consultations, and personal interviews with old and experienced tribal people from different tribal settlements in Chhattisgarh. During the study, prior informed consent was obtained from the information providers. Fresh fruitbodies were collected during the monsoon season and brought to the laboratory for identification. Collected specimens were identified using standard literature and published keys for macromorphological characters (Purkayatha and Chandra, 1985; Singer, 1986; Tibuhwa *et. al*., 2010; Semwal *et al*., 2014). All samples were deposited in the Mushroom Research Laboratory, Department of Plant Pathology, College of Agriculture, Indira Gandhi Krishi Vishwavidyalaya, Raipur, Chhattisgarh for future reference.

Table 1: Questionnaire used for data collection of Macrofungi

1	Survey-Date and season :
2	Survey Area :
3	Ethnic group/ tribe name:
4	Local Name/Vernacular name of mushroom:
5	Mushroom habitat:
6	Substrate on which mushroom appears:
7	Extent of mushroom:
8	How much is the mushroom preferred?
9	Any special consideration for identification?
10	Is the mushroom seasonally, or is it available all year?
11	Does the mushroom appear in the same place often?
12	Any specific efforts to grow mushroom *ex-situ*?
13	Any specific method for sampling and transport?

[Table Contd.

Contd. Table]

14	Which method was followed for processing?
15	Any medicinal use of mushrooms?
16	Any peculiarities of mushrooms or habitat?
17	What is the marketing strategy for financial benefits?

In Chhattisgarh's southern region, Bastar division has dense forests covering 60% of its territory. The primary native tribes of Bastar are the Gond, Bhatra, Halba, and Dorla. Even though Bastar is part of Chhattisgarh, it maintains its own cultural identity and culinary traditions. Bastar's inhabitants have inherited the purest type of cuisine culture and culinary techniques. The tribes in this area depend heavily on the forest. The forest produces a variety of goods throughout the year, depending on the season. Mahua appears in the spring, Tendu and Chaar (Chironji) appear in the summer, and bamboo shoots and wild mushrooms appear in the monsoon. Bastar is covered with Sal and Teak forests. In the monsoon season, when vegetables start becoming expensive in the market, then the tribals of Bastar change their taste. In this season, colorful mushrooms including Boda (*Astraeus hygrometricus*), Phutu (*Termitomyces heimii*) found in the Sal forest and crab found in the fields, Charota Bhaji (*Cassia tora*) and Munga Bhaji (Moringa leaves) growing around the house is the main part of their diet, which do not have to be bought. Mushrooms are an important part of Bastar's culture. It is a forest-dominated region, and here in the rainy season, many types of phutu grow on the ground and old trees. Here, almost every type of mushroom is eaten under the popular name phutu. Many species of mushroom (phutu) are eaten in Bastar.

Amanita

Mushrooms are essential for both human health and the preservation of the forest environment. Many *Amanita* species are edible, while others are very toxic. *Amanitas* are characterized by a ring on the stem below the cap, a veil (volva) that tears as the cap develops, and a cup from which the stalk emerges. A huge number of new *Amanita* species have been discovered in India, where they are establishing their ectomycorrhizal association with Sal trees (Bhatt *et al.*, 2003). Some edible *Amanita* species, *Amanita caesarea, Amanita loosii* (Figure-1a) and *Amanita vaginata* var. *fulva* found ectomycorrhizal association with sal trees in the forests of Bastar.

Astraeus:

Astraeus hygrometricus (Figure-1b) is an ectomycorrhizal mushroom locally known as Boda. This mycorrhizal fungus emerges from the various chemicals released from its roots in the Sal forests. The mutualistic relationship between tree roots and fungal mycelium aids the trees in extracting nutrients (especially phosphorus) from the ground in return for carbohydrates from photosynthesis. The form of the fruitbody is roughly spherical. The fruit body is covered in a smooth white mycelial layer. The mycelial layer tears away as the fruit body matures, and the exoperidium, the outer tissue layer, bursts apart in a star-shaped pattern to create numerous irregular "rays." Inside the exoperidium, the spore case was surrounded by a thin papery endoperidium. The rays respond to the amount of moisture in the air, opening in high humidity and closing in dry weather. The particular epithet hygrometricus refers to the way the rays of this earthstar react to variations in atmospheric humidity in the same manner as a barometer reacts to changes in atmospheric pressure. The fruitbodies develop just below the surface of the soil and emerge as they expand and the outer layer of fruitbody tissue splitting open in a star-like manner so called earthstar. If you cut through a young fruitbody the interior (gleba) is white and black, but it gradually turns into a brown powdery mass as the spores mature.

Boletus

The genus *Boletus* contains many members which are edible. These fungi form ectomycorrhizal associations with the roots of forest trees. They are having hymenial pores instead of gills. Locals gave names to them according to their associated tree or habitat, like Jam Phutu, Parsa Phutu, etc. (Figure 1c).

Cantharellus

The genus *Cantharellus* is a popular edible mushroom. *Cantharellus* species (Figure- 1d) appears to be host-specific, locally known as Baans Phutu, since it is most commonly associated with bamboo plantations showing symbiotic mycorrhizal relationship making it very challenging to cultivate. The mushroom's meaty and solid fruit bodies, which are frequently funnel-shaped, make it easy to identify.

Lactarius

Lactarius have almost similar characteristics to *Russula*, but they also have milky latex while breaking gills. During the study, several *Lactarius* species were collected many of which seem to be ectomycorrhizal. Among all *L. piperatus* (Figure-1f) was the edible one, found in an ectomycorrhizal relationship with Sal tree. Many species of *Lactarius* were found edible among tribals and residents of the Chhattisgarh.

Lycoperdon

Lycoperdon is a genus of a puffball mushroom. *L.Pyriforme* (Figure- 1g) pear shaped puffball mushroom found in the forests of Bastar. Tribals of Bastar eat this puffball mushroom in the early stage and if the gleaba is found yellow or other than white then it is not considered for consumption.

Russula

The genus *Russula* includes ectomycorhizal members. *Russula* showed large diversity and appear in a wide variety of colors. An abundance of *Russula* in Bastar might be due to the richness of Sal trees, a most important partner of ectomycorrhizal mushrooms species. Many species were also seen in the local markets as well as on the roadside for sale. Some edible *Russula* species, *Russula emetica, Russula rosea* (Figure 1e), *Russula albonigra,* and *Russula xerampelina* were found ectomycorrhizal relationship with Sal tree in the forest of Bastar.

Termitomyces

Termitomyces is a genus of agarics that is closely associated with termites. The presence of fungi in their nests can be seen as brain-like structures called fungus combs or fungus gardens. The mycelium forms white nodules on the fungus comb. These nodules eventually develop into fruiting bodies. The fungus comb is made from a special substrate maintained by the continuous addition of fecal pellets and pre-digested plant material by the termites. It grows as a single-strain monoculture. *Termitomyces* species are usually characterized by association with termites. The pileus bearing a prominent perforatorium and pseudorhiza subterranean in nature connects the comb in the termite nest. *Termitomyces* had many species and display a broad variety of macro-morphology, from the tiny *T. microcarpus* (Figure- 1h) (locally known as Kanki phutu) to big *T. titanicus*. *T. microcarpus* grows on ejected comb material and barely reaches 2 cm in cap diameter, and *T. titanicus* with a cap diameter up to 1 m making it the world's biggest gilled mushroom fruit body which is found in Africa (Piearce, 1987).

Fig. 1. a) ***Amanita loosii*** b) ***Astraeus hygrometricus*** c) ***Boletus edulis*** d) **Cantharellus spp.** e) ***Russula rosea*** f) ***Lactarius piperatus*** g) ***Lycoperdon pyriformis*** h) ***Termitomyces microcarpus***

Ethnomedicinal Uses

The ethnic information from tribals especially Baiga and Maria tribes revealed the use of different *T. heimii* (Raja phutu) which are having bigger heads normally appearing during Dusshera, to treat persons suffering from sickle cell anemia. *T. titanicus* was used as a tonic for various gastrointestinal ailments *e.g,* abdominal pain, constipation stomach ache, and ulcers, and also used with lemongrass to treat stomach complications in Camroon, Tanzania (Tibuhwa, 2012 and Kinge *et al.*, 2017). The powders of *Termitomyces* species are also to treat measles, viral fever, diarrhea, bleeding, constipation, wound, muscular pain, arthritis, skin disease, cough, cold, indigestion, *etc.* Lahiri *et al.* (2010) and Tibuhwa (2012) found that people from India and Tanzania used *T. tylerianus* for prophylaxis throughout the year for general protection against disease. People from Nepal and Tanzania used powder of *T. umkowaan* as a mouthwash for buccal cavity infection and relief from arthritis. It is also reported to be used as a tonic for various gastrointestinal ailments *e.g,* abdominal pain, constipation stomach ache, and ulcers (Tibuhwa, 2012; Aryal and Budathoki, 2016). It was also found that fruiting bodies of *T. microcarpus* are burned and inhaled to speed up the occurrence, maturation, scaling of chickenpox. Tayung and Panda (2015) reported that in India *T. reticulatus* is used in rheumatism and lowering high blood pressure. In Bastar, fruiting bodies of several *Termitomyces* species are wrapped along with ajwain seeds in banana leaves and kept over low flames the fumes coming from it help the tribal women and after pregnancy to reduce pain. Harsh *et al*. (1999) and Tibuhwa (2012) reported that the spore mass of *Astraeus hygrometricus* (Boda) is mixed with mustard seed oil and used as a burn salve. Assist in the clotting of blood also. Sal boda is used in traditional medicine to treat constipation. Children below 8 years are not given Boda to eat frequently.

Tribal Gastronomy and Culinary Status

Termitomyces are also cooked along with drumsticks which is another signature combination of vegetable recipes during the monsoons. *Amanita loosii* and *Amanita vaginata* are boiled for a few minutes before cooking. They are never cooked in the mushroom broth arising from it. *Termitomyces* and *Russula* are sun-dried and kept by the tribals of Bastar. Even other vegetables are also dried and kept for future use. Tripathy *et al.* (2014) reported that to lessen the bitter scent, *Russula brevipes* are fried with ginger and sugar. Because boiling decreases the mucosal component in the cap and stipes, *Amanita* mushrooms are first boiled before being fried.

Boda rice: The tribal food preparations of Boda rice are a signature recipe of Bastar. Sal Boda, Futka, or rugdra are cleaned thoroughly to remove the adhered soil particles with water. They are cut into two halves. If the center portion is white, it is used as such but if the center portion (gleba) is black, then that portion is removed. However, if the center portion is yellow then that entire fruiting body is not used in preparations. The fruiting bodies of boda are boiled in water with a little salt and lemon juice. Then they are fried along with raw spices, curry leaves onions, garlic, and tender ginger stems. Rice is washed and added to the fried mushroom and spices and it is also fried along. Thereafter coconut milk and water are used to cook the Boda rice recipe. Most of the tribal food preparations are prepared in earthen utensils.

Boda masala: Before using it, the boda is kept immersed in water for a while and thoroughly cleaned so that the taste of the vegetable does not deteriorate due to soil. It is used only after washing it with water four to five times. After this, it is cut into two parts. In a little more oil than common vegetables, garlic and boda are fried until golden. After this, add ginger, onion, and garlic paste and fry them all for a while. Thereafter, water is added and heated till it comes to a slight boil. Lastly, chopped coriander is added so that it can be tasted. This vegetable is eaten with fervor with roti and rice. Its taste is many times more wonderful than that of artificially prepared mushrooms and non-veg. Harsh *et al.* (1999) reported that boda is washed and cut into pieces like poatao, cooked in mustard oil with spices and salt, then eaten with cornflour chapaties.

Phutu fry: *T. heimii* is cooked just like chicken or meat but the only difference is that they are cooked in the mushroom broth itself. The mushroom broth or the water that comes out during initial cooking is separated and mushrooms are fried a little along with spices, onions, chili, garlic, and curry leaves. Once fried or sauteed the mushroom broth is added slowly and the entire preparation is cooked in that. Tibuhwa (2012) reported that Termitophilous mushrooms are boiled with salt and spices such as onion, cooking oil, and tomatoes are used in the preparation.

Fig. 2. Phutu fry (*T.heimii*)

Chutney: Partially fried *T. microcarpus* and Red Ants Paste (Chapda) are used as chutney in Bastar tribal recipe.

Rasam: Fresh *Termitomyces* and *Russula* spp. are used in Rasam preparations (Seasonal vegetables, tubers, horse gram pulse, and seasonal wild mushrooms) in Bastar. These Rasam preparations are garnished with fried cumin and *T. microcarpus* tadka.

Sociobiological Aspects and Wild Edible Mushroom Market

The local people of Chhattisgarh call it Bhodu phutu. Bhodu means the home of termites, where this mushroom automatically grows in the rain. Phutu is quite famous, it has many species. Here, phutu with long stalks (*Termitomyces* spp.) is more preferred. These are collected in tonnes in Chhattisgarh from mid-June to September.

Mushrooms are an important part of Bastar's culture. Bastar is a forest-dominated area, here in the rainy season, many types of phutu grow on the ground and old trees. Many species of phutu are eaten in Bastar. In Bastar, almost every type of mushroom is eaten under the popular name phutu. Here people are crazy about boda and phutu mushroom, white umbrella like (*Amanita loosii, Calocybe indica*), mace-like (*Lycoperdon pyriforme*), button (*Agaricus biosporus*), paira futu (*Volvariella volvacea*), dengur futu (*Termitomyces* spp.), hardulia, manzoordhundha, and tendu chhati, jam phutu (*Boletus* sp.) are the food of common people. Dengur phutu (*Termitomyces* sp.) which grows near termite mound (bambi) is considered the best. It comes in the market after Raksha Bandhan festival till Navratri. The phutu grown during Dussehra festival is called Dussehra phutu. It is liked by everyone. Which is considered to have health and medicinal properties. They are also eaten dried.

Surprisingly, the people who know the most about wild mushrooms are those who rely on traditional wisdom passed down from generation to generation rather than science. As the first showers arrive, the young and elderly walk out early in the morning into the deep woods to pick mushrooms that appear as the rains wash the forest. The locals believe that if it thunders while raining, they will get a bountiful harvest of mushrooms the next morning. For ages, ethnic groups and tribal societies have penetrated the heart of forests, knowing exactly when and where to hunt for a certain kind of mushroom. Shepherds who feed cattle also have a good knowledge of where phutu can be found. However, tribal families roaming the forest are also skilled in this work. Especially women bring mushrooms (phutu) from the forest and family members do the work of selling. For boda collection, tribal women make groups and leave in the morning. Even after taking risks in these forests full of wild animals and poisonous snakes, these women keep collecting this mushroom. These women can easily be seen scraping the moist ground near the Sal trees and storing the hidden sacks in it and saving them in the basket. Where the land looks a bit high and soft, the tribals of Bastar dig the ground and take it out. Being under the soil, it contains a lot of soil attached to it. This special mushroom (Boda) of Chhattisgarh is sold only for a few months of the year. The boda vegetable found here is not sold in the shopping mall of Metro City but the local markets. It comes in the market for a very short time. Those who belong to Chhattisgarh know the taste and benefits of this mushroom. Therefore, as soon as its season comes, people try to taste it at least once. When the arrival of boda starts, its price remains more than 2000 Rs/kg. Despite the expensive price, people buy it easily. Some people buy small quantities just for the taste. By August, the boda has vanished. Many people think that when lightning hits, the sal blooms on the ground transform into mushrooms. There are two types of boda: white (rakhadi) and black (jaat). Initially, the black colored sack has come in the market, called Jaat boda. It is very soft and delicious thus

People prefer the black one over white. Apart from this, when the upper layer turns white and becomes hard, it is called Rakhdi boda.

After foraging wild mushrooms like, *Astraeus hygrometricus* (Boda), Termitophilous mushrooms (Bhodu phutu, Dengur phutu, Kanki phutu, Patera phutu, Dussehra phutu), *Cantharellus* sp. (Baans phutu), Edible *Amanita* (Sugamunda), *Boletus* sp. (Jam phutu), villagers go out to sell it collectively on the side of the road or sell them in the mandis. In Tanzania, edible species of *Termitomyces, Cantharellus, Lactarius, Russula*, and *Amanita* were sold at roadside markets (Harkönen, 2002). Some middlemen buy it from rural areas at a lower price and sell it in the city market at a higher price. *Termitomyces* species are a highly prized delicacy collected both for home consumption and for sale in local markets or along roadsides where they command a high price (Pegler and Vanaecke, 1994). People of tribal society earn good money by selling it during this period. If you buy Termitomyces mushroom from the local mark*et al*so, then it will get at least Rs 600-800 kg but as the end of monsoon (October or Dussehra time) price got decline to 100-200 Rs/kg. Merchants collected on an average 2000kg of *Termitomyces mammiformis* from kodugu district of Karanataka state with an approximate cost of 1.5-2$ (mature fruitbody) to 4-5$ (Young fruitbody) (Pahlevanlo and Janardhana, 2012).

Every panchayat holds a haat on a specific weekday. These haats attract traders from all across the region (Bastar). Villagers from the panchayat's neighboring villages go to the haats to purchase and sell things. Local haats are usually managed by traders from the villages or towns along the route. Non-adivasi settlers from other states *viz*. Uttar Pradesh, Bihar, Rajasthan, and Punjab constitute the majority. These activities are increasingly being undertaken by the Sahu community as well. These small business owners are linked to larger business owners who operate huge godowns and cold storage facilities. Some local traders also operate truck fleets and influence transportation in their communities. They retain control over smaller traders operating in the local haats by paying advance cash, which the smaller traders use to acquire forest goods from the villages in advance; in turn, they come and sell the aggregated produce to the 'seth,' or larger trader. Hundreds of quintals of such forest goods are hoarded by the 'Seths,'

Payli and Soli are the vessels made by the tribals to weigh the grains. Soli is of small scale which is about 8 cm. diameter and 11 cm. deep container which contains about 500 grams of grain. The payli is about 14 cm in diameter and 15 cm in diameter deepens in which 4 soli grains come very easily, 4 soli is considered as 1 payli (Almost 2kg). The same vessel was also used to weigh the boda. About one kg boda in Payli and 250 grams in Soli comes. Its demand is increasing year after year in the Border States including Bastar. Boda is being supplied from Bastar to Andhra Pradesh, Telangana and Odisha. It is also in great demand in

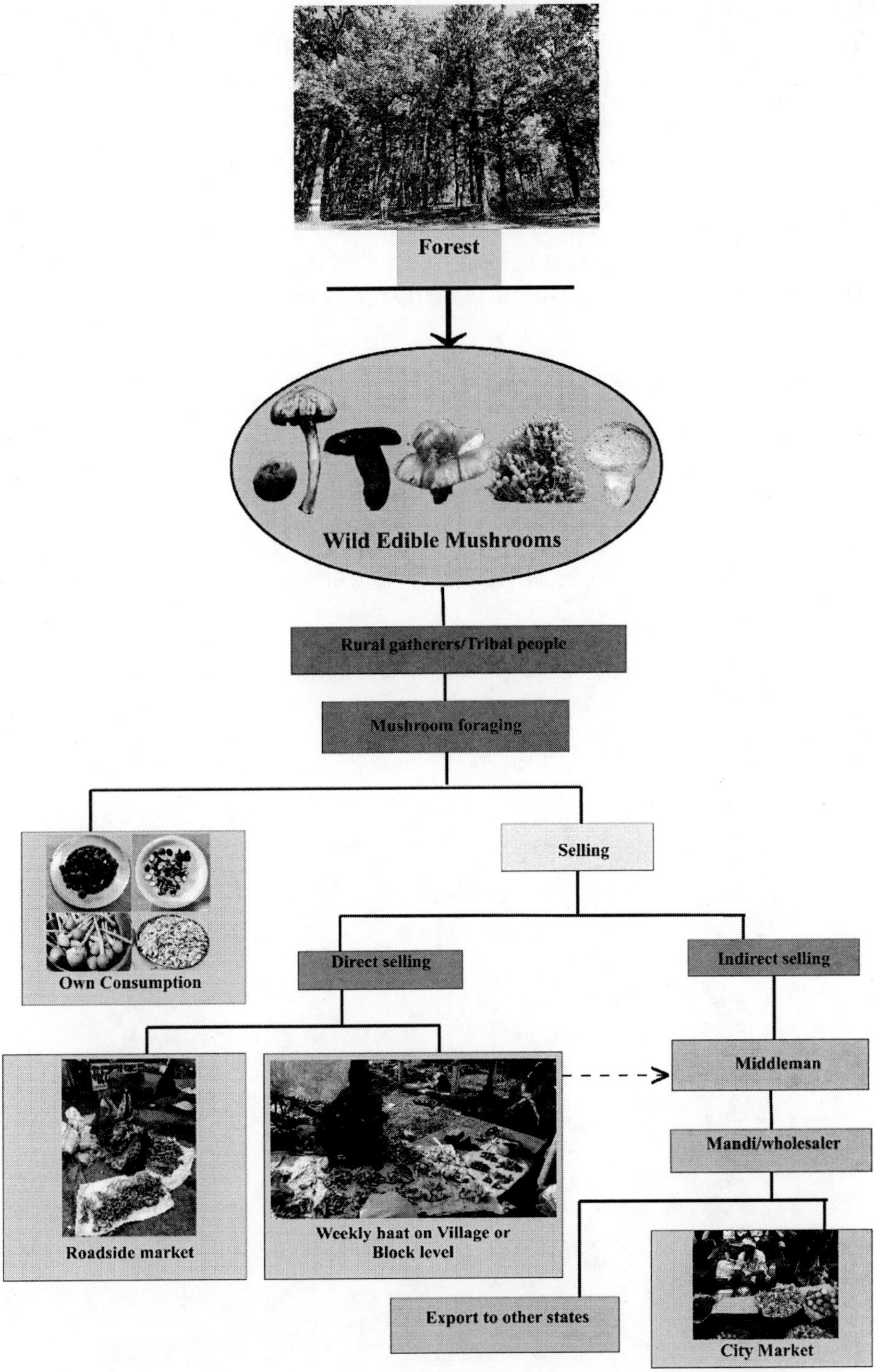

Marketing Channel of wild edible mushrooms in Chhattisgarh

the capital Raipur. At present, its market price is up to Rs 3,000 per kg. Its demand is high and supply is less. Due to this also the prices have increased. Sal trees in Bastar have also decreased a lot in the past few years.

However, an annual turnover of Rs 25 lakhs to 30 lakhs through sale of forest mushrooms especially *T. heimii and A. hygrometricus* (Sal boda, Sarai boda), *Cantharellus* and edible *Amanita* spp. Seasonal collection of *T. heimii and Astraeus hygrometricus* (Sal boda, Sarai boda) have been accounted as 300 to 350 tonnes from Dantewada, Bijapur, Bastar, Narayanpur Gariyaband, Kanker, Nagri, Banjari (Kawardha), Korba, Ambikapur, Jashpur and Tamnar forests of Chhattisgarh.

Table 2: Price list of some mushrooms sold on the markets of Chhattisgarh

Name of the Mushroom	Month	Price/kg	Place
Termitomyces microcarpus, Termitomyces heimii, Termitomyces clypeatus	August to September	Rs.1000/-	Raipur, Kanker, Gariyaband, Jagdalpur, Kawardha, Ambikapur, Korea, Jagdishpur
Astraeus hygrometricus	August to September	Rs. 2000-3000/-	Raipur, Kanker, Gariyaband, Jagdalpur
Cantharellus	August to September	Rs. 300/-	Korba, Kanker, Jagdalpur
Volvariella	June to October	Rs. 300/	Raigarh, Janjgir Champa, Jagdishpur, Saraipalli, Mahasamund, Kanker

Fig. 3. a)-d) Showing *Astraeus hygrometricus* from markets of Bastar

Fig. 4. Wild edible mushrooms at local market in Bastar a) b) c) & d) *Astraeus hygrometricus* e) f) & g) *Russula albonigra*, *Astraeus hygrometricus*, *Amanita caesarea* (Haldi phutu) and *Amanita loosii*

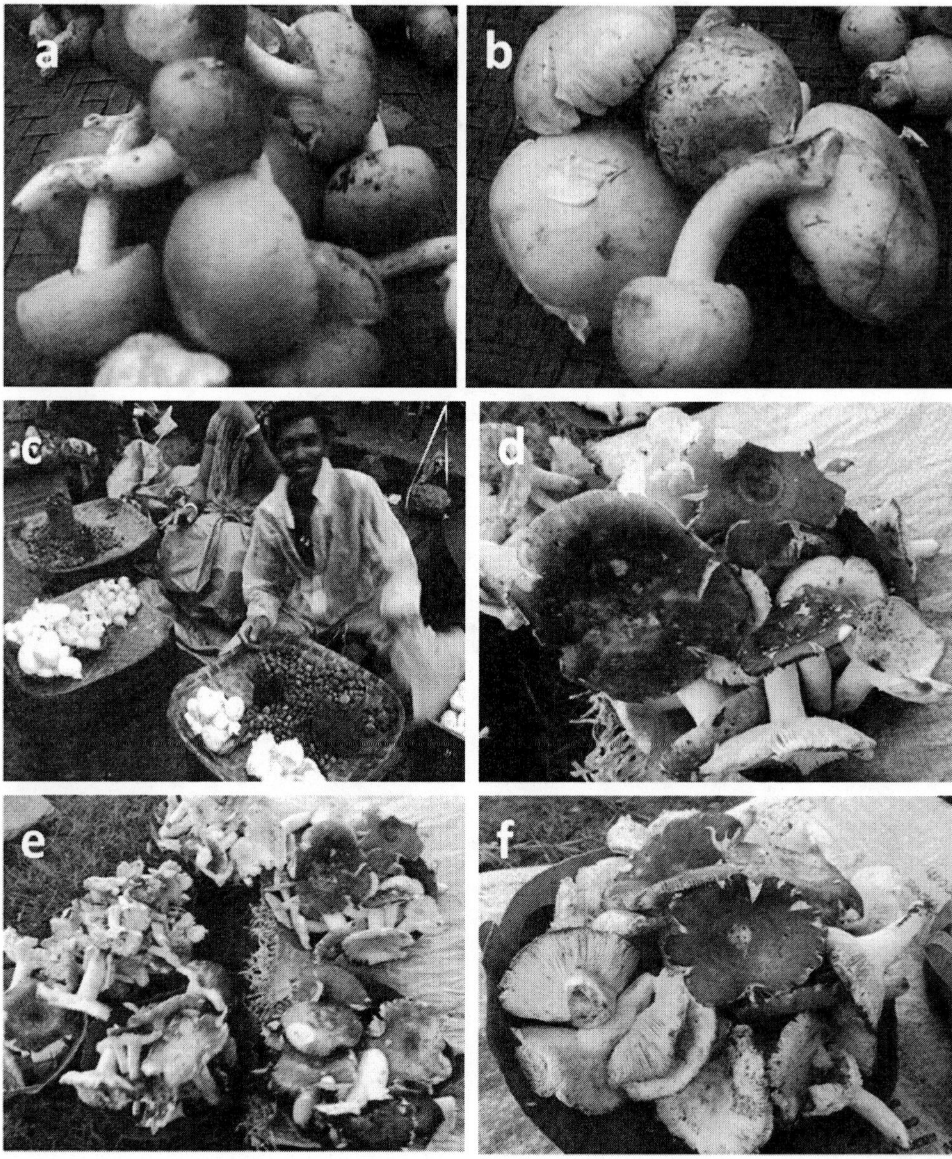

Fig. 5. **Showing Mushrooms at local markets a)** ***Amanita caesarea*** **(Haldi phutu) b)** ***Amanita loosii*** **c)** ***Astraeus hygrometricus*** **d) & e)** ***Russula xerampelina*** **f)** ***Russula rosea***

Fig. 6. a)-c) *T.heimii* (mature stage) d)-g) Young *T. heimii* with long pseudorhiza at local market

Fig. 7. a)-f) *Termitomyces* mushrooms at local markets

Fig. 8. a) *Termitomyces* mushroom at Sanjay market, Jagdalpur b) *Termitomyces heimii* at roadside market Bilaspur c) d) & e) *Termitomyces* mushroom at Shashtri market Raipur f) *Termitomyces heimii* at roadside market Bilaspur

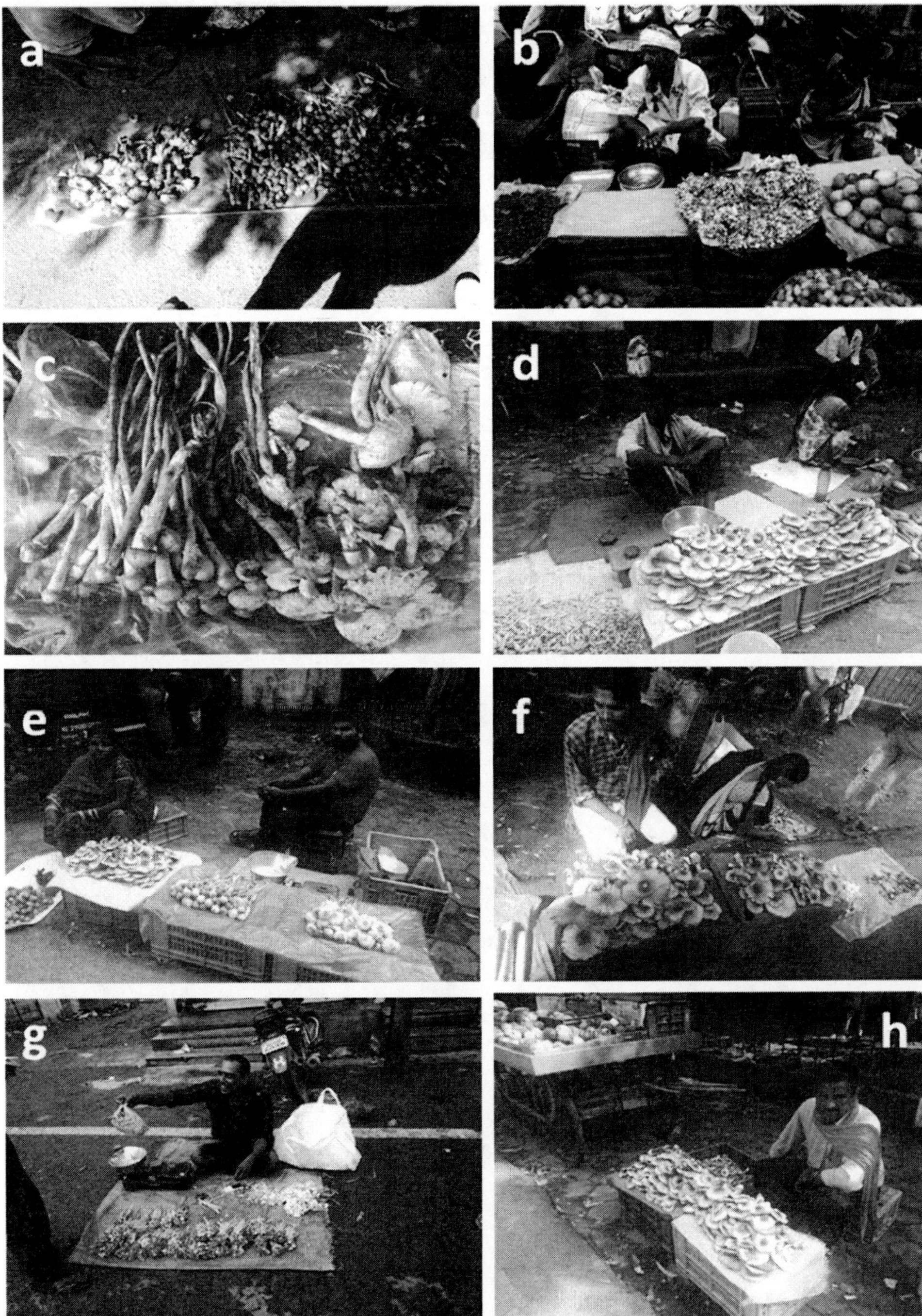

Fig. 9. a) & c) *T. heimii* (Greyish) at roadside market in Bilaspur b) & g) *Termitomyces* mushroom at shastri market Raipur d) e) f) & h) *Volvariella volvacea* at Raigarh

Fig. 10: a) - f) *Volvariella volvacea* in local market

Fig. 11. Showing Mushrooms in a market a) *Tricholoma* spp. b) *Pleurotus* spp. c) *Cantharellus* spp. from Rampur, Korba d) Different species of edible *Russula* from Bastar e) & f) *Cantharellus* spp. g) *Volvariella volvacea*

Fig. 12. ***Termitomyces heimii*** **(Dussehra phutu) on market for sale**

Table 3: List of mushrooms collected by locals with their vernacular name

Sl. No.	Scientific Name	Local / Vernacular name
1	*Amanita vaginata*	Sugamunda
2	*Amanita caesarea*	Haldi Phutu, Hardula phutu
3	*Amanita loosii*	Sugamunda, Safed anda,
4	*Amanita vaginata* var. *fulva*	Munjar dhunda
5	*Astraeus hygrometricus*	Boda, Sal boda, Sarai Boda, Rugda, Chharkeni phutu, Jaat boda, Rakhdi boda, Sargi phutu, Patras phutu, Gohiya, Bastariya phutu, Phutka
6	*Boletus* sp.	Jam Phutu
7	*Bovista dermoxantha*	Dhela phutu
8	*Cantharellus subalbidus*	Bans phutu (khukhadi)
9	*Ganoderma lucidum*	Lakdi phutu
10	*Macrocybe gigantea*	Dudhiya phutu
11	*Macrolepiota procera*	Bhiden phutu
12	*Pluerotus ostreatus*	Pihiri, phutu
13	*Russula emetica*	Paan Phutu
14	*Russula rosea*	Murga Chundar, Paan Phutu, Paan ki begum, Sargi phutuLal badar Khukhdi, Patar Chhati
15	*Sparassis* sp.	Dimag phutu
16	*Trametes versicolor*	Lakdi phutu
17	*Termitomyces. eurrhizus*	Desi Chhattisgarhiya Phutu
18	*T. heimii*	Bhadwahi/ Bhodu phutu/Dusherra phutu, Bhimbhora phutu, Goncha phutu (appearing during Dusshera), Dengur phutu (appearing in termitarium), Khukhdi, Manya Chhati/Badedhunda
19	*T. clypeatus*	Patera Phutu,
20	*T. robustus*	Kumha Khukdi/phutu
21	*T. microcarpus*	Kanki Phutu, Chowk Phutu, Chapat phutu
22	*T. umkowaan*	Patera Phutu, kodri Chhati
23	*T. mammiformis*	Phutu, Dengur phutu
24	*T. aurantiacus*	Patera Phutu
25	*Volvariella volvacea*	Paira phutu, Chhati, Paira ke phool

Wild Mushrooms Need to be Conserved

As part of the State's rich traditional culture, the local committees have played a significant role in preserving its bio-resources by developing informal but participatory mechanisms. Due to the population pressures and diminishment of natural resources, perceptions of local communities have changed over time. As a result, local communities have become ecologically poor and alienated from natural resources. The result of this has been the conversion of many of these communities into actors with solely economic motives, thus leading to the destruction or over extraction of these bio-resources.

The adivasi society's structure reflects their symbiotic relationship with the forest and its flora and wildlife. The Gonds, Marars, Halba, Dhurwas, Murias, and Marias are the forest dwellers. 'Koyatur' is a term used by Gonds and Murias to describe themselves. In Gondi, Mahua is known as 'Koya,' and 'tur,' which signifies people. Marias take their name from the 'Mar,' or hills and woodlands, where they live. Whatever they gather from the forest and produced on their land, they first present to their ancestors, their Devi-Devta! They think that their deos are with them at all times. They represent the species that a certain group of Gonds (identified by their surnames) will defend. The groups are referred to as gotras. Whatever they gather from the forest and create on their property, they first present to their deos, their forefathers! They think that their deos are with them at all times. Lingo, their progenitor, helped the adivasis establish their society inside the jungles of Bastar, and thus the Gotras were born. He developed this strategy to safeguard both his tribe and the flora and creatures of the forest, which had been their predecessors' realm. To maximize resource utilization, Gonds are split into 750 gotras, a system of categorization and social structure. Because each gotra is entrusted with saving 3 species, it is believed that Gonds together protect and preserve virtually all of the species that live in the forest. The Gonds have such a strong ethic of protecting totemic wildlife that they neither injure nor hunt them, nor do they reveal their whereabouts. It would keep the balance and allow for consumption restriction, avoiding exploitation of specific animals, plants, insects, birds, and fungi if the entire society foraged, hunted, and gathered them. The gotras have kept their traditions up to this day. For them, Forest products are essentials that require care, protection, and compassionate usage. They frequently refer to these fruits, flowers, tree sap, roots, and mushrooms as manifestations of their forefathers, imbuing these natural components with spiritual and ecological worth and connecting them with life in nature. The adivasi communities, on the other hand, have begun their shift. They're making their way towards the roadside villages and towns, steadily infiltrating the capital's circuits. People also worship the Mushrooms (*Macrocybe* spp.) that grow out near Bhooteshwarnath Temple in Gariyaband District of Chhattisgarh.

Sal tree is also called Kalpa tree of tribals and Bastar is also called the island of Sal forests. To say that the sal is the state tree of Chhattisgarh, but it is revered by the tribals because more than half a dozen products are found from this one tree. The irony is that year after year, due to encroachment in forests and illegal logging, the forests of the sal are ending. The popularity of edible wild mushrooms is turning into an ecological threat for the forests of Chhattisgarh. Locals who live on the outskirts of the forest scour the forest for edible mushrooms to forage. These locals are accused of removing all the mushrooms, leaving not a single one to develop and release spores, resulting in a rich harvest the next season, despite their lack of awareness of the symbiotic link between the mushrooms and the termites. Some foragers pull out the mushrooms with the whole stock since it sells for more money than the one cut at ground level. Mushrooms, like other fungi, reproduce via spores. When they are harvested before they mature, generally for human consumption, no spores may form, resulting in fewer mushrooms of that species in subsequent years. Furthermore, not all spores will live to see another season. Overexploitation, which is usually the result of ignorance, has resulted in many productive termite mounds becoming barren over the years. Land-use changes, destruction of termite hills, overexploitation due to the growing market, wildfire, and building of roads, and development of projects pose a combined threat to this biodiversity. Overexploitation of the natural resource in the shape of delectable mushrooms may result in the extinction of these species from Chhattisgarh's forests permanently. Mushroom diversity and density are threatened by unrestricted harvesting. The older notion of preserving the gene pool by leaving some mushrooms for God has faded. Termitomyces can only develop with the aid of termites owing to their intricate mutualistic nature, making artificial cultivation difficult. Another major reason why this fungus is becoming increasingly endangered is that there are no rules governing its harvesting. As the mushroom becomes increasingly endangered, the price rises, enticing mushroom hunters to continue looking for and selecting such rare fungus. There appears to be a strategy in the works to pass legislation prohibiting the harvesting of *Termitomyces*. The '*Termitomyces*' mushrooms will be gone forever unless greater awareness is raised and tougher restrictions are implemented in forested regions. Conserving Mushroom biodiversity is as important as conserving Rhinoceros, Turtles, and Tigers but these big species got more attention. Conservation is only possible when the owners of the bioresources, the local people, are convinced, involved, and can sustainably derive the benefits by ensuring that varieties are not overexploited or lost. By creating awareness about the conservation of *Tertmitomyces, Astraeus,* and other wild edible macrofungi.

CONCLUSION

Several tribal groups are greatly dependent on forest agriculture or forest resources. There are still many mushroom species whose medicinal and nutritional properties are unknown to us. However, indigenous people or tribals are aware and know how to use wild macrofungi as nutritional supplements and as a treatment for various diseases. Though knowledge is typically passed down from generation to generation in the baiga tribe, traditional food values and ethnomedicinal knowledge of macrofungi are at risk of disappearing in the community since they are confined to the elderly. Hence, it is important to document traditional knowledge of mushrooms for future generations. Also, the way wild edible mushrooms are being exploited recklessly, these indigenous mushrooms need to be conserved and preserved for posterity under competent supervision; otherwise, they would go extinct like many other species that have already disappeared or are on the verge of disappearing forever.

ACKNOWLEDGMENTS

The authors appreciate the technical assistance provided by the Department of Plant Pathology, College of Agriculture, Indira Gandhi Krishi Vishwavidyalaya, Raipur, Chhattisgarh, India. We'd also want to thank the tribal healers of Chhattisgarh for their help with the ethnomycological study.

REFERENCES

Adhikari MK (2000) Mushrooms of Nepal. Kathmandu: P. U. Printers.

Ao T, Deb CR and Khruomo N (2016) Wild edible mushrooms of Nagaland, India: a potential food resource. Journal of Experimental *Biology and Agricultural Sciences*, 4(1): 59-65.

Aryal HP and Budathoki U (2016). Ethnomycology of *Termitomyces* R. Heim in Nepal. *Journal of Yeast and Fungal Research*, 7(4): 28–38.

Averill C, Turner BL and Finzi AC (2014) Mycorrhiza-mediated competition between plants and decomposers drives soil carbon storage. *Nature*, 505: 543-45.

Bhatt RP, Tulloss RE, Semwal KC, Bhatt VK, Moncalvo JM and Stephenson SL (2003) *Amanita*ceae from India. A critically annotated checklist. *Mycotaxon* 88: 249–270

Boa E (2004) Wild edible fungi: a global overview of their use and importance to people. NonWood Forest Products, No. 17, FAO, Forestry Department, Rome, Italy.

Borah N, Semwal RL and Garkoti SC (2018) Ethnomycological knowledge of three indigenous communities of Assam, India. *Indian Journal of Traditional Knowledge*, 17(2): 327-35.

Chang, ST and Miles, PG (1992) Mushroom biology - a new discipline. *Mycologist*, 6: 64–65.

Chang, ST and Miles, PG (2004) Mushrooms: Cultivation, Nutritional Value, Medicinal Effects and Environmental Impact. USA: CRC Press, 480.

Cowan A (2001) Fungi—life support for ecosystems. Essential ARB, 4: 1-5.

Devkota S (2006) Yarsagumba; Traditional utilization in Dolpa district, Western Nepal. *Our Nature*. 4:48-52.

García-Lafuente A, Moro, C Villares A, Guillamón E, Rostagno MA, D'Arrigo M, Martínez JA (2010) Mushrooms as a source of anti-inflammatory agents. Anti-Inflammatory & Anti-Allergy Agents in Medicinal Chemistry. 9: 125–141.

Härkönen M (2002) Mushroom collecting in Tanzania and human (Southern China): inherited wisdom and folklore of two different cultures. In Watling, R, Frankland JC, Ainsworth AM, Isaac S and Robinson CH (Eds.) *Tropical Mycology, Vol. 1, Macromycetes* 1: 149-165.

Harsh NSK, Tiwari CK and Rai BK (1996) Forest fungi in the aid of tribal women of Madhya Pradesh. *Sustainable Forestry*, 1: 10-15.

Harsh NSK, Rai BK and Ayachi SS (1993). Forest Fungi and Tribal Economy-A Case Study in Baiga tribe of Madhya Pradesh (India). *Journal of Tropical Forestry* 9: 270-279.

Harsh NSK, Rai, BK and Soni VK (1999) Some ethnomycological studies from Madhya Pradesh, India. In *From ethnomycology to fungal biotechnology*. Springer, Boston, MA. pp. 19-31.

Kinge RT, Nkengmo AA, Theobald MN, Ache NA and Afui MM (2017) Species Richness and Traditional Knowledge of Macrofungi (Mushrooms) in the Awing Forest Reserve and Communities, Northwest Region, Cameroon. *Journal of Mycology*, 1-9.

Koné, NA, Yéo K, Konaté, S and Linsenmair KE (2013) Socio-economical aspects of the exploitation of Termitomyces fruit bodies in central and southern Côte d'Ivoire: raising awareness for their sustainable use. *Journal of Applied Biosciences*, 70: .5580-5590.

Lahiri SS, Shukla MD, Shah MB and Modi HA (2010) Documentation and Analysis of Certain Macrofungal Traditional Practices from Western-India (Gujarat). *Ethnobotanical Leaflets*, 14(5): 626-641.

Matsumoto T and Abe T (1979) Studies on the distribution and ecological role of Termites in a lowland rain forest of West Malaysia. (2) Food and Feeding Habits of Termites in Pasoh Forest Reserve. *Japanese Journal of Ecology*, 29: 121-135.

Pahlevanlo A and Janardhana GR (2012) Diversity of *Termitomyces* in Kodagu and need for conservation. *Journal of Advanced Laboratory Research in Biology*, 3(2): 54-57.

Pegler, DN and Vanhaecke, M (1994) *Termitomyces* of Southeast Asia. *Kew Bulletin*, 49(4): 717-736.

Piearce GD (1987) The genus *Termitomyces* in Zambia. *Mycologist*, 1(3): 111-116.

Pradhan P, Dutta AK, Roy A, Basu SK and Acharya K (2012). Inventory and Spatial ecology of macrofungi in the Shorea robusta forest ecosystem of lateritic region of West Bengal. *Biodiversity*, 13: 88-89.

Purkayastha RP and Chandra A (1985) Manual of Indian Edible Mushrooms. New Delhi, India: Jagendra Book Agency.

Rai BK, Ayachi SS and Rai A (1993) A note on Ethno-mycomedicines from central India. *Mycologist*, 7: 192-93.

Rammeloo J and Walleyn R (1993) The edible fungi of Africa South of the Sahara: a literature survey. *Scripta Botanica Belgica*, 5: 1-62.

Sarma TC, Sarma I and Patri BN (2000). Wild Edible Mushrooms used by some ethnic tribes of Wester Assam. *The Bioscan*, 3: 613-625.

Semwal, KC, Stephenson SL, Bhatt VK and Bhatt RP (2014) Edible mushrooms of the Northwestern Himalaya, India: a study of indigenous knowledge, distribution and diversity. *Mycosphere*, 5(3): 440-61.

Shukla K, Giri B and Shukla RV (2017) Occurrence and Distribution of Mushrooms in Semi-evergreen Sal (*Shorea robusta*) Forest Chhattisgarh, Central India. In: Developments in Fungal Biology and Applied Mycology, edited by T. Satyanarayana, S. Deshmukh & B. Johri, (Springer, Singapore), p. 501-523.

Singer R (1986) *The Agaricales in Modern Taxonomy*, (Bishen Singh Mahendra Pal Singh, Dehradun, India), 981-982.

Tanti B, Gurung, L and Sarma GC (2011) Wild edible fungal resource used by the ethnic tribes of Nagaland, India. *Indian Journal of Traditional Knowledge*, 10(3): 512- 15.

Tayung K and Panda KM (2015). Documentation and Ethnomedicinal knowledge on wild edible mushrooms among ethnic tribes of Northern Odhisa, India. *Asian Journal of Pharmaceutical and Clinical Research*, 8(4): 139-43.

Tibuhwa DD (2012) Folk taxonomy and use of mushrooms in communities around Ngorongoro and Serengeti National Park, Tanzania. *Journal of Ethnobiology and Ethnomedicine*, 8(1): 1-9.

Tibuhwa DD, Kivaisi AK and Magingo FSS (2010) Utility of the macro-micromorphological characteristics used in classifying the species of *Termitomyces*. *Tanzania Journal of Science*, 36.

Tidke G and Rai MK (2006) Biotechnological potential of mushrooms: drugs and dye production. International Journal of Medicinal Mushrooms, 8(4).

Tiwari SK (1994) Encyclopaedia of Indian tribals, vol 1. Rahul Publishing House, New Delhi.

Tripathy SS, Rajoriya A and Gupta N (2014) Wild Indegenous Mushrooms as a Source of Food. *Journal of Pharmaceutical Biology*, 4(3): 138-147.

Wood TG and Sands WA (1978) Production Ecology of Ants and Termites. In: The role of termites in ecosystems (Eds. M. V. Brian). Cambridge University Press, UK. 245 – 292.

Wood TG and Thomas RJ (1989) The mutualistic association between Macrotermitinae and *Termitomyces*. In: Insect- Fungus Interaction (Eds. Wilding N., Hammond, P.M. and Webbe, J.F). Academic Press, London, 69-92.

Yamada A, Inoue T, Wiwatwitaya D, Ohkuma M, Kudo T, Abe T and Sugimoto A (2005) Carbon mineralization by termites in tropical forests, with emphasis on fungus combs. *Ecological Research*, 20(4): 453-460.

Sustainable Agriculture and Biodiversity: Innovations and Insights, Pages: 158–181
Edited by: Bibhuti Bhusan Sahoo, Harvinder Kumar Singh, Niranjan Chinara and Amitabh Panda

CHAPTER - 6

COMPREHENSIVE DISEASE MANAGEMENT STRATEGIES FOR MAXIMIZING APPLE CROP HEALTH AND YIELD

Upma Dutta[1*] Arashdeep Kour[1] and Deepti Jha[2]

[1]*Division of Microbiology,*
Sher-e-Kashmir University of Agricultural Sciences & Technology of Jammu-180009, Jammu & Kashmir
[2]*Department of Plant Pathalogy, College of Agriculture, IGKV, Raipur-492012, Chattisgarh*
**Corresponding email: upmaduttamicro@gmail.com*

Apple (*Malas domestica*) is a major temperate fruit crop valued for its economic and nutritional importance. However, its yield and quality are often threatened by various diseases. This chapter outlines integrated disease management strategies for maximizing Apple crop health and yield.

1. SCAB DISEASE

Introduction/Economic Importance: The first reports of scab disease dates back to 1819 in Sweden, and it spread to other apple-growing regions of the world. However, in India, reports of scab disease on the local cultivar Ambri date back to 1935 in the Jammu and Kashmir region (Nath, 1935). Nearly all of India's developing states have reported cases of the illness. In 1973, scab disease broke out as an epidemic in the Jammu and Kashmir area, infecting over 70,000 acres of apple orchard and causing an estimated loss of rupees of approximately 55 lakhs (Joshi *et al.*, 1975). Apple scab plant disease has a significant economic impact. Every year, it results in considerable financial losses in yields. Fruit infections are the primary cause of losses, which lead to low yields, poor fruit quality, and poor.

Causal Organism: (*Venturia inequalis*) (Cke) Went.

Imperfect stage: Spilocaea pomi Fr.

The fungal pathogen *V. inaequalis* is the cause of the sickness. The pathogen makes it through two separate stages of its life cycle. The pathogen's perfect stage, or sexual stage, lives on sick fallen leaves during the winter, while its imperfect stage, or asexual stage, lives as a parasite on leaves and fruits during the growth season. Blooming crab apples (*Malus* spp.), hawthorn (*Crataegus* spp.), and loquat (*Eriobotrya japonica*) are the corresponding pathogen host plants. Similar symptoms are caused by *Venturia pirina*, a similar fungus, which infects pears (*Pyrus* spp.) trees and associated plant components (Gupta, 1985).

Symptoms: The apple tree's upper ground plant components, such as the leaves, petioles, blooms, sepals, pedicels, juvenile shoots, bud scales, and fruits, displayed signs of the scab disease. The majority of bloom indications are tiny, dull green lesions near the base of the blooming. Fruit drop results from it occurring on the stem pedicel and sepals both before and during flowering. Usually at the underside of growing leaves, the lesions are first noticed in the spring. Subsequently, it also spreads to the top leaf surfaces. Because older leaves are frequently more resistant to infection, lesions may look modest on enlarged and older leaves and fairly big (more than 1 cm in diameter) on young leaves. In the early stages, the lesions are tiny and round. It becomes larger over time and acquires an olive and velvety colour from the asexual spores (conidia) that are produced. Tissues that are impacted progressively get puckered and blurry. The leaf damage are frequently ripped and shattered. Damage to the foliage and fruits appears as huge, rankled, "scabby," and edge-like lesions. Velvety, green to olive-brown lesions are the symptom on fruits in water-soaked places (Gupta, 1990, 1985). Fruit deformity may appear in newly infected fruit. Fruits and leaves from the tree that are severely diseased often fall off. When

Fig. 1.1 Scab lesions on fruits

Fig. 1.2 Secondary scab lesions on leaves

a scab disease infection causes leaves to get defoliated for two or three years in a row, the tree becomes more susceptible to diseases, insect damage, and freezing damage. The symptoms of scab disease in fruits and leaves are depicted in Figures 1.1 and 1.2.

Disease Cycle and Epidemiology: The apple scab organism (*V. inaequalis*) inhabits the space in living leaves that exists between the host cuticle and the epidermis. The infectious mycelium completely colonises damaged leaves once they have fallen and propagated. The majority of pseudothecial initials in leaf detritus appear a month or so after leaves fall and enter a dormant phase. Conidia that serve as necessary inoculums will be produced throughout development in infected buds in temperate environments. In spring, there is mating among the leaves. The asci develop through the top level of the pseudothecium when the pseudothecia get moist. Persuasively released into the air, the ascospores spread out with the wind and breeze. Ascospores infect apple blooms and young leaves, causing lesions if there is sufficient humidity from rain or dew (Gupta and Lele, 1976). Nine to thirty days after the initial lesions on apple blossoms and young unfolding leaves occur, conidia are produced. These are dispersed to other leaves and fruits, resulting in secondary contaminations. Up to 100,000 conidia can be formed from a single lesion. The temperature and humidity levels, as well as the tissue host, genotype, and age features of the plant, influence the rate of lesion infection by ascospores or conidia. Lesions may be smaller in size and grow more slowly in resistant apple varieties or during a cold season, which reduces the number of secondary cycles. The formation of conidia is influenced by moisture, and the generation of spores requires 60% to 70% moisture. Ascospore and conidia contamination is strongly influenced by the average temperature and the degree of moisture in the leaves and fruits. Conidia contamination is determined using the Mills table, which links the term and temperature of the leaf moisture. For example, an infection will be minimal if leaves are wet for nine hours at an average temperature of eighteen degrees Celsius; but, if leaves are wet for eighteen hours or more at the same temperature, the infection will be quite serious. According to Gupta and Lele (1976), lesions with an average temperature of 18 °C would develop conidia after 9 days, whereas at lower temperatures (8 °C, it takes 17 days).

Integrated Disease Management

Cultural Control:

- Cultural initiatives, planting malady-safe assortments, using synthetic chemicals, and using natural control agents can all be used to manage the management of apple scab disease.

- Adhere to cultural practices; cleaning up and getting rid of contaminated leaves that have fallen to the ground will help prevent sickness. However, large orchards cannot use it (Gupta, 1985).
- Applying a urea solution splash to apple trees shortly before leaf drop or on diseased leaves that have fallen to the ground lowers the quantity of ascospore inoculum that develops during the upcoming growing season
- Pruning the trees on a regular basis is a common method of controlling apple scab illness.
- Proper air circulation and enough sunshine reaching the ground will help in the quick drying of fallen diseased leaves, which will inhibit the formation of scab ascospore. This procedure will aid in increased sunlight penetration to the soil surface in orchards.
- Excessive watering should be avoided, particularly in climates where scab generation is encouraged (Gupta, 1985).

Choose Resistant Cultivars and Rootstocks:

- There are vast amounts of apple varieties that are protected from scab infections. The most effective treatment for the condition is to plant scab-safe assortments.
- Thc apple types that don't scab are free of the following: Jonafree, Liberty, Pure, Red, Gold Rush, Independence, and Capitalism. Farmers and customers alike welcome the variety. Crab apples with scab resistance to blossom include David, Harvest Gold, Mary Potter, and Prairie Fire. Because there is no chance of defoliation due to apple scab infection and these kinds make lovely landscape plants, they have a higher potential for flowering.
- In arid locations where scab disease is not a barrier to apple production, customers are advised to cultivate the vulnerable types of apples, which include Delicious, Fuji, Gala, Golden Delicious, McIntosh, and Winesap (Gupta and Lele, 1976).

Chemical Control:

- Preventive chemical fungicides are sprayed to apple leaves and fruits before to the onset of the illness, while curative fungicides are used to treat scab sickness in apples.
- In the unlikely event that ascospores or conidia are present on plant surfaces, the application of fungicide prevents the parasite spores from germinating and infiltrating the host tissue.
- To avoid infection on freshly developing tissues in the early season, chemicals must be sprayed at regular intervals of seven days (Gupta and Lele, 1976).

- During the crop-growing season, seven non systemic and systemic sprays are applied to apple trees in protective mode.
- Chemicals used in therapy have limited fundamental activity. When applied, even soon after contamination, they have the ability to penetrate the leaf and regulate further development.
- Growers only use curative fungicides when absolutely necessary. Fungicide use during the dry spring season can help manage early infection. Fungicide residues are either nonexistent or present at a safe level when apples are harvested (Gupta and Lele, 1976).

Biological Control: In the future, Apple scab disease can be effectively managed by using biological control agents, also known as natural microbial antagonists, as an alternative to chemical fungicides (Ellis *et al.*, 1997).

2. POWDERY MILDEW

Introduction/Economic Importance: The first reports of powdery mildew disease on apple seedlings date back to 1871 in Iowa, the United States, and thereafter in apple-growing regions worldwide (Bessey, 1877). Nearly all of India's developing states have reported cases of the illness. The main victims of this illness are nurseries and seedling supplies. All in all, Victorian apples provide a significant issue. Winter squash, Golden Delicious, Buckingham, White dotted Red (Maharaji), Granny Smith, and Ambary varieties are particularly susceptible to the illness. The powdery mildew disease can result in stunted tree development, dead buds, smaller leaf areas, slower growth of the shoots, and worse fruit quality. Plants with severe mildew infections might have an 80% reduction in yield (Glawe, 2008).

Causal Organism: *Podosphaera leucotricha* (El. Ev.) Salmon Imperfect stage: *Oidium farinosum* (Cooke) Powdery mildew is caused by *P. leucotricha* [(El. Ev.) Salmon], fungal organism The sickness may also be brought on by the presence of *Podosphaera oxycanthae* and *Erysiphe heraclei*. The host plants are pears (*Pyrus* spp.) and crab apples (*Malus* spp.).

Symptoms: The freshly formed shoots, leaves, both young and old, flowers, and fruits all displayed symptoms. Generally, leaves and fruits are where symptoms are most visible. A 3-5 day delay in the emergence of new buds is the first sign of the illness that emerges in the spring. The illness manifested as a white to light-dim powdered material covering foliage, flowers, and fruits; this powdery mass is made up of pathogen spores (Glawe, 2008). Plant components are quickly affected by the illness since it quickly spreads from leaves. The pathogen's dark brown or earthy-colored fruiting bodies cover the whole white to light brown powdery mass on apple tree branches. It turns out that the infected foliage is brittle and rigid.

White patches on the upper and lower surfaces of the newly emerging shooting leaves are indicative of the illness. Some leaves may twist, shatter, or develop disfigurement as a result of the illness. Pathogens that infect succulent leaves and developing shoots can cause illness, which can diminish the size of the entire shoot and cause the leaves and shoots to become brown in the middle of summer. According to Hickey and Yoder (1990), flowers that do not develop properly become greenish-white and cease to yield fruits. Figures 2.1–2.3 depict the symptoms of powdery mildew on dormant shoots, blossoms, leaves, and fruits.

Fig. 2.1 Symptoms on dormant shoot.

Fig. 2.2 Symptoms on blossom and leaves.

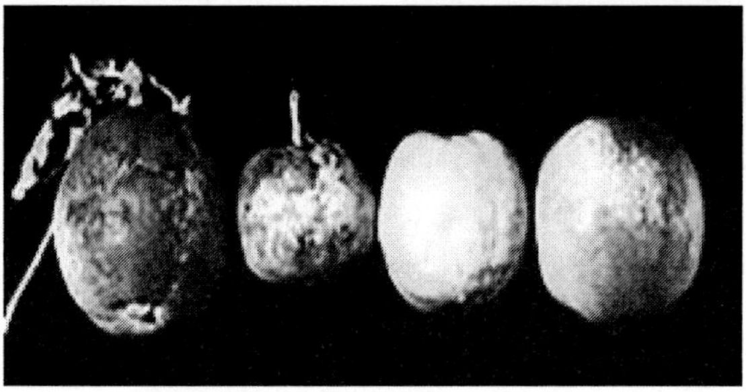

Fig. 2.3 Symptoms on fruit.

Infection in Apple Shoots: The development of terminal buds begins in the spring. Young green tissue is colonised by the fungus. In addition to their silver-gray colour, these diseased flag shoots may have displayed dieback signs, reduced growth, and defoliation. In the early stages of the developing season, the main contaminations of the shoots generate inoculums that lead to secondary contamination on leaves, flowers, and even fruits. Trees that are heavily infected become weak, and in order to infect secondary inoculums, mycelium and different coloured fruiting bodies (ascocarps) are needed (Hickey and Yoder, 1990).

Infection in Leaves and Fruits: Powdery mildew spores carried by the wind cause secondary infection on newly enlarged leaves. On the surface of leaves, fungal colonies made up of hyphae, mycelium, and spores exhibit chlorotic and white spots. The margins of diseased leaves twist, wrinkle, or fold lengthwise. As the illness progresses, affected tissues take on the powdery, silver-grey look typical of powdery mildews. Net resetting and discolouration during maturing fruits are caused by infections on the bloom receptacle or immature fruit. According to Hackett and Yoder (1990), deformed and small-sized fruits result in inferior quality and reduced production.

Infection in Apple Blossoms: After healthy buds, infected blooming buds become silver-gray in colour and open 5-8 days later. Petals have a light green or pale green colour that is blurry. Blossoms begin to wither and struggle to produce fruit. The buds don't wake up until the next spring. According to Turechek *et al.* (2004), significantly infected buds that have pathogens at the time of opening may completely destroy the yield for the next season.

Disease Cycle and Epidemiology: *P. leucotricha's* mycelium infected apple blossom and shoot buds the year before. In winter, the mycelium remains dormant. With the arrival of spring, dormancy is broken and fungal growth resumes, feeding on the growth of developing leaf shoots and surrounding tissue. These first infections on conidiophores result in the production of asexual conidia, which are carried by the wind. At high humidity (>70%) and temperatures between 10 °C and 25 °C, pathogen conidia germinated. Conidia expand to enclose hyphal outgrowths, which pass through the leaf's surface, develop, and finally straighten into the form of an oppressor. The chemical released by these structures facilitate the penetration of parasite contamination pegs into the plant's epidermal cells, enabling them to expand and frame haustoria. Haustoria are particular organs that take on the structure of live plant cells, absorbing nutrients and continuing to be parasites. Until helpless tissue is no longer accessible at this point, the contamination routine (hyphal outgrowth > appressorium > contamination peg > haustorium) is repeated as the mould status worsens or as optional illnesses cause fresh settlement production. Late-season development might result in an unanticipated rise in mould activity. Auxiliary disease cycles not only aid in the quick inoculum formation but also contaminate the terminal and horizontal buds that will carry the organism through the winter. (Turechek *et al.*, 2004; Heck and Yoder, 1990).

Integrated Disease Management:

Cultural Control:

- The earliest inoculum sources that is, the diseased flower and shoot buds from the previous year are also eliminated in order to control the primary contamination.
- It is necessary to cut out infected white terminal shoots in the winter or early spring. Large orchards are not economically viable for pruning. There will be a significant benefit if there is inexpensive labour available (Yoder, 1992).

Choose Resistant Cultivars and Rootstocks:

- Cultivars that are resistant to accumulation and require management only in situations of elevated illness tension are available. Jonafree, Prima, and Enterprise are resistant apple cultivars, albeit they are not widely cultivated.
- Rather than illness blockage, the choice of cultivar is influenced more by business bid, fruit size, fruit appeal, and pollination features.
- Although cultivars like Granny Smith and Golden Delicious are commonly cultivated, they can require the treatment of chemical diseases because to their moderate to high susceptibility to mildew (Hickey and Yoder, 1990). According to Verma and Gupta (1988), cultivars of Golden Chinese and Maharaja Chunth are resistant to powdery mildew.

Chemical Control:

- Fungicide foliar sprays can be used to manage fruit infection and secondary infections. Fungicides are virtually always used in commercial orchards to manage apple diseases like mildew.
- Fungicides are typically sprayed every 7–10 days starting in the tight-cluster stage and continuing until the conclusion of the terminal shoot development in the middle of summer. Chemical sprays work best in the postbloom phase, while the leaf is still growing. In extremely sensitive cultivars, this may occasionally entail 18 sprays for disease management (Yoder *et al.*, 1997).
- The most effective agents for controlling mildew include strobilurins, myclobutanil, fenbuconazole, and sulphur. Apple producers use several activity groups of fungicides in alternate applications rather than a single fungicide. Sulphur, for example, is sprayed during a low risk period to cultivars that are less vulnerable. Although benzomidazoles were successful in combating mildew, their use was curtailed because to the extensive resistance that V. inaequalis (apple scab) acquired.
- The long-term consequences of powdery mildew include the trees' lifetime and profitability. It also has a chronic influence on tree vigour and productivity. The understanding of apple growers on management plan is also necessary for the successful management of mildew.

3. PHYTOPHTHORA COLLAR ROT

Introduction/Economic Importance: Apples are frequently afflicted with Phytophthora collar rot. The "Grimes Golden" variety of apples is especially prone to this illness. On apple trees grown on Malling Merton 106 (MM.106) rootstocks, phytopthora collar rot is extremely severe. Apple farming without pollution and using contaminated surface water for overhead irrigation are two further practices that contribute to the disease's development. Numerous other tree fruits, especially nut trees, are the host plants of the causative bacterium.

Causal Organism: *Phytophthora cactorum* is the causative agent of this disease; however, *Pythium ultimum, Phytophthora megasperma, Phytophthora syringae*, and *Phytophthora cambivora* can also occasionally contribute to the development of collar rot (Rana and Gupta, 1984).

Symptoms: The disease phytopthora starts around the crown, travels to the rootlets, and eventually results in a lesion at the soil's surface. The illness can also spread to the bud's association if the scion cultivar is helpless. The canker forms a characteristic shape and callus tissues accumulate at the margins as the bark dries out. Likewise, susceptible cultivar fruits are susceptible to damage by the Phytophthora pathogen. Apple trees that have been contaminated experience rot and a solid, light tan spoilage on or near the ground (Rana and Gupta, 1984). Apple trunks are attacked by phytophthora collar rot up to 76 cm above the soil's surface. There is contamination in the disease between the root-crown layer and the soil line. At the tree's lateral root crossing, contaminations frequently start. At the ground level, define the new canker problem. They grow larger until they are dark, gloomy, and roughly oval in form with an odd border. The symptoms of contaminated bark are earthy in colour and frequently thin, fragile, or delicate. The presence of Phytophthora is indicated by the earthy to reddish earthy coloured wood discoloration and a sticky exudate beneath the ulcer's dead bark. Additionally, there are dark streaks that start close to the shift and continue past the canker's edge.. In the unlikely event that an infection spreads for a considerable amount of time, only the smaller areas exhibit the typical colour and texture of freshly killed tissue. The canker spreads swiftly in both vertical and even bearings. The support of the affected appendage roots, or trunk, leading to the demise of the affected organs or the entire tree, is a conclusive effect of neckline spoilage. Declining vigour, poor terminal bud development, and the early production of tiny, sparse yellow leaves in the summer or rosy bronze leaves in the early harvest are the key indicators of the infection. Symptoms of the illness include early fruit maturity and the growth of additional coloured fruits. Only branches above the canker can experience the foliar symptoms, and badly afflicted trees finally perish. According to Rana and Gupta (1984), phytopthora collar rot is typically linked to malignancies brought on by sporadic or secondary invaders. Figures 3.1 and 3.2 show the contamination of collar rot disease in apple fields and apple trees, respectively.

Fig. 3.1 Phytophthora collar rot at the base of apple tree. Fig. 3.2 Apple orchard infected with collar rot.

Disease Cycle and Epidemiology: The pathogen mycelium can persist in trunk cankers throughout the winter and in the soil for several years as thick-walled oospores or chlamydospores. The highest concentrations of oospores and chlamydospores are found in the soil of ancient apple plantings. These oospores with strong walls may withstand harsh conditions, such as drought and cold temperatures, for an extended duration. They are also mainly resistant to treatment with chemicals. Late winter is the ideal time for oospores to develop in bark holes, or the mycelium can infiltrate the bark near the soil's surface. Sporangia in the soil are the source of the many zoospores that develop in the late spring or early summer. Zoospores, which swim in irrigation or rainfall-fed water, reach the bottom reaches of trees. Rain splash also contaminates fruits that have grown close to the ground, infecting new branches and causing fruit deterioration. The Phytophthora fungus may develop a huge number of spores with the aid of a small number of zoospores (Anonymous, 1989).

For the growth and infection of fungi, naturally occurring temperatures and moist soil are ideal. The pathogen *Phytophthora cactorum* develops at 13 °C and remains active at 21 °C. The most common times for apple "Midwest" sensitive cultivars to be impacted are during the flowering cycle, which occurs again in late August or early September, and in October, when dormancy begins. During these periods, the host is most vulnerable and the fungus is most active in the soil.

Integrated Disease Management: The disease known as phytopthora collar rot is extremely difficult to trace and control. Before blossoming, draw out a lengthy, chilly season and moist environment. The illness can have a major impact on plants. It takes defensive tactics to contain the sickness. If an illness is discovered after it is too late to treat, the apple tree may still be rescued (Anonymous, 1989).

Cultural Control:

- Soils with a high clay content or those that are moist and poorly drained tend to be more susceptible to phytophthora collar rot.
- Apples should be replanted on slopes with adequate drainage. In fields, the creation of a modest slope should come before planting apples on hills. It will support the water's smooth flow.
- Orchard wetlands with well-drained tiling enhance internal soil drainage and manage the illness.
- It is best to prevent flooding and dirt saucer around the stems of apple trees.
- Apple trees should be planted shallowly, much as at a nursery.
- The majority of disease contamination will result from high-density planting.
- Late summer or early autumn is not the best time to fertilise orchards. Fertiliser use during these times results in low cold tolerance and delayed development. We need to provide the young apple trees with assistance.

Choose Resistant Cultivars and Rootstocks:

- In areas with thick, poorly drained soil, sensitive rootstocks should not be planted.
- Rotating collar disease poses a serious threat to varieties of apples, Midwest, Dutch, Grimes Golden, and Lodi.
- Rome Beauty, Jonathan, McIntosh, Golden Beautiful Reliability of cultivars seedling rootstocks against collar rot disease is rather high.
- The disease known as collar rot is largely resistant to the cultivars Melba, Red Wonderful, Rich, and Winesap.
- Since contaminated nursery stock is a major source of illness, disease-free nursery stock need to be planted.
- There are differences in the Dwarf rootstocks' vulnerability to Phytophthora collar rot.

- It is acknowledged that the rootstocks M-4 and M-9 of East Malling (M) are quite resistant. It has also been established that Ottawa-3 is comparatively resistant to Canadian rootstock, viz.
- Extremely sensitive rootstocks include M-25, M-26, Malling-Merton (MM) 103, MM-104, MM-106, MM-107, MM-109, MM-110, MM-111, MM-113, and MM-115.
- The rootstocks of the M-2, M-7, and MM-112 are resistors of the intermediate type. However, M-9 and M-26 rootstocks have a considerable susceptibility.
- Offers recommended methods for managing the generation of resistant rootstocks on apple cultivars (Anonymous, 1989).
- MM-106 and other dwarf rootstocks are particularly susceptible to winter damage.

Chemical Control:

- Blitox/fytolon/Blue Copper @0.1% or Dithane M45 @0.03% can be sprayed in a 30-cm radius around the apple tree trunk to suppress the disease (Agrwal, 1970).
- Fosetyl–aluminum foliar treatments totally eradicate the disease while simultaneously promoting tree growth and increased fruit production (Utkhede and Smith, 1995).
- If collar rot infection develops, the grower may have some success in stopping and curing the illness if they adhere to the following guidelines:
- One-year-old inarch-resistant varietal whips into the trunk much above the cankered area. This therapy is beneficial if the tree's trunk circumference is infected to a depth of 25% to 50%.
- Remove dirt from a late-contaminated tree's base to expose the whole cancer zone.
- Take out any diseased tissue and keep the storage compartment area open to let the contaminated area dry. It will prevent the disease from getting better. Fill up the area surrounding the storage compartment with excellent fresh soil in late autumn. Agrwal (1970).
- When planting or in the spring, a fungicide designed specifically for phytopthora should be used before growth starts.
- It has been demonstrated that using tolerant rootstock planting and sound cultural techniques in conjunction with chemical application in the autumn after harvest can be successful (Byrde and Jordan, 1977).

4. SOOTY BLOTCH AND FLYSPECK DISEASES

Introduction/Economic Importance: Sooty blotch and flyspeck are two distinct illnesses that typically coexist on the same fruit as superficial discolorations or blemishes with dark pigmentation. Fruit prices are negatively impacted by both illnesses. They frequently made simultaneous surface appearances on the same fruit. Fruits with flyspeck and sooty blotch illnesses have lower market prices and values. These illnesses become widespread in the "Midwest" apple cultivar. The two conditions are often treated concurrently and are controlled in the same manner.

Causal Organism of Sooty Blotch: *Gloeodes pomigena, Peltaster fructicola, Geastrumia polystigmatis,* or *Leptodontium elatius*

Causal Organism of Flyspeck Sooty blotch is a complex disease caused by the fungi *Schizothyrium jamaicensis*, also known as *Zygophiala pomi* and *G. pomigena* (Schw), while flyspeck is caused by the fungus *S. pomi* (Mont and Fr.) (previously *Microthyriella rubi; anamorph Z. jamaicensis*) (Saha, 2002). Based on their morphology, three fungus species were identified as the primary culprits behind the dirty smudge infection on North Carolina apples in the latter half of the 20th century. According to Johnson *et al.* (1997), they are *P. fructicola, G. polystigmatis,* and *L. elatius.* Ash, blackberry, citrus, crab-apple, dogwood, grape, hawthorn, leatherwood, magnolia, maple, mountain ash, oak, orchids, peach, pear, persimmon, prickly-ash, raspberry, orange, tulip poplar, willow, and wintergreen are the host plants of sooty blotch and flyspeck. Flyspeck mostly feeds on the following plants: carnation, apple, grapes, banana, blackberry, and American elderberry.

Symptoms:

A. **Sooty Blotch:** Sooty blotch symptoms manifested as a "smudge" or sooty look on the apple fruit surface (Figure 1.8). On the surface of fruits, sooty blotch symptoms shown as dark or discoloured patches. The olive green spots have indiscriminate edges. can can be grouped together to cover a significant area of the fruit, and can measure a quarter of an inch or more wide. From late May/early June until October, this fungus produces pycnidia, which are huge spores that travel by air or rain splash. The pycnidia are characterised by their strong walls, connected dull hyphae, and breach-like pieces that resemble cells. The existence of "smudge" is caused by a mass of free, linked dull hyphae that connects several moments-long, dim pycnidia. Usually, just the outermost layer of the fingernail's epidermis is affected by the development of the discoloured spot. Rarely, the hyphae penetrate through the nail's epidermis and the epidermal cell dividers. The temperature range in which the fungus grows and develops is 18 °C to 27 °C.

B. **Flyspeck:** Flyspeck symptoms first showed up as glossy, black, spherical spots on the apple fruit's surface that resembled fly excrement (Figure 4.1). Comparing the individual "fly spots" to the sooty blotch pathogen pycnidia, the former are bigger and more widely dispersed. The pathogen flyspeck is joined by tiny hyphae and generates sexual fruiting structures, or pseudothecia. Excessive scouring or blanching will get rid of the flaws. Figure 4.1 show symptoms of flyspeck and sooty blotch illnesses.

Fig. 4.1 Sooty blotch and

Disease Cycle and Epidemiology: Both S. *pomi* and *G. pomigena* overwinter as pseudothecia on woody reservoir hosts and diseased apple twigs. The optimal temperature and humidity levels that both illnesses require for infection are involved in the disease cycles. A. Sooty Blotch: Massive numbers of spores, or conidia, are produced by the pathogen pycnidia on diseased plants. These conidia run out of illnesses and collect in a coagulated mass. From late May or early June until harvest time, conidia are spread through plantations by wind-blown raindrops and the atmosphere. The dark, thick-walled hyphae that encased apple apples and other twigs often broke apart like cells. After spreading, they infect spherical fruit thereafter. The temperature range where the pathogen development occurred was 18 °C to 27 °C. The heated conditions in May and June were followed by a cool climate in July and August, resulting in a sooty blotch since no filthy smudge had formed on the spherical fruits. When summer showers and preharvest low temperatures combine with blustery, chilly spring weather, illness bouts are typically severe. Under ideal circumstances, the incubation period between the infection and the development of symptoms may be as brief as 5 days. In any event, this time on the plantation often lasts between twenty and more days. Fruit contamination can happen at any time when a petal falls, although it is more common in the intermediate to prefall stage.

B. Flyspeck: In late April, this fungus produced both conidia and ascospores. Air currents transfer conidia and ascospores from one plant to another. In cold conditions (18 °C), the fungus takes around 15 days to incubate.

Integrate Disease Management: On apple plantations, sooty blotch and flyspeck are common in wet, shady places. The best management techniques for controlling disease in orchards are those that promote air flow and provide a dry environment in plantation regions. Applying fungicides is a useful disease control method; it should begin soon after petal fall (calyx) and continue until cooler weather arrives in the fall before harvest. The majority of producers apply preventative fungicide to control flyspeck/sooty blotch disease at intervals of 10–14 days. However, producers have a wide range of integrated pest control techniques at their disposal to reduce the need of pesticides.

Cultural Control:

A. Pruning: Pruning the trees can help control flyspeck and sooty blotch. Ideally, the disease may be controlled by annual pruning. Pruning dries out the fruit's surface, reducing the likelihood of disease infestation. The flyspeck/ sooty blotch pathogen grows best in damp or moist conditions.

Well-pruned trees also improve fungicide penetration. In the event that fruits are located in the inner canopy of improperly trimmed trees, even regular applications of fungicides are insufficient to mitigate the damage caused by sooty blotch/flyspeck.

B. Bramble Management:

- Various rubus species, such as raspberries and blackberries, are a major source of flyspecks and sooty blotches. As a result, growing berries close to apple planting regions should be avoided or planted far away, with berries and apple orchards separated by a hedgerow or windbreak.
- In the spring and summer, wild brambles should be planted around the edges of orchard blocks. This aids in controlling the amount of flyspeck and sooty blotch spores.

C. Thinning:

- The fruit surface developed soothy blotch smudges or flyspecks in the early season. The apples are too little to consider interacting with one another.
- High-density planting favours the two sooty blotch smudge/flyspeck spots and provides a wet environment.
- When fruit is thinned properly, cluster development is reduced, which affects the effectiveness of fungicide treatment in controlling disease.

D. Mowing:

- Mowing is a practice used in plantation regions to promote faster dry conditions and improved wind currents.
- Dew can remain moist on tall, grassy apple trees with low-hanging branches into the late hours of the morning.
- Regular mowing inhibits the growth of pathogenic sooty blotch/flyspeck and enables low-draping apples to dry rapidly.

Choose Resistant Cultivars and Rootstocks:

- Apple cultivars that are resistant to both flyspeck and sooty blotch disease do not exist.
- Physiological traits of certain cultivars lessen the likelihood of sooty blotch/ flyspeck damage.
- Cultivars with shorter maturation periods exhibit less damage because the sooty blotch/flyspeck pathogen has less opportunity to proliferate on the apple surface. Sooty blotch and flyspeck damage can be lessened by planting short-duration maturing varieties, such as red apple cultivars.

Chemical Control:

- Flyspeck and sooty blotch may be safely and successfully managed using chemicals.
- The flyspeck/sooty blotch infection warning system was implemented recently. Between the first cover spray and the second cover fungicide treatment, there must be a 175-hour period during which the moisture from rain or dew is evaluated. Fungicides are sprayed as a preventative measure (every 10–14 days) in second-cover spraying till harvest. In primary and second cover sprays, a combination of benzimidazole fungicide (i.e., thiophanate-methyl) and a contact fungicide (captan) should be employed with a disease warning system.
- Additionally, the moisture of trees may be determined using a personal weather sensor. The sensor is installed in the lower shelter of a delegate tree in the plantation for information assortment. Data is downloaded onto a PC at a frequency determined by the system (usually once a week initially, increasing in frequency when the limit is approached). Information gathering with the wetness estimate framework begins following the primary distributed shower. Every day up to 175 full hours are included in the wetness hours. After 175 hours, the second spread splash is administered. For the remainder of the term, a typical interim timeline of 14 days is adhered to (Agrwal, 1967).
- To control for sooty blotch and flyspecks, two strobilan chemicals—kresoxim-methyl and trifloxystrobin—were investigated. For the purpose of managing sooty blotch/flyspeck disease, alternate sprays of kresoximmethyl + trifloxystrobin and thiophanate-methyl (Topsin-M) + captan were just as successful as full-season applications of thiophanate methyl plus captan. Fungicides containing strobilan are less harmful to the environment and human health. In addition, it offers the same degree of control as conventional sprays (Agrwal, 1967).

- Potassium bicarbonate ($KHCO_3$) may be a novel, somewhat successful chemical for managing the flyspeck/sooty blotch disease in apples. However, it is the less complicated remedy for managing apple diseases, especially mildew in organic growing zones.
- Flyspecks are fruiting bodies of sexual fungi called pseudothecia, which are joined by extremely thin hyphae. Often, the flaws are eliminated by vigorously rubbing or bleaching.

5. BLACK FRUIT ROT DISEASE

Introduction/Economic Importance: *Diplodia seriata* and *Botryosphaeria obtusa* (Schwein.) are the disease's causative organisms. Shoemaker infections frequently result in branch canker, dieback of shoots in apple trees, frog-eye leaf spot, and black fruit rot. Regarding summer and autumn variations, the illness is severe. By ablation of the limb cankers and adherence to the spray schedules, the disease can typically be entirely managed inside the orchard itself.

Causal Organism: *B. obtuse* Schweinitz (1832) provided the pathogen's first detailed description, naming it Sphaeria obtusa Schwein. Cooke (1892) determined that *Physalospora obtusa* (Schwein.) was the pathogen. Shoemaker (1964) proposed the new combination *Botryosphaeria obtuse* and was thought to be a species of the fungus *Botryosphaeria Ces.*

Symptoms: Fruit, leaves, and bark all had indications of the *B. obtusa* pathogen that causes black fruit rot disease. Start inspecting your apple trees a few weeks after the petals fall off for any indication of illness.

(a) Symptoms as Fruit Rot: Large sections of the fruit may be covered with intense brown and black rings made of rotting tissue. On older infections, Pycnidia fungal spores appear as little black dots. Certain fruits stick on the tree after mummifying. Fruits infected early on ripen one week faster than they would normally. The centre of infected fruits is rotting.

(b) Symptoms as Frogeye Leaf Spot: Infected apple leaves displayed frogeye leaf spot symptoms, which were followed by the development of black rot disease. The round lesions with light tan centres and violet or crimson exterior borders are observed on leaves.

(c) Symptoms as Branch Cankers: A depressed, reddish-brown area that occasionally developed with rough-looking or cracked bark was the sign of an infected limb. Those cankers might not be immediately noticeable. It's important to pay attention to the canker on shoots if the plants have fruit rot or frogeye leaf spot. Figures 5.1, 5.2 and 5.3 show symptoms of branch cankers, frogeye leaf spot, and fruit rot in apples.

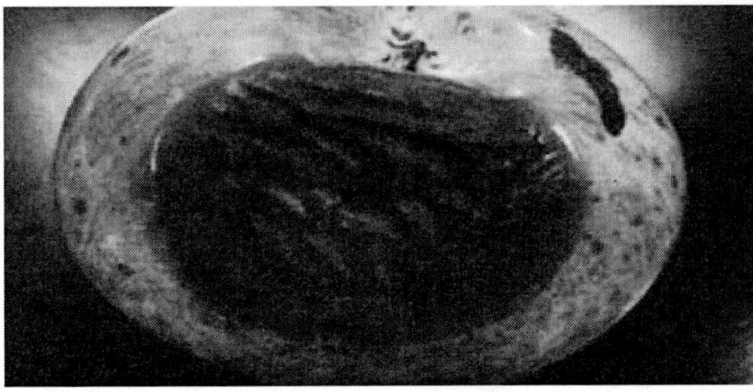

Fig. 5.1 Fruit rot.

Fig. 5.2 Frogeye leaf spot

Fig. 5.3 Branch cankers.

Disease Cycle and Epidemiology: The fungus is prevalent in deadwood bark, mummified fruits, and cankerous throughout the winter. Conids and ascospores are released by the black fungal fruiting bodies (perithecia and pycnidia) over the season. The disease is spread by these spores into wood, fruit, and healthy leaves. The greatest release of spores happens around flowering, however during rainy summer months, conidia might continue to assemble. For a minimum of a year, the conidia will stay viable. Typically, leaf infection happens during the petal-fall phase. Conidia are affixed to the leaf and, in a moist environment, should germinate in five to six hours. After conidia germination, they enter the leaves through natural apertures under the surface, as well as through wounds caused by insects, hail, or other factors. Furthermore conducive to the propagation of the pathogen and disease are relative humidity (RH) and precipitation. At around 32 °C, spore germination and infection are most favourable.

Integrated Disease Management:

- The most effective method of controlling apple disease is to combine cultural activities with fungicidal treatment.
- Removing preserved apples off the tree and cutting down deadwood on the plantation's ground surface are two ways to lessen the inoculum.
- Twigs and punched shoots might harbour illness for the following growing season.

- Annual tree pruning and the use of modified composts on soil test sites reduce the amount of disease transmission.
- Stress and weak trees are known to cause cancer. Maintaining sturdy apple trees requires adhering to specific social customs and behaviours.

6. FIRE BLIGHT DISEASE

Introduction/Economic Importance: The fire blight infection was initially brought to northern Europe in the 1950s and 1960s through fruit containers imported from Northern America that included bacterial slime. The bacterium *Erwinia amylovora* was initially identified as the illness's aetiology (Burrill, 1883). The causes of apple fireside blight were finally identified in Hungary, Romania, Northern Italy, and Northern Spain between 1995 and 1996. More than 40 nations have reported *E. amylovora* as a disease cause.

Causal Organism: *E. amylovora.* The causative organism's host plants include hawthorn, cotoneaster, pyracantha, loquat, pears, and quince. Host plants have comparable symptoms that are easy to recognise.

Symptoms Brownish coloration and a necrotic look of the leaves, twigs, and flowers were the symptoms. Because the symptoms also resembled fire or a charred look, the term "fire blight" was coined. The distinctive "shepherd's crook" of terminal shoots is caused by the assembly of exudate, diseased twigs or branches, and infected leaves that exhibit a brown to black colour. In light of this, the illness causes limb or trunk blight, collar or rootstock blight, fruit blight, and blossom blight (Van der Zwet and Keil, 1979; Van der Zwet and Beer, 1995).

(a) Symptoms as Blossom Blight and Spur Blight: Blistering blight and blight symptoms appear in the spring. Bacteria can infiltrate the trees through the flowers in addition to new shoots. The signs of infected flowers include browning, wilting, and switching. Bacteria migrate quickly down the spur and into other flowers within the cluster. According to Van der Zwet and Keil (1979) and Van der Zwet and Beer (1995), infected spurs on pears and apples turn brown.

(b) Symptoms as Shoot Blight: Shoot curse starts at the tips of the shoots and spreads swiftly to the more established segments of the twig. At first, it was seen that the infected branches were drenched in water, resulting in a dull, earthy or dark hue. The branches curl at the development point of diseased shoots, like the evildoer of a shepherd or a "J." Throughout the middle of the year, tainted leaves are still clinging to dead twigs. In warm, humid climates, beads of silky white microscopic creatures will overflow (Van der Zwet and Keil, 1979; Van der Zwet and Beer, 1995).

(c) Symptoms as Twig/Branch/Stem Cankers: Bark cracks along the surface of contaminated places often, resulting in a visible canker, if bacteria that cause illness spread from infected twigs to the most branches. A water-saturated environment and dark brown to purple cankers are visible in the bark of younger trees. Most tree limbs are susceptible to cankers, and the trunk produces additional dieback (Van der Zwet and Keil, 1979; Van der Zwet and Beer, 1995).

(d) Symptoms as Fruit Blight: Apple fruits might have a fire blight infection. Fruit rot was caused by a pathogen, and the rotten sections became dark to black and were covered in droplets of whitish-colored bacterial slime. In the end, apple fruits shrivel and dry up to become mummies.

(e) Symptoms as on Rootstock: Symptoms of fire blight that resemble stem cankers growing close to the graft union. By girdling the rootstock, fire blight infections in rootstocks can quickly destroy the tree (Van der Zwet and Keil, 1979).

Disease Cycle and Epidemiology: The fire blight bacteria survive the winter as tumours in living tissue on the trunk, major limbs, and even on fruit that has been mummified. Primary infection caused by raindrop splashing, pollinating insects (bees, wasps carrying pollen, flies, ants), or pruning during the growing season (Paulin, 2000). Bacteria proliferate and cankers become active in the spring. A single active canker can create enough germs to infect an entire orchard. Secondary infections are caused by these bacteria, which grow quickly in the nectar of the bloom and spread to the spurs (twigs that carry flowers), new shoots, and branches. Additionally, freezing or frost damage, wind, wind-driven rain, hail, and wounds from sucking insects (aphids, leafhoppers, or tamed plant bugs) can result in shooting diseases. The heart blight bacteria reproduce swiftly after a shoot becomes infected, and within three days, oozing droplets are frequently visible. Until vegetative development ceases and the final bud forms, the shooting is still extremely susceptible to infection (Thomson, 2000). At 27 °C, the pathogen grows at its best. Additionally, it can grow in temperatures between 5 and 31 °C. Unfavourable conditions cause illnesses to grow more slowly and lesions to spread more slowly at high temperatures (5 °C or less, 31 °C and above) (Thomson, 2000).

Integrated Disease Management: Controlling the apple disease fire blight is a challenging task. The use of integrated management techniques may reduce fire damage and losses, notwithstanding the complexity of the management. The farmer may find that the following suggested actions greatly aid in reducing the fire blight disease.

(a) Cultural Control:

- Pruning cankers and twigs afflicted with blight over the winter (dormant season) clears the orchard of fire blight bacteria and monitors the infection in the spring.
- When sick branches are properly pruned in the summer, the orchard's bacterial population is lowered.
- Only in the dry season can contaminated shoots be trimmed. Wood with sick good tissue below 12 to 15 inches cannot be cut. Sanitising pruning tools is necessary following each cut.
- A common method for sanitising tools is to dip them into a 10% bleach solution (1 volume of bleach to 9 volumes of water) that also contains a few drops of soap. Preparing a fresh bleach solution is necessary when the solution appears unclean.
- It is best to gather and burn or discard portions of the contaminated plant that were removed from the orchard.
- Good hygiene habits have a significant impact in preventing illness infestation.

(b) Chemical Control:

- Sprayed with a pesticide containing copper to rescale the bacterial accumulation on the bark and buds. Green tips can receive copper additions when the buds begin to swell near the silver tip, but not after.
- The antibiotic streptomycin eliminates pear blight-causing bacteria before they have a chance to infect the fruit tree.

(c) Control Sucking Insects: Insects that feed on blood produce wounds that allow fire blight germs to infiltrate. It is best to control these pests throughout the growth season. During bloom, insecticides should not be used to safeguard bees.

REFERENCES

Agarwal RK (1961) Problem of root rot in Himachal Pradesh and prospect of its control with antibiotics. *Himachal Hortic.*, 2: 171–178.

Agrwal RK (1967) Relative efficacy of fungicides for the control of Gloeodes pomigena (Schw) causing Sooty blotch disease of apple. *Proc. Acad. Sci.* 37: 171–178.

Anonymous (1989) Pakage and Practices for fruit crops. Dr Y. S. Parmar University of Horticulture and Forestry, Nauni, Solan, Himachal Pradesh.

Bessey CW (1877) On Injurious Fungi: The Bligh (Erysiphe). Iowa State College of Agriculture. Bienn. Rep., pp. 185–204.

Burrill TJ (1883) New species of Micrococcus (bacteria). Am. Nat., 17: 319.

Byrde RJW and Jordan VWL (1977) Systemic Fungicides (R. W. Marsh, Ed.) Longman, London and New York. 280 pp.

Ellis MA, Chatfield J and Draper E (1997) Scab of Apple and Crab Apple. The Ohio State University Extension Factsheet. HYG-3003-94.

Glawe DA (2008). The powdery mildews: a review of the world's most familiar (Yet poorly known) plant pathogens. *Annu. Rev. Phytopathol.*, 46: 27–51.

Gupta GK (1985) Apple Scab (Venturia inequalis) (Cke) Went. E. Merck (India) Ltd., Bombay, 36 pp

Gupta GK (1990) Apple scab – a review. *Indian J. Agric. Sci.* 18; 157–165.

Gupta GK and Lele VC (1976). Morphology, physiology and epidemiology of the apple scab fungus (Venturia inequalis) (Cke) Went., in Kashmir valley. *Indian J. Agric. Sci.* 18; 157–165.

Hickey KD and Yoder KS(1990) Apple powdery mildew. Pages 9–10 in A. L.

Jones and Aldwinckle HS Eds (1990) Compendium of Apple and Pear Diseases. American Phytopathological Society, St. Paul, MN, USA.

Johnson EM, Sutton TB, Hodges CS (1997) Etiology of apple sooty blotch disease in North Carolina. *Phytopathology*, 87: 88–95.

Joshi NC, Malik AG, Kaul ML and Anand SK (1975) Some observation on epidemics of scab disease of apple in Jammu and Kashmir during 1973. *Indian Phytopathol.*, 28: 288–289.

Nath P (1935) Studies on the diseases of Apple in northern India. II A short notes on apple scab due to Fusicladium dendriticum Puck. *J. Indian Botany*, 14: 121–124.

Paulin JP (2000) Erwinia amylovora: general characteristics, biochemistry and serology. In J. Vanneste (Ed.) Fire Blight: The Disease and Its Causative Agent, Erwinia amylovora, pp. 87–116. Wallingford, UK, CABI. 370 pp.

Pfeiffer DG, Bergh JC, Fell RD, Yuan R, Walsh CS, Yoder KS, Biggs AR, Kotcon JB, Derr JF, Chandran RS, Weaver MJ, Baniecki JF, Brown A and Parkhurst J (2010) Spray Bulletin for Commercial Tree Fruit Growers. Virginia Cooperative Extension Publication.

Rana KS and Gupta VK (1984) Occurrence of pythiaceous fungi in collar rot affected apple soil of Himachal Pradesh. *Indian Phytopathol.*, 37: 39–42.

Saha LR (2002) Hand Book of Plant Diseases. Kalyani Publisher, Ludhiana, pp. 260–270.

Shoemaker RA (1964) Conidial states of some Botryosphaeria species on Vitis and Quercus. *Can. J. Botany*, 42: 1297–1301.

Thomson SV (2000) Epidemiology of fire blight. In J. Vanneste, Ed. Fire Blight: The Disease and Its Causative Agent, Erwinia amylovora, pp. 9–36. Wallingford, UK, CABI. 370 pp.

Turechek WW, Carroll JE and Rosenberger DA (2004) Powdery mildew disease of apple. Tree Fruit Factsheet, Cornell University.

Utkhede, Raj and Smith, Emmerson (1995) Control of Phytophthora crown and root rot of apple trees with fosetyl-aluminium in new plantings. *Pest Manage. Sci.*, 45 (2): 117–122.

Van der Zwet, T and Beer, S (1995) Fire blight: its nature, prevention and control. A practical guide to integrated disease management. USDA Agricultural Information Bulletin, No. 631.

Van der Zwet T and Keil HL (1979). Fire blight: a bacterial disease of rosaceous plants. United States Department of Agriculture (USDA) Handbook 510. Washington, DC, USDA.

Verma KD and Gupta, GK (1988) Field reaction of apple germplasm to powdery mildew. *Indian J. Agric. Sci.*, 58: 223–234.

Yoder KS (1992) Powdery mildew of apple. In: J. Kumar, H. S. Chaube, U. S. Singh, and A. N. Mukhopadhyay (eds.) Plant Diseases of International Importance, pp. 66–89, Vol. 3. Diseases of Fruit Crops, Prentice Hall, NJ.

Yoder KS, and Biggs AR (1997) Apple cultivar susceptibility to the powdery mildew fungus. *Biol. Cult. Tests Control Plant Dis.*, 12: 42–43.

Sustainable Agriculture and Biodiversity: Innovations and Insights, Pages: 182–194
Edited by: Bibhuti Bhusan Sahoo, Harvinder Kumar Singh, Niranjan Chinara and Amitabh Panda

CHAPTER - 7

MECHANISM OF PATHOGENESIS AND BIOLOGY OF ENTOMOPATHOGENIC FUNGI: A NOVEL BIOLOGICAL INSECT MANAGEMENT SYSTEM

Vinod Kumar Nirmalkar[1*], Narendra Lakpale[2], Harvinder Kumar Singh[2] and Rajkumar Singh Tiwari[1]

[1]*Section of Plant Pathology, BTC College of Agriculture & Research Station, Sarkanda-495001, Chattisgarh*
[2]*Department of Plant Pathology, College of Agriculture, IGKV, Raipur-492012, Chattisgarh*
**Corresponding email: vinod.nirmalkar01@gmail.com*

INTRODUCTION

Fungi which control the insect pest population are associated with agricultural crops are called as entomopathogenic fungi. There are naturally occurring organisms, such as bacteria, viruses and fungi for the control of crop pests, which can act as a parasite of insects and kills or seriously disables them. In recent years, microbial pathogens like viruses, bacteria, fungi and protozoa have been recognized for the biological suppression of many insect pests. About 1.5 million species of fungi alone are known to occur worldwide out of which nearly half of the species have been identified. Amongst these, several asexual stages of fungi are associated with insect infection. Entomopathogenic fungi are naturally occurring organisms which perceived as less damaging to the environment (Tahira *et al.,* 2014). Among biopesticdes, entomopathogenic fungi played a significant role in insect pathology and especially in microbial control (Steinhous, 1956). Comprise a diverse group of over 90 genera with approximately 750 species, only 10 species have identified for potential biocontrol activity. Hajek and Leger (1994) reported from different species of insects and classified them in to different phylums *i.e. Oomycota*, *Ascomycota*, *Chytridiomycota* and *Zygomycota* includes several genera like *Coelomomyces, Entomophaga, Entomophthora, Ascomycotina, Cordyceps, Septnbasidium, Uredinella, Stereum, Aschersonia, Aspergillus, Beauveria, Cephalosporium, Hirsutella, Metarhizium, Nomuraea,*

Paecilomyces, Penicillium, Trichoderma, Verticillium etc. Most of the entomopathogenic fungi belong to either entomopthorales (Zygomycotina) or hyphomycetes (Deuteromycotina). Prominent genera that have been exploited for the pathogenic properties include *Nomuraea, Beauveria, Metarhizium, Verticillium, Hirsutella, Aspergillus, Coelomomyces, Lagenidium, Paecilomyces* and *Tolypocladium* of which the first five are the best known cosmopolitan insect pathogens. These are potentially the most versatile biological control agents due to their wide host range, higher virulent, higher host specificity, economical, easy production, shelf life, safe, minimal effect to natural enemies, infect insect by contact, persist in environment for long time, safer alternative and compatibility with other methods (Rai *et al.,* 2014). Globally, the most commonly used genders are *Metarhizum anisopliae* (Metsch.) Sorokin (33.9%), *Beauveria bassiana* (Balsam) Vullemin (33.9%), *Isaria fumosorosea* (before *Paecilomyces fumosoroseus*) [(Wize) Brown and Smith] (5.8%) and *Beauveria brongniartii* (Sacc.) Petch (4.1%).

Fungi, which induce disease symptoms in insects, include fungi from quick killers to absolute parasites that provide disease symptoms in the host. Absolute parasites are considered to be fungi which live in association with a host and benefit at the host's expenses. Entomopathogenic fungi cause lethal infections and regulate insect and mite population in nature by epizootics. They are host specific with a very low risk of attacking non-target organisms or beneficial insects. They are reported to infect a very wide range of insects including lepidopterous larvae, aphids and thrips, which are of great concern in agriculture worldwide (Nirmalkar *et al.*, 2020).

OCCURRENCE

Entomopathogenic fungi naturally occur in insect hosts as infections; they can be collected in the field and grown in the laboratory for the documentation of the fungus. *B. bassiana* have been reported to occur naturally in more than 700 species of hosts. Studies on the prevalence of fungi in insects have usually been limited to species that are pests or are important non-target species such as certain predators and parasitoids. However, it is likely that almost any major insect taxon collected intensively will be found to be a natural host for *B. bassiana* in temperate regions. Infections in hosts are considered the only part of the fungal life cycle in which the fungi can build up significant population sizes by producing vast numbers of conidia. By considering environmental manipulation strategies, the contribution to the availability of susceptible hosts for fungal population increase is a key component.

BIOLOGY OF ENTOMOPATHOGENIC FUNGI

The entomopathogenic fungi have life cycles that synchronize with insect host stages and environmental conditions. In broad terms, the differences between Hyphomycetes and Zygomycetes, particularly Entomophthorales. However, as with all general principles, there are exceptions and features of the two groups form a continuum. Species and sometimes isolates within a species can behave very differently. For example, insect host range, infection levels, germination rates and temperature optima can vary between species and isolates. Members of the Hyphomycetes are generally considered to be opportunistic pathogens infecting many species in a range of insect orders, and host death is commonly associated with toxin production which overwhelms the host defense responses (Samson *et al.*, 1988). In contrast, other groups of fungi are thought to have evolved into higher parasitic forms. For example, infection and host death by Entomophthorales tends to occur due to tissue colonization with little or no use of toxins. One of the best examples of a highly evolved insect pathogenic fungus is Strong well *sea castrans* that infects flies. Infection does not interfere with insect feeding and movement but conidia are discharged and thereby dispersed from a cavity in the abdomen of infected insects over a long period of time prior to death. In general, Entomophthorales have biotrophic relationships with their insect hosts with little or no saprophytism, while Hyphomycetes can be hemibiotrophic with well-defined parasitic phases within insect hosts and saprophytic phases on the death of their hosts.

CLASSIFICATION OF ENTOMOPATHOGENIC FUNGI

Entomopathogenic fungi are found in the divisions Zygomycota, Ascomycota and Deuteromycota (Samson *et al.*, 1988), as well as Chytridiomycota and Oomycota, which were previously classified within Fungi. Many of the genera of entomopathogenic fungi currently under research belong either to the class Entomophthorales in the Zygomycota or to the class Hyphomycetes in the Deuteromycota. It is important to mention that fungal infections occur in other arthropods as well as insects and/or species that are not pests of cultivated crops. For example, *Gibellula* species infect spiders and several species of *Cordyceps* and *Erynia* infect ants.

Division- Ascomycota,

Class- Sordariomycetes,

Order- Hypocreales,

Family- Clavicipitaceae

FUNGAL INFECTION PROCESS AND TRANSMISSION OF ENTOMOPATHOGENIC FUNGI

Mechnism

Differing from bacteria and viruses, fungi can infect insects not only through the gut, but also through spiracles and particularly through the surface of the integument, infecting insects independently of their feeding activity. Mode of infection by a fungus have often resulted in contradictory positions because of the fact that all possible ways of infection are not necessarily exploited by a fungus for a given insect: for instance. artificial contamination of the intestinal tract of larvae of *Melolontha melolontha* by *Beauveria brongniartii* does not necessarily bring about infection of the host , whereas contamination of the integument or of the buccal appendages causes the development of the white muscardine. Although the integument seems to be the main pathway of infection, the importance of contamination through the mouth parts or the intestinal tract cannot be neglected. If the mycelial penetration through the integument can be favored by the spores, several workers also described that involvement of both physical and enzymatic mechanisms, with or without the development of appressorial structures. This penetration is frequently followed by a melanic reaction of the integument at the sites of infection, probably because of changes in phenoloxydase activity caused by the fungus. It is important to remember that the insect integument is essentially composed of proteins and chitin associated with lipids and phenolitic compounds. The very thin outer layer or epicuticle contains lipids (fatty acids and paraffin) whose antifungal activities have been demonstrated, but in higher concentrations than those present in the insect integument . Furthermore, it was demonstrated that entomogenous fungi such as *Beauveria bassiana*, *B. brongniortii*, *Metarhizium anisopliae*, *Cordyceps militaris*, *Aspergillus flavus* and *Entomophthora muscae* have lipolytic enzymes. The originality of the enzymatic equipment of entomogenous fungi is due, there-fore, to systems capable of attacking the proteo-chitin complex. It is estimated that only 10% of the total proteins of the integument are not linked. Therefore, it is necessary that a hydrolysis of the proteins takes place before the action of chitinases begins. It is known that in addition to the lipases, proteases and chitinases are effectively produced by these fungi and that lipolytic and proteolytic activity precede chitin breakdown. The production of proteases was also reported in 10 species of *Entomophthorales*.

Electron microscope studies on the histopathology of fungal infections of wire-worms by *Metarhizium anisopliae* show that the fungus develops through at least six identifiable stages: external infective spores, which produce the germ tube; appressorial cells formed from germ tube against the surface of the cuticle, which produce the penetration peg through the epicuticle; subepicuticular penetrant plate, which produces hyphae that give rise to the penetrate hyphal bodies; irregular-walled and smooth-walled penetrate hyphal bodies that give rise to hyphae that penetrate the procuticle to the coelom; detached coelomic hyphal bodies produced by the penetrant hyphae, which circulate in the hemolymph, germinate to produce new hyphae and thus spread the fungus in the body cavity and chlamydospores produced after death of the host, which can maintain the fungus in a viable state within the host cadaver. These spores subsequently germinate to form emergent hyphae that sporulate on the surface of the host to produce new external infective spores. It is important to emphasize that because of arthropod moulting the infection can be considered successful only when the fungus has reached the hypodermis. This phenomenon is even more important if

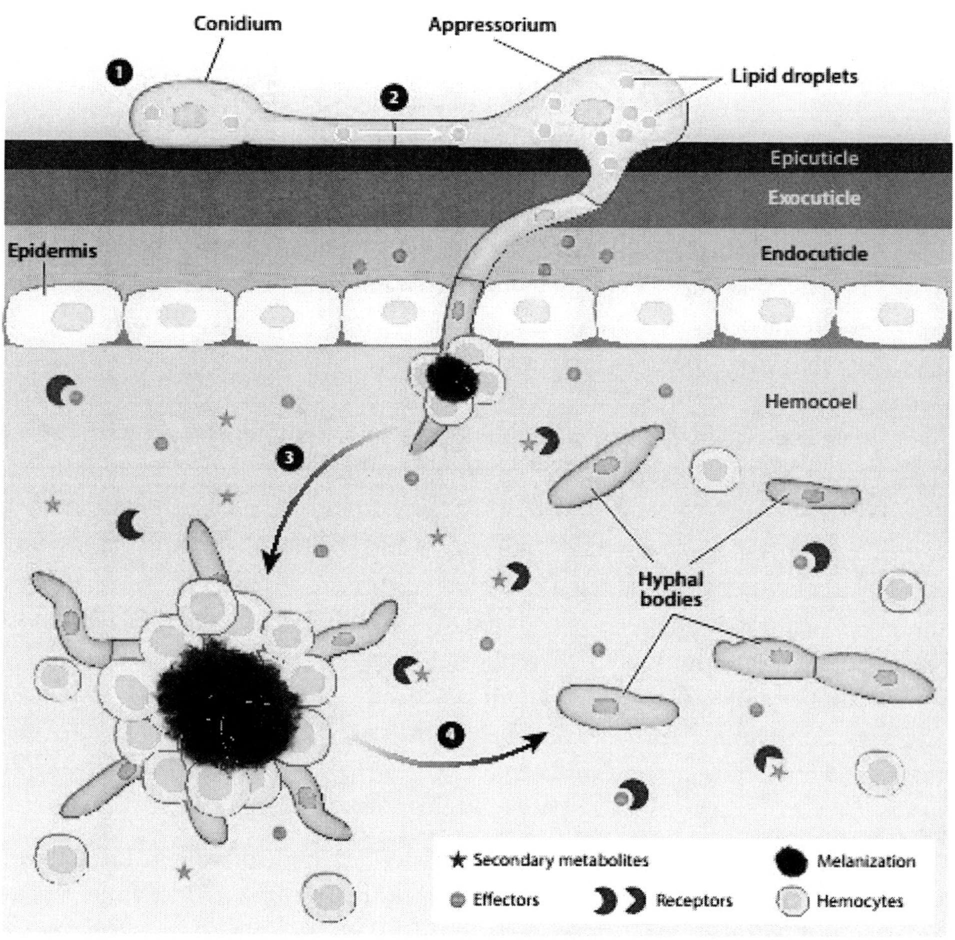

Fig. Infection process of entomopathogenic fungi (Wang and Wang, 2016)

the insect under consideration undergoes frequent moults. In fact, this infectious process may vary as a result of the coincidence between fungal penetration and the moulting stage. Therefore, *B. bassiana* can cross through the integument of third instar larvae of *Leptinotarsa decemlineata* (Colorado potato beetle) just prior to moulting. In this instance the hyphae and blastospores invade the exuvial fluid and infect the newly formed integument, which cause localized lysis of the hypodermis by the usual process of primary infection of larvae between moults. In the times the fungal action is more advanced when ecdysis occurs and the new cuticle apparently is wounded during ecdysis because of the adhesion of the two cuticles caused by the mycelium. These wounds are an open door not only to mycelial elements of the exuvial fluid, but also to bacteria that can cause lethal septicemia. However, if the penetration into the old integument is only uperficial during the moult, the host insect can escape infection by rejecting the infectious inoculums.

Process: The virulence of fungal entomopathogens involves four steps: adhesion, germination, differentiation and penetration. Each step is influenced by a range of integrated intrinsic and external factors, which ultimately determine the pathogenicity. A successful infection is achieved by the attachment or adhesion of spores to the host. Adhesion is necessary and normally achieved through the secretion of mucilage. However, enzymes, lectins, as well as hydrophobic and electrostatic forces also play a role. The virulence of an entomopathogenic fungus is recognized first by adhesion to an insect body. The failure of a pathogen to adhere with the epidermis is considered a feature of virulent strains. After adhesion, the next factor for the virulence of a strain is the enzymes that hydrolyze the epidermis of the insect. The most important enzymes secreted by entomopathogenic fungi are lipases, proteases and chitinases, which are produced sequentially, reflecting the order of the substrates they encounter. A wide range of factors such as water, ions, fatty acids and nutrients on the cuticle surface and the physiological state of the host, influence spore germination and behavior. Successful germination requires the assimilation of utilizable nutrients and a tolerance to any toxic compound present on the surface. After germination, appresoria appear at the end of short germ tubes, subterminally or on the side branches. Appressorium morphology is reported to be found either in swollen clavate or spherical structures to slightly swollen terminal hyphae. Penetration of the cuticle is accomplished by the germ tube itself or by the formation of an appressorium that attaches to the cuticle and gives rise to a narrow penetration peg (Humber, 1981). Penetration is both a mechanical and an enzymatic process. Most terrestrial pathogens are known to penetrate directly, rarely via wounds, sense organs or spiracles. The penetration process is considered to be a combination of enzymatic and mechanical forces. The exact mechanism for entry is usually peculiar to the species. A range of cuticle-degrading enzymes is produced during penetration into the host. In forestry, horticulture and agriculture, entomopathogens are considered as primary candidates for mycoinsecticides. Fungal insect pathogens are important natural control agents for many insects and other arthropods, which significantly reduces host insect populations. The significance of fungi in regulating insect population was noted early in recorded history by the ancient Chinese. Approximately 750 species of entomopathogenic fungi are known from 85 genera found throughout the classes of fungi.

DEVELOPMENT OF MYCOSIS

After crossing the cuticular barrier, the fungus comes up against the cellular defense reactions of the host. Plasmatocytes, normally dispersed in the hemolymph, accumulate around the fungus and give rise to melanization. This hemocytic pseudotissue becomes lamellae . By using invertebrate tissue culture techniques, it was noted that the hemocytes are attracked from a distance by the mycelium . After a few days, these hemocytic formations may become large and visible to the naked eye. Either the infection is blocked and the insect continues its normal development or the mycelium may also overcome this hemocytic barrier and invade the rest of the organism. The degree of cyst formation is related to the virulence of the fungus. With less virulent species such as *Aspergillus niger*, the infection seems to be easily overcome, but with wound pathogens, such as *Mucor hiemalis*, cyst formation is not very effective because of the rapid development of the hyphae and the secretion of toxins. Finally, in true pathogens such as *Beauveria*, alteration of cyst formation is probably affected by the production of toxins, in which case the mycelium may rapidly resume its normal growth and invade the hemocoel by means of blastospores. The role of toxins, consequently, is considered to be very important. In most of elaterid larvae infected by *Metarhizium anisopliae* , only the hypodermis at the sites of fungal penetration showed signs of histolysis before death.

Hyphal bodies were produced by abstriction from the original hypodermal hyphae, by budding from pre-existing cells, by the separation of cells comprising free-floating filaments, which themselves developed from free cells, and by abstriction from terminal or lateral pegs on cells of free filaments. Phagocytes were seen congregated around the hyphal bodies and hyphae, but no phagocytosis was observed; indeed, phagocytosis was clearly observed in an early stage of the formation of hypha bodies of *Paecilomyces fumororoseus* artificially infecting the silkworm and in tissue culture infected with *Metarhizium anisopliae* . Normally hyphal growth and hyphal production within the hemocel were not very extensive at death; most of the hyphal bodies tended to be concentrated in the colonies at the point of penetration. Just before and after death of the host, hyphal bodies proliferate and are distributed throughout the body cavity. Symptoms of sluggishness and lack of coordination are seen just before death. The histolytic coloration of the nervous system of Eurygaster integriceps infected by *Beauveria bassiana* was modified by toxic substances produced in the hemolymph by the fungus before death. Similar results were observed in the case of an infection by *B. bassiana* of the tick Ixodes ricinus (Castor bean tick). For these reasons, the role played by entomogenous toxins in the lethal process of the disease is particularly important . Several toxic compounds were isolated and identified from cultures of *Beauveria* , *Cordyce* , *Entomoph-thora, Metarhizium*, and *Paecilomyces*. Cyclic depsipeptides, such as destruxins A, B, C, and D and desmethyl destruxin B, were isolated from cultures of *Metarhizium anis pliae* and their chemical structure was determined and subse-quently synthesized. Another cyclodepsipeptide, beauvericin, was found in *B. bassiana* and in *Paecilomyces* . A different compound of the same family was identified in *Beauveria brongniartii* . Production of toxic substances by the Entomophthorales was also reported. *Cordycepin* was isolated from cultures of *Cordyceps militaris* and non-identified toxic substances were reported in many entomogenous *Aspergillus*. The cytological effect of *beauvericin* was studied by using *in vitro* culture techniques for insect cells. In addition to the influence on the form, reactions, and migrations of the cells, the most striking effect concerned the nuclear changes. A correlation between toxin production from different strains of *B. bassiana* and their virulence to *Galleria melongella* was pointed out. Histological studies of insect tissues attacked by entomogenous fungi and especially from elaterid larvae infected by *Metarhizium anisopliae* , showed that the toxins produced by the fungus incite progressive degeneration of the host tissues, with no appreciable swelling or shrinkage of host cells. The primary effects appeared to be on the structural integrity of the membranes of the various cytoplasmic organelles. The serious dehydratation of tissue cells through fluid loss, perhaps also incited by the toxins, seemed to be primarily responsible for host mortality. Perturbations of the electric activity of the nervous circuit of insects experimentally infected by fungi was shown , as also was an increase of the oxygen consumption . A perturbation of moulting or metamorphosis can occur . The death of the insect marks the end of the parasitic phase of the fungus development, but the lethal result of a disease is only one aspect of an infection. In fact, knowing that the lethal development of a mycosis is closely connected with the number of infecting spores, it is important to examine what eventually happens to insects infected by an inoculum at a level distinctly lower than the optimum, as occurs frequently in the case of field application as well as in epidemiological development of the disease in nature.

SAPROPHYTIC DEVELOPMENT OF THE FUNGUS

When the host insect dies, the fungus proceeds to grow saprophytically and spreads through virtually all tissues of the insect. Competition between the fungus and the intestinal bacterial flora occurs. Nevertheless, in the majority of cases, the cadaver is transformed into a mummy resistant to bacterial decay apparently because of antibiotics produced by the fungus. Cordycepin is known to have an antibacterial activity; also in *Beauveria*, oosporein, a red pigment responsible for the color changes of cadavers, has a bactericidal activity. In addition to oosporein, two yellow pigments called *bassianin* and *tenellin* also were identified. After the complete invasion of the cadaver, further development of the fungus on the mumified insect depends on the relative humidity of the environment. It is only when the atmosphere reaches saturation that the mycelium, after emerging through the integument, develops conidiophores, otherwise the cadaver remains as a hard and brittle mummy. Observations show that some entomogenous agents are conserved in the form of chlamydospores, other forms of conservation are sporanges in Coelomomycetaceae, resting spores in Entomophthoraceae and sclerotes in Cordyceps. Other than the active dispersion of spores by projection, as in the case of Entomophthorales, spores are carried away by the air, rain drops, and even insects and mites. Some researchers also reported a possible saprophytic development of Deuteromycetes in nature on plant debris and in the soil. .

FAVORABLE CONDITIONS FOR MYCOSIS DEVELOPMENT

All stages of insect development, eggs, larvae, pupae, and imagos, are generally susceptible to mycosis. Contradictory data were found, however, with regard to the susceptibility of the egg , but it frequently concerns a stage of short duration that only later manifests the disease. For example, eggs of Eurygaster integriceps infected by *Beauveria bassiana* or *Aspergillus flavus* show only a low mortality (2 1 and 33%, respectively, against 8% in the control), whereas from the first instar the disease attacks 100% of larvae in the case of *Beauveria* and 54% in the case of *Aspergillus* . This phenomena of postponed mortality is quite frequent in the case of contamination by entomogenous fungi for *Leptinotarsa decemlineata* infected *by B. bassiana*. On the other hand, all insect species are not susceptible to the same fungi. It is well known that the families of fungi are characterized by a certain specificity. For example, Coelomomyces parasite aquatic mosquito larvae, Entomophthorales develop principally on aphids, Diptera, and some Lepidoptera, and Moniliaceae infect primarily Coleoptera and Lepidoptera . However, a strict systematic view of the problem should be avoided, because for example there is a *Beauveria brongniartii* pathogenic on mosquito larvae and an Entomophthorale pathogenicn thrips, mosquito larvae Isoptera, and mites. Furthermore, in the same species of fungus, the different strains can have a very different activity spectra. Studies with *Scarabaeida*e larvae infected by different strains of *Metar-hizium anisopliae* showed that there was a very strict adaptation to the original host . The study of this specificity shows the importance of mechanisms acting at the level of the integument and in the body cavity. For example, two strains of *Metarhizium ani sopliae* tested both on their original hosts, *Oryctes rhinoceros* and *Cetonia aurata* and on the heterogenous host are subjected to similar hemocytic reactions, but eventually only the specific pathotype passes through the granuloma wall.

In the presence of a virulent strain and a susceptible host insect, the development of the infection is closely connected with the dose of infecting spores, the physiological state of the host, and abiotic conditions. Until recently, physical phenomena of the environment, especially hygrometry, were considered as predominant factors to the extent that, for certain workers, propsectives for the practical utilization of entomogenous fungi seemed to be dependent on specific, if not exceptional, ecologi-cal conditions. Influence of temperature and humidity. In effect, water, either liquid or vapor, has long been recognized as essential for the germination of spores of most fungi, and furthermore, high atmospheric humidities are known to favor development of epizootic mycosis. At present, various information permits us to distinguish between the role of atmospheric humidity during the process of inf ection and on the various epizootic phenomena. It is known, indeed, that the development of the fungus on the cadaver, and therefore the sporulation, is associated with humidity close to the saturation point. Under these conditions the auto-multiplication of the inoculum results in a heavier contamination of healthly insects, favoring an epizootic development of mycosis. The infection of insects can be obtained independently of the environmental humidity. This would lead to the supposition that physical phenomena of the boundary layer of the insect integument facilitates spore germination even if the atmosphere is practically dehydrated. The caterpillars of *Carpocapsa pomonella* held at 40% RH show physiological disturbances characterized by an alteration of the hemocytes, which modified the host susceptibility to infection and thereby interfered with the study of the phenomenon. For this reason, the stored product pest, *Acanthoscelides obtectus*, which is able to support conditions of extreme humidity apparently with-out difficulty, seem to support these findings. It is evident that spore longevity depends upon the relative humidity, conidia of *Beauveria bassiana* and *Paecilomyces farinosus* lived longer at a RH of 0 or 34% than at 75%; on the other hand, conidia of *Metarhizium anisopliae* survived the longest at high and low humidities and the least at humidities near 45%. In the case of Entomophthora, the conidia could not survive exposure to humidities below 75% RH. The rapidity of mycelial development and therefore the rapidity of the evolution of infection, depends on temperature.

In general, optimum values fall between 20 and 30^{0}C (for example, 23°C for *Beauveria brongniartii*, 24°C for *Entomophthora obscura* and *Entomophthora exitialis* 25°C for *Beauveria bassiana* and *Nomuraea rileyi*, 27° -28°C for *Metarhizium anisopliae*, 30^{0}C for *Entomophthora virulenta*), with the limits between 5° and 35°C. Temperatures lower than the optima distinctly retard the development of mycosis without necessarily affecting the total mortality. Furthermore, the enzymatic activity of the fungi whose importance in the penetration through the cuticle was reported, does not always appear to follow strictly the development of the mycelia as a function of temperature. On the other hand, it is important to point out that some fungal strains have a behavior different than that expected for the species, thereby offering pros-pects of utilization not yet studied. The optimum temperature for the development of the fungus is not necessarily the same for the development of the disease. However, the influence of temperature on the host insect must be taken into consideration, since very short periods between moults resulting from a high temperature may reduce, for example, the duration of the instar to an extent that penetration of the fungus through the integument is impeded.

CONIDIA DISPERSAL

Conidia of Hyphomycetes such as *Metarhizium* and *Beauveria spp.* are hydrophobic and passively spread from infected cadavers. Entomophthoralean conidia are actively discharged under hydrostatic pressure with the exception of *Massospora spp.* After discharge, entomophthoralean conidia are carried on wind or by co-occurring insects. If the primary conidia of entomophthoralean species fail to find suitable host substrates on which to germinate, then most form higher order conidia which can either be actively discharged, e.g. *Erynia* or *Pandora spp.* or form passively held secondary conidia (capilli conidia) borne on long stalks (capilli conidiophores), e.g. *Zoophthora spp.* With *Neozygites spp.*, the primary conidia are not infective but serve to disperse the infective secondary conidia produced on capilli conidiophores. The conidia of some entomophthoralean species are discharged while the host insect is still alive, e.g. flies infected with *S. castrans* and thrips infected with Entomophthora thripidum.

FUNGUS-PLANT INTERACTIONS

The entomopathogens have the potential to engage in fungus-plant interactions. The large majority of investigated higher vascular plants have been found to host fungal endophytes including species in *Clavicipitaceae* contained within Hypocreales. *B. bassiana* has also been included in this spectrum of fungi with endophytic activity by infecting corn (Zea mays) . Endophytic fungi are often regarded as plant-defending mutualists and the presence of *B. bassiana* in internal plant tissue has been discussed as an adaptive protection against herbivorous insects. Besides its natural occurrence in the leaf tissue of corn, *B. bassiana* exhibited endophytic activity in cacao, poppy (*Papaver somniferum*) , coffee (*Coffea spp.*) and tomato (*Lycopersicon esculentum*). In temperate regions the inoculum of *B. bassiana* has been isolated from the phylloplanes of various plants in hedgerows. This occurrence was hypothesized to be a consequence of deposition from the surroundings but was also suggested to act as a natural infection pathway of endophytic activity. These new findings open exciting perspectives for the understanding of the ecology of *B. bassiana*. However, no knowledge is currently available about the natural endophytic activity and host plant range of *B. bassiana* in temperate regions or of the significance of as an endophyte for fungus or plant fitness. Plant association has also recently been documented for *M. anisopliae*, but this association occurred below ground in the rhizosphere . The rhizosphere is the layer of soil immediately surrounding the root and many interactions between plants and other organisms occur in this interface by releasing a recombinant isolate of *M. anisopliae* to the soil in an experimental cabbage field, the released isolate persisted better in the soil immediately surrounding the cabbage roots as compared to the bulk soil. Factors in the rhizosphere seemed to promote the persistence and biological activity of *M. anisopliae* , from bean roots and on a nutrient rich medium, while the fungus when growing on insect cuticle and in insect hemolymph expressed different genes. This indicates that *M. anisopliae* has developed different adaptations to function as a pathogen and to grow saprophytically in the rhizosphere. The implication for biological control with *M. anisopliae* exploring the rhizosphere competence. Inoculated conidia of *M. anisopliae* persisted significantly better (up to one year) in the rhizosphere compared to the bulk soil. Survival outside the host may thus be critical for the ability of *M. anisopliae* to control insect pests in the soil. Whether, the rhizosphere of plants generally provides a "refuge" (where the fungus can survive outside insect hosts) for *M. anisopliae* in the soil, remains to be investigated. Perhaps associations with plants are important in the life cycle of both *B. bassiana* and *M. anisopliae* in temperate regions

LIFE CYCLE

The life cycle of entomopathogenic fungi on the insect cuticle begins with spore germination and penetration, followed by a rapid proliferation of fungal cells which ultimately results in the death of the host. Host death may be followed by the production of infective spores that can insert immediately to repeat the cycle, or by the production of resting spores or other resistant structures that requires a period of dormancy.

REFERENCES

Ferron P (1978) Biological control of insect pests by entomopathogenic fungi. *Ann. Rev. Entomol.* 23:409-42.

Hajek AE and Leger RGS (1994) Interactions between fungal pathogens and insect hosts. *Annu. Rev. Entomol.* 39: 293-322.

Humber R (1998) Fungi: identification. In: Manual of Techniques in Insect Pathology. Lacey, L.A. (Ed.), Academic Press, San Diego. pp 153–185.

Nirmalkar VK, Lakpale N and Tiwari RKS (2020) Natural occurrence and distribution of entomopathogenic fungi from Chhattisgarh. *Int. J. Curr. Microbiol. App. Sci.* 9(1): 1990-1998.

Nirmalkar VK, Tiwari RKS and Lakpale N (2020) Efficacy of different carbon and nitrogen sources against mycelial growth and sporulation of Beauveria bassiana and Metarhizium anisopliae. *J. Soils and Crops* 30 (2): 206-2012.

Rai D, Updhyay V, Mehra P, Rana M and Pandey AK (2014) Potential of entomopathogenic fungi as biopesticides. *Ind. J. Sci. Res. and Tech.* 2(5):7-13.

Samson RA, Evans HC and Latge JP (1988) Atlas of entomopathogenic fungi. Spröner-Verlag, Berlin, Heidelberge, New York. p. 187.

Shahid A, Gayyum A, Bakhsh A and Husnain T (2012) Entomopathogenic fungi as biological controllers :new insights into their virulence and Pathogenicity. *Arch. Biol. Sci., Belgrade* 64 (1) 21-42.

Tahira G, Shafqat S and Khan A (2014) Entomopathogenic fungi as effective insect pest management tactic: A review. *Applied Sci. and Business*. I (1): 10-18.

Wang C and Wang S (2016). Insect pathogenic fungi: Genomics, molecular Interactions and genetic Improvements. *Annu. Rev. Entomol.* 62:73–90.

Sustainable Agriculture and Biodiversity: Innovations and Insights, Pages: 195–227
Edited by: Bibhuti Bhusan Sahoo, Harvinder Kumar Singh, Niranjan Chinara and Amitabh Panda

CHAPTER - 8

VIRAL DISEASES OF HORTICULTURAL CROPS AND THEIR MANAGEMENT

Devendra Kumar Choudhary[1], Harvinder Kumar Singh[1*] and Sunil Kumar Suhani[2]

[1]*Department of Plant Pathology, College of Agriculture, IGKV Raipur-492012 Chhattisgarh*
[2]*Scientist (Plant Pathology) ICAR-IIPR, Regional Station, Khordha-752055, Odisha*
**Corresponding email: harry.0452@gmail.com*

INTRODUCTION

Horticulture in India is a significant agricultural sector encompassing the cultivation of fruits, vegetables, flowers, spices, and medicinal plants (Singh *et al.*, 2021). With diverse agro-climatic zones, India boasts a rich variety of horticultural crops grown across the country. Fruits play a vital role in horticulture, with India being one of the largest producers of mangoes, bananas, and citrus fruits globally (Kumar *et al.*, 2020). The country also cultivates a wide range of other fruits such as apples, grapes, guavas, and papayas. Vegetable cultivation is another integral part of Indian horticulture. India produces an array of vegetables including potatoes, tomatoes, onions, and leafy greens (Singh *et al.*, 2021). With increasing demand for fresh and nutritious produce, there has been a growing emphasis on organic farming and the adoption of modern cultivation techniques (Kumar *et al.*, 2020). Floriculture is gaining traction in India with the rise in demand for flowers both domestically and internationally. Roses, marigolds, jasmine, and chrysanthemums are some of the prominent flowers grown in the country (Kumar *et al.*, 2020). The floriculture industry not only caters to the domestic market but also exports flowers and flower products to various countries. Spices are an essential component of Indian cuisine and culture. India is renowned for its diverse range of spices such as cardamom, pepper, turmeric, and cumin, which are not only used for culinary purposes but also have medicinal and therapeutic properties (Rahman *et al.*, 2019). Additionally, medicinal plants are cultivated extensively in India, given the country's rich biodiversity (Kumar *et al.*, 2020).

Ayurveda, an ancient system of medicine, heavily relies on medicinal plants for various treatments and therapies. India exports medicinal plants and herbal products to numerous countries, contributing to the global herbal market (Rahman *et al.*, 2019).

Horticulture production during 2013-2022						**Production in million Tonne**			
2013-14	2014-15	2015-16	2016-17	2017-18	2018-19	2019-20	2020-21	2021-22	2022-23
277.4	280.9	286.2	300.6	310.7	311.1	320.5	334.6	347.2	351.9

Source: https://pib.gov.in/

The production trend of horticultural crops in India is expected to continue on an upward trajectory in the coming years, driven by increasing demand from domestic and international markets, favorable government policies, and ongoing efforts to enhance productivity, sustainability, and value addition in the sector (Singh *et al.*, 2021). However, continued investment in research and development, infrastructure, market linkages, and skill development will be crucial to sustainably harness the full potential of India's horticulture sector and ensure food security, income generation, and rural development (Kumar *et al.*, 2020). The cultivation of horticultural crops in India faces various production constraints, with diseases caused by fungi, bacteria, and viruses being significant challenges (Rahman *et al.*, 2019). These pathogens pose serious threats to crop health, yield, and quality, thereby impacting agricultural productivity and farmer livelihoods. Fungal diseases are prevalent in horticultural crops, affecting fruits, vegetables, and ornamental plants. Examples include powdery mildew, anthracnose, and fruit rot, which can lead to significant yield losses if not managed effectively. Bacterial diseases also pose significant challenges to horticultural crop production. Bacterial wilt, bacterial blight, and soft rot are among the common bacterial diseases affecting crops like tomatoes, potatoes, and leafy greens (Singh *et al.*, 2021). These diseases often spread rapidly under favorable environmental conditions, causing wilting, leaf lesions, and fruit rot, thereby reducing marketable yields. Viruses represent another group of pathogens that cause considerable damage on horticultural crops (Kumar *et al.*, 2020). Viral diseases such as mosaic viruses, leaf curl viruses, and necrosis viruses can lead to stunted growth, leaf distortion, and reduced fruit quality. Viruses are typically transmitted through vectors like insects, leading to widespread dissemination within crop populations. To address these challenges, farmers employ various disease management strategies including cultural practices, chemical treatments, and biological control methods (Rahman *et al.*, 2019). Integrated pest management (IPM) approaches, which combine multiple management strategies, are increasingly adopted to minimize reliance on synthetic pesticides and reduce environmental impacts. Research and extension efforts play a crucial role in developing resistant crop varieties, understanding disease epidemiology, and disseminating best management practices to farmers. Collaborative initiatives between agricultural universities, research institutions, and extension agencies are essential for fostering innovation and promoting sustainable disease management strategies. Despite the persistent threat of diseases, advancements in biotechnology, precision agriculture, and digital farming offer promising avenues for enhancing disease surveillance, early detection, and control measures (Singh *et al.*, 2021). By integrating cutting-edge technologies with traditional knowledge systems, India can mitigate the impact of diseases on horticultural crop production and ensure food security and economic sustainability for farmers.

IMPORTANT FAMILY OF VIRUS INFECTING HORTICULTURAL CROPS

List of Important Crops and their Virus Diseases:

DS-DNA Virus

Caulimoviridae: *Caulimoviridae* is a family of plant viruses characterized by their unique double-stranded DNA (dsDNA) genome and icosahedral virion structure. The virions of *Caulimoviridae* viruses are non-enveloped and possess an icosahedral capsid, typically with a diameter ranging from 50 to 60 nanometers. These viruses have a circular dsDNA genome, which ranges in size from approximately 7 to 9 kilobases. The genome encodes several proteins involved in viral replication, transcription, movement, and encapsidation (Hohn *et al.*, 2008).

In terms of host range within horticultural crops, *Caulimoviridae* viruses infect a variety of plants including vegetables, fruits, and ornamental plants. Among the most important genus within the *Caulimoviridae* family is *Caulimovirus*, which includes *Cauliflower mosaic virus* (CaMV). CaMV is known to infect a wide range of cruciferous plants, including cauliflower, cabbage, broccoli, and other *Brassicaceae* species, causing significant losses in yield and quality (Hohn *et al.*, 2008).

Symptoms induced by *Caulimoviridae* infections in horticultural crops can vary depending on the specific virus and host plant. Common symptoms include leaf mottling, chlorosis (yellowing) of leaves, stunting, leaf curling or distortion, and reduced vigor or productivity of plants. In some cases, infections may also lead to the development of necrotic lesions on leaves or other plant tissues. Severe infections can result in plant death, particularly in young plants or under conditions conducive to virus spread.

Transmission of *Caulimoviridae* viruses primarily occurs through mechanical means, such as contaminated tools or hands during cultivation practices. These viruses can also be transmitted through vegetative propagation, infected seeds, or grafting, although these modes of transmission are less common. In some cases, *Caulimoviridae*viruses may be transmitted by aphids in a non-persistent manner.(Hohn *et al.*, 2008).

SS-DNA Virus

Geminiviridae: *Geminiviridae* is a family of plant viruses characterized by their unique geminate particle structure and a single-stranded, circular DNA genome. The name "*geminivirus*" is derived from the Latin word "gemini," meaning twin, referring to the twinned particles that are characteristic of this virus family. The virions typically consist of two incomplete icosahedral particles joined together, forming a characteristic twin structure. These particles are usually around 18-30 nm in diameter (Harrison, 2013).

The genome of *Geminiviridae* viruses is composed of one or two components of circular, single-stranded DNA, ranging from approximately 2.5 to 3.0 kilobases in size. The genome encodes several open reading frames (ORFs) that are involved in viral replication, transcription, movement, and encapsidation. Some *geminiviruses* have a bipartite genome, meaning they are composed of two separate DNA components, while others have a monopartite genome (Heyraud-Nitschke *et al.*, 2013).

Geminiviridae viruses infect a wide range of host plants, including many important agricultural crops such as tomatoes, peppers, beans, cassava, cotton, and cereals. The family is divided into several genera, with the most prominent ones being *Begomovirus*, *Curtovirus*, *Mastrevirus*, and *Topocuvirus*. *Begomoviruses*, in particular, are responsible for devastating diseases in tropical and subtropical regions, affecting crops like tomatoes, peppers, and beans (Zerbini *et al.*, 2017).

Symptoms induced by *Geminiviridae* infections in horticultural crops can vary depending on the virus species and the host plant. Common symptoms include leaf curling, stunting, yellowing (chlorosis), vein thickening, and the development of characteristic leaf mosaic patterns. Severe infections can lead to reduced yields, distorted fruits, and even plant death, significantly impacting agricultural productivity.

Transmission of *Geminiviridae* viruses primarily occurs through insect vectors, particularly whiteflies and leafhoppers. These vectors acquire the virus while feeding on infected plants and then transmit it to healthy plants during subsequent feeding. *Geminiviruses* can also be transmitted through vegetative propagation, infected seeds, and mechanical means such as contaminated tools or hands, although vector-mediated transmission is the most efficient and common route (Heyraud-Nitschke *et al.*, 2013).

Nanoviridae: *Nanoviridae* is a family of plant viruses known for their unique multipartite genome structure and their ability to infect a wide range of horticultural crops. These viruses possess small, non-enveloped, isometric particles with a diameter of around 18-20 nm. The most distinctive feature of *Nanoviridae* viruses is their multipartite genome, which consists of multiple, circular single-stranded DNA molecules, typically ranging from 1.0 to 1.1 kb in size each. Each DNA component encodes one or more viral proteins essential for replication, movement, and encapsidation (Hohn *et al.*, 2008).

Within the family *Nanoviridae*, the genus *Nanovirus* is particularly important in horticulture. *Nanoviruses* infect a variety of horticultural crops including legumes, vegetables, and ornamental plants. Examples of crops susceptible to *nanovirus* infections include beans, peas, cucurbits, and petunias.Symptoms induced by *nanoviruses* in horticultural crops can vary depending on the specific virus-host combination. However, common symptoms include stunting, leaf yellowing, vein clearing, leaf distortion, and necrosis. In some cases, *nanovirus* infections can lead to severe yield losses due to reduced plant growth and fruit development.

Transmission of *nanoviruses* primarily occurs through insect vectors, particularly aphids. These viruses are phloem-limited and are efficiently transmitted by aphids during feeding activities. *Nanoviruses* are often transmitted in a circulative, non-propagative manner, where the virus particles are ingested by the aphid, circulate within its body, and are subsequently excreted onto healthy plants during feeding. Additionally, *nanoviruses* can also be transmitted through vegetative propagation materials such as infected plant cuttings or seeds (Timchenko *et al.*, 1999).

SS-RNA Virus

Alphaflexiviridae: *Alphaflexiviridae* is a family of plant viruses known for their flexuous, filamentous particles and single-stranded RNA genome. The virions of *Alphaflexiviridae* viruses typically have a helical structure, with lengths varying between 500 and 800 nanometers. These viruses contain a positive-sense, single-stranded RNA genome, which ranges in size from approximately 8 to 9 kilobases.

The genome encodes several proteins involved in viral replication, movement, encapsidation, and suppression of host defenses (Adams *et al.*, 2012).

In terms of host range within horticultural crops, *Alphaflexiviridae* viruses infect a wide variety of plants, including vegetables, fruits, and ornamental plants. One of the most important genera within the *Alphaflexiviridae* family is *Potexvirus*, which includes *Potato virus X* (PVX) and *Narcissus mosaic virus* (NMV). These viruses are known to infect a broad range of horticultural crops, such as potatoes, tomatoes, peppers, cucumbers, and ornamental plants like narcissus, causing significant yield losses and quality reduction (Adams *et al.*, 2012).

Symptoms induced by *Alphaflexiviridae* infections in horticultural crops can vary depending on the specific virus and host plant. Common symptoms include mosaic patterns on leaves, leaf distortion, yellowing (chlorosis) of leaves, stunting, and reduced vigor or productivity of plants. In some cases, infections may also lead to the development of necrotic lesions on leaves or other plant tissues. Severe infections can result in plant death, particularly in young plants or under conditions conducive to virus spread.

Transmission of *Alphaflexiviridae* viruses primarily occurs through mechanical means, such as contaminated tools or hands during cultivation practices. These viruses can also be transmitted through vegetative propagation, infected seeds, or grafting, although these modes of transmission are less common. In some cases, *Alphaflexiviridae* viruses may also be transmitted by insect vectors, such as aphids or whiteflies, although this mode of transmission is not as common compared to other plant virus families (Adams *et al.*, 2012).

Bromoviridae: Bromoviridae is a family of plant viruses known for their non-enveloped, spherical particles and segmented, positive-sense, single-stranded RNA genome. The virions of Bromoviridae viruses typically have a diameter ranging from approximately 25 to 35 nanometers. These viruses possess a genome consisting of three segments, referred to as RNA1, RNA2, and RNA3. Each segment encodes one or more proteins involved in viral replication, movement, encapsidation, and suppression of host defenses (Adams *et al.*, 2017).

In terms of host range within horticultural crops, Bromoviridae viruses infect a variety of plants, including vegetables, fruits, and ornamental plants. The most important genus within the Bromoviridae family is Bromovirus, which includes Cucumber mosaic virus (CMV) and Tomato mosaic virus (ToMV). These viruses are known to infect a wide range of horticultural crops, such as cucumbers, tomatoes, peppers, lettuce, and ornamental plants, causing significant yield losses and quality reduction (Adams *et al.*, 2017).

Symptoms induced by Bromoviridae infections in horticultural crops can vary depending on the specific virus and host plant. Common symptoms include mosaic patterns on leaves, leaf distortion, yellowing (chlorosis) of leaves, stunting, and reduced vigor or productivity of plants. In some cases, infections may also lead to the development of necrotic lesions on leaves or other plant tissues. Severe infections can result in plant death, particularly in young plants or under conditions conducive to virus spread.

Transmission of Bromoviridae viruses primarily occurs through mechanical means, such as contaminated tools or hands during cultivation practices. These viruses can also be transmitted through vegetative propagation, infected seeds, or grafting, although these modes of transmission are less common. In some cases, Bromoviridae viruses may also be transmitted by insect vectors, such as aphids or thrips, although this mode of transmission is not as common compared to other plant virus families (Adams *et al.*, 2017).

Closteroviridae: *Closteroviridae* is a family of plant viruses that belong to the order *Martellivirales*. These viruses are known for their flexuous filamentous particles and their single-stranded, positive-sense RNA genomes (Karasev & Bar-Joseph, 2012). The virions typically range from 650 to 2000 nm in length, making them one of the largest plant viruses. The viral genome is approximately 15–20 kb in size and encodes a polyprotein that is processed into various functional proteins. These proteins are involved in viral replication, movement, and encapsidation, among other functions. The genome also contains a number of non-structural proteins that play crucial roles in virus-host interactions.

Within the family *Closteroviridae*, the important genus relevant to horticultural crops is the genus *Closterovirus*. *Closteroviruses* have a wide host range, infecting many horticultural crops including citrus, grapevine, and various ornamental plants. They are responsible for causing significant economic losses in agriculture due to the diseases they induce.

Symptoms induced by *closteroviruses* in horticultural crops can vary widely depending on the specific virus-host combination. However, common symptoms include leaf chlorosis, mosaic patterns, leaf curling, stunting, and fruit disorders such as reduced size, deformities, or premature ripening (Bar-Joseph *et al.*, 1989). In citrus, for example, *Closterovirus* infections can lead to diseases like *citrus tristeza virus* (CTV), which causes decline in tree vigor, fruit quality deterioration, and eventual death of the infected trees.

Transmission of *closteroviruses* typically occurs through propagation materials such as infected plant cuttings, grafting, or budding. Insects, particularly phloem-feeding insects such as aphids, whiteflies, and mealybugs, play a crucial role in the transmission of these viruses. *Closterovirusesare* often phloem-limited, and they are efficiently transmitted from plant to plant by these insect vectors during feeding activities. Additionally, some *closteroviruses* can also be mechanically transmitted through contaminated tools or sap.

Comoviridae: Comoviridae is a family of plant viruses characterized by their non-enveloped, spherical particles and positive-sense, single-stranded RNA genome. The virions of Comoviridae viruses are typically 30-35 nanometers in diameter and composed of two capsids, each containing one RNA molecule. The RNA genome of Comoviridae viruses ranges in size from approximately 8.3 to 9.4 kilobases and encodes several proteins involved in viral replication, movement, encapsidation, and suppression of host defenses (Adams *et al.*, 2012).

In terms of host range within horticultural crops, Comoviridae viruses infect a wide variety of plants, including vegetables, fruits, and ornamental plants. The most important genus within the Comoviridae family is Comovirus, which includes Bean pod mottle virus (BPMV) and Cowpea mosaic virus (CPMV). These viruses are known to infect leguminous crops such as beans, peas, cowpeas, and other related species, causing significant yield losses and quality reduction (Adams *et al.*, 2012).

Symptoms induced by Comoviridae infections in horticultural crops can vary depending on the specific virus and host plant. Common symptoms include mosaic patterns on leaves, leaf distortion, yellowing (chlorosis) of leaves, stunting, and reduced vigor or productivity of plants. In some cases, infections may also lead to the development of necrotic lesions on leaves or other plant tissues. Severe infections can result in plant death, particularly in young plants or under conditions conducive to virus spread.

Transmission of Comoviridae viruses primarily occurs through mechanical means, such as contaminated tools or hands during cultivation practices. These viruses can also be transmitted through vegetative propagation, infected seeds, or grafting, although these modes of transmission are less common. In some cases, Comoviridae viruses may also be transmitted by insect vectors, such as aphids or beetles, although this mode of transmission is not as common compared to other plant virus families (Adams *et al.*, 2012).

Luteoviridae: Luteoviridae is a family of plant viruses known for their non-enveloped, quasi-spherical particles and their single-stranded, positive-sense RNA genomes (Brault & Uzest, 2014). These viruses typically have particles that are about 25-30 nm in diameter. The RNA genome of luteoviruses is approximately

5.5-6.0 kb in size and contains a single open reading frame (ORF) encoding a polyprotein (Miller *et al.*, 1997). This polyprotein is processed into various functional proteins involved in viral replication, movement, and encapsidation. Additionally, the genome also contains untranslated regions (UTRs) at the 5’ and 3’ ends, which play crucial roles in viral RNA replication and translation.

Within the family Luteoviridae, the genus Luteovirus is particularly important in horticulture. Luteoviruses infect a wide range of horticultural crops including cereals, vegetables, and ornamental plants (Brault & Uzest, 2014). Examples of crops susceptible to luteovirus infections include wheat, barley, oats, potatoes, and roses.Symptoms induced by luteoviruses in horticultural crops can vary widely depending on the specific virus-host combination. However, common symptoms include yellowing or reddening of leaves, stunting, leaf curling, and reduced yield (Brault &Uzest, 2014). In cereals, luteovirus infections can lead to diseases such as barley yellow dwarf, which can cause significant reductions in grain yield and quality.

Transmission of luteoviruses typically occurs through insect vectors, particularly aphids (Brault & Uzest, 2014). These viruses are phloem-limited and are efficiently transmitted by aphids during feeding activities. Luteoviruses are often transmitted in a persistent, circulative manner, where the virus particles are ingested by the aphid, circulate within its body, and are subsequently excreted onto healthy plants during feeding. Additionally, luteoviruses can also be transmitted through vegetative propagation materials such as infected plant cuttings or grafting.

Potyviridae: Potyviridae is a family of plant viruses that encompasses a vast array of species with significant economic importance due to their ability to infect a wide range of crops, causing severe losses in agricultural production. The viruses in this family are characterized by flexuous, filamentous particles with helical symmetry, typically ranging from 680 to 900 nm in length and 11 to 20 nm in width. These virions are non-enveloped and contain a single-stranded, positive-sense RNA genome (Adams *et al.*, 2017).

The genome of Potyviridae viruses is approximately 10,000 to 12,000 nucleotides in length and encodes a single polyprotein that undergoes proteolytic processing to yield functional proteins. This polyprotein includes proteins involved in viral replication, movement, and encapsidation, as well as proteins responsible for modifying host physiology to facilitate infection and spread (Chung *et al.*, 2008). *Potyviridae* viruses exhibit a broad host range, infecting numerous plant species across different families, including many important crops such as potatoes, tomatoes, peppers, beans, cucurbits, and cereals. Among the most significant genera within the *Potyviridae* family are *Potyvirus*, the type genus, which includes several economically important viruses like *Potato virus Y* (PVY), *Soybean mosaic virus* (SMV), and *Zucchini yellow mosaic virus* (ZYMV) (Adams *et al.*, 2017).

Symptoms induced by *Potyviridae* infections in horticultural crops can vary widely depending on the virus species and the host plant. Common symptoms include mosaic patterns on leaves, leaf distortion, stunting, yellowing (chlorosis), necrosis, and reduced yield or quality of fruits. In some cases, infections can lead to systemic effects, affecting the entire plant and causing wilting or even plant death.

Transmission of *Potyviridae*viruses occurs through various means, including aphids, which are the primary vectors for many species within this family. These viruses are typically transmitted in a non-persistent or semi-persistent manner, meaning that the virus can be acquired and transmitted by the vector within a short time frame. Other modes of transmission include mechanical transmission through contaminated tools or hands, seed transmission, and transmission through infected plant material during propagation or grafting.

Secoviridae: Secoviridae is a family of plant viruses characterized by their non-enveloped, spherical particles and their single-stranded, positive-sense RNA genomes (Martelli *et al.*, 2002). These viruses typically have particles that are about 28-30 nm in diameter. The RNA genome of secoviruses is approximately 8-10 kb in size and contains a single open reading frame (ORF) encoding a polyprotein. This polyprotein is processed into various functional proteins involved in viral replication, movement, and encapsidation (Dietzgen *et al.*, 2003). Additionally, the genome also contains untranslated regions (UTRs) at the 5' and 3' ends, which play crucial roles in viral RNA replication and translation.

Within the family Secoviridae, the genus Nepovirus is particularly important in horticulture. Nepoviruses infect a wide range of horticultural crops including fruit trees, vegetables, and ornamental plants (Martelli *et al.*, 2002). Examples of crops susceptible to nepovirus infections include grapevines, tomatoes, peppers, and roses. Symptoms induced by nepoviruses in horticultural crops can vary widely depending on the specific virus-host combination. However, common symptoms include leaf mottling, mosaic patterns, leaf distortion, stunting, and necrosis. In some cases, nepovirus infections can also lead to reduced fruit quality, such as deformities, discoloration, or malformation.

Transmission of nepoviruses typically occurs through mechanical means, such as contaminated tools or sap, or via insect vectors. Nematodes are often important vectors for the transmission of nepoviruses in horticultural crops. These viruses can also be transmitted through vegetative propagation materials such as infected plant cuttings or grafting (Martelli *et al.*, 2002).

Tombusviridae: *Tombusviridae* is a family of plant viruses that possess a non-enveloped, icosahedral capsid structure. The virions are small, spherical particles with a diameter of approximately 28-30 nm. These viruses have a positive-sense, single-stranded RNA genome that ranges from 4.8 to 6.4 kilobases in size. The genome typically contains five open reading frames (ORFs), which encode proteins involved in viral replication, movement, encapsidation, and suppression of host defenses (Hyodo *et al.*, 2019). Host range in horticultural crops, *Tombusviridae* viruses infect a variety of plants including tomatoes, peppers, cucumbers, and other vegetables and ornamental plants. Among the most important genera within the *Tombusviridae* family is *Tombusvirus*, which includes *Tomato bushy stunt virus* (TBSV) and *Cucumber necrosis virus* (CNV). These viruses can cause significant losses in horticultural production due to the diseases they induce in susceptible crops.

Symptoms induced by *Tombusviridae* infections in horticultural crops can vary depending on the specific virus and host plant. Common symptoms include stunting, chlorosis (yellowing) of leaves, mosaic patterns on leaves, leaf distortion, and reduced yield or quality of fruits. In severe cases, infections may lead to plant death, particularly in young plants or under conditions conducive to virus spread.Transmission of *Tombusviridae* viruses typically occurs through mechanical means, such as contaminated tools or hands during cultivation practices. Some viruses within this family may also be transmitted through seed, although this mode of transmission is less common. Additionally, some *Tombusviridae* viruses have been reported to be transmitted by insect vectors in experimental settings, although natural transmission by insects is not widely documented for this family (Martelli *et al.*, 2016).

Tymoviridae: *Tymoviridae* is a family of plant viruses characterized by their non-enveloped, spherical particles and single-stranded RNA genome. The virions of *Tymoviridae* viruses typically have a diameter of approximately 28-30 nanometers. These viruses contain a positive-sense, single-stranded RNA genome, which varies in size depending on the virus species but typically ranges from around 6 to 8 kilobases. The genome encodes several proteins involved in viral replication, movement, encapsidation, and suppression of host defenses (King *et al.*, 2012).

In terms of host range within horticultural crops, *Tymoviridae* viruses infect a variety of plants, including vegetables, fruits, and ornamental plants. The most important genus within the *Tymoviridae* family is *Tymovirus*, which includes

Turnip yellow mosaic virus (TYMV) and *Odontoglossum ringspot virus* (ORSV). These viruses are known to infect a broad range of horticultural crops, such as turnips, cucumbers, tomatoes, and orchids, causing significant yield losses and quality reduction (King *et al.*, 2012).Symptoms induced by *Tymoviridae* infections in horticultural crops can vary depending on the specific virus and host plant. Common symptoms include mosaic patterns on leaves, leaf distortion, yellowing (chlorosis) of leaves, stunting, and reduced vigor or productivity of plants. In some cases, infections may also lead to the development of necrotic lesions on leaves or other plant tissues. Severe infections can result in plant death, particularly in young plants or under conditions conducive to virus spread.

Transmission of *Tymoviridae* viruses primarily occurs through mechanical means, such as contaminated tools or hands during cultivation practices. These viruses can also be transmitted through vegetative propagation, infected seeds, or grafting, although these modes of transmission are less common. In some cases, *Tymoviridae* viruses may also be transmitted by insect vectors, such as aphids or leafhoppers, although this mode of transmission is not as common compared to other plant virus families (King *et al.*, 2012).

Virgaviridae: *Virgaviridae* is a family of plant viruses characterized by their non-enveloped, rod-shaped particles and single-stranded RNA genome. The virions of *Virgaviridae* viruses are typically elongated and filamentous, ranging from approximately 600 to 800 nanometers in length. These viruses possess a positive-sense, single-stranded RNA genome, which varies in size depending on the virus species but typically ranges from around 5 to 9 kilobases. The genome encodes several proteins involved in viral replication, movement, encapsidation, and suppression of host defenses (Martelli *et al.*, 2002).

In terms of host range within horticultural crops, *Virgaviridae* viruses infect a variety of plants including vegetables, fruits, and ornamental plants. Among the most important genera within the *Virgaviridae* family is *Tobamovirus*, which includes *Tobacco mosaic virus* (TMV) and *Tomato mosaic virus* (ToMV). These viruses are known to infect a wide range of horticultural crops, including tomatoes, peppers, cucumbers, lettuce, and other solanaceous and cucurbitaceous species, causing significant yield losses and quality reduction (Martelli *et al.*, 2002). Symptoms induced by *Virgaviridae* infections in horticultural crops can vary depending on the specific virus and host plant. Common symptoms include mosaic patterns on leaves, leaf distortion, yellowing (chlorosis) of leaves, stunting, and reduced vigor or productivity of plants. In some cases, infections may also lead to the development of necrotic lesions on leaves or other plant tissues. Severe infections can result in plant death, particularly in young plants or under conditions conducive to virus spread.

Table : Important Horticultural Crops Susceptibility to Viral Diseases and Associated Information

SN	Crop	Disease	Virus	Genus	Family	Vector
1	Fruit					
2	Apple	Apple chlorotic leafspot	Apple chlorotic leafspot virus (ACLSV)	Trichovirus	Betaflexiviridae	Unknown
3		Apple dwarf (Malus platycarpa)	Apple stem pitting virus (ASPV)	Foveavirus	Betaflexiviridae	Transmitted through grafting and possibly by nematodes.
4		Apple flat apple	Cherry rasp leaf virus (CRLV)	Nepovirus	Secoviridae	Unknown
5		Apple mosaic	Apple mosaic virus (ApMV)	Ilarvirus	Bromoviridae	Transmitted through grafting and vegetative propagation.
6		Apple stem grooving = Apple decline of Virginia crab	Apple stem grooving virus (ASGV)	Capillovirus	Betaflexiviridae	Transmitted through grafting and possibly by nematodes.
7		Apple stem pitting = apple Spy 227 epinasty and decline	Apple stem pitting virus (ASPV)	Foveavirus	Betaflexiviridae	
8		Apple union necrosis and decline	Tomato ringspot virus (ToRSV)	Nepovirus	Secoviridae	Transmitted by dagger nematodes.
9	Banana	Bract mosaic	Banana bract mosaic virus	Potyvirus	Potyviridae	Unknown
10		Bunchy top	Banana bunchy top virus	Babuvirus	Nanoviridae	Transmitted by banana aphids.
11		Mosaic	Cucumber mosaic virus	Cucumovirus	Bromoviridae	Transmitted by aphids in a non-persistent manner.
12		Streak	Banana streak virus	Badnavirus	Caulimoviridae	Unknown
13		Banana mild mosaic	Banana mild mosaic virus	Potyvirus	Potyviridae	Unknown
14		Banana virus X	Banana virus X	Potexvirus	Alphaflexiviridae	Unknown

[Table Contd.

Contd. Table]

SN	Crop	Disease	Virus	Genus	Family	Vector
15	Papaya	Apical necrosis	Papaya apical necrosis virus	Potyvirus	Potyviridae	Transmitted by aphids.
16		Droopy necrosis	Papaya droopy necrosis virus	Potyvirus	Potyviridae	Transmitted by aphids.
17		Leaf curl	Papaya leaf curl virus (PaLCuV)	Begomovirus	Geminiviridae	Transmitted by whiteflies.
18		Mosaic	Papaya mosaic virus	Potyvirus	Potyviridae	Transmitted by aphids.
19		Papaya ringspot	Papaya ringspot virus	Potyvirus	Potyviridae	Transmitted by aphids.
20		Papaya lethal yellowing	Papaya lethal yellowing virus	Bobemovirus	Virgaviridae	Transmitted by aphids.
21		Spotted wilt	Tomato spotted wilt virus	Tospovirus	Peribunyaviridae	Transmitted by thrips.
22		Sticky disease ('meleira')	Papaya meleira virus complex	Closterovirus	Closteroviridae	Unknown
23		Terminal necrosis and wilt	Tobacco ringspot virus	Nepovirus	Secoviridae	Transmitted by nematodes and possibly by seed.
24	Citrus	Citrus mosaic	Citrus mosaic virus	Potyvirus	Potyviridae	Unknown
25		Bud union crease	Citrus tristeza virus (CTV)	Closterovirus	Closteroviridae	Transmitted by aphids, grafting, and possibly by seed.
26		Citrus leaf rugose	Citrus leaf rugose virus (CLRV)	Ilarvirus	Bromoviridae	Unknown
27		Citrus yellow mosaic	Citrus yellow mosaic virus	Badnavirus	Caulimoviridae	Unknown
28		Crinkly leaf	Citrus tatter leaf virus (CTLV)	Capillovirus	Betaflexiviridae	Unknown
29		Infectious variegation	Citrus variegation virus (CVV)	Ilarvirus	Bromoviridae	Unknown
30		Navel infectious mottling	Citrus tatter leaf virus (CTLV)	Capillovirus	Betaflexiviridae	Unknown
31		Psorosis	Citrus psorosis virus (CPsV)	Ophiovirus	Aspiviridae	
32		Satsuma dwarf	Citrus tatter leaf virus (CTLV)	Capillovirus	Betaflexiviridae	
33		Tatter leaf = citrange stunt	Citrus tatter leaf virus	Capillovirus	Betaflexiviridae	

[Table Contd.

Contd. Table]

SN	Crop	Disease	Virus	Genus	Family	Vector
34		Tristeza = decline and stem pitting seedling yellows	Citrus tristeza virus (CTV)	Closterovirus	Closteroviridae	
35	Grapevine	Alfalfa mosaic	Alfalfa mosaic virus	Alfamovirus	Bromoviridae	Transmitted by aphids in a non-persistent manner.
36		Arabis mosaic	Arabis mosaic virus	Nepovirus	Secoviridae	Unknown
37		Artichoke Italian latent	Artichoke Italian latent virus	Macluravirus	Tymoviridae	Unknown
38		Bratislava mosaic	Bratislava mosaic virus	Ilarvirus	Bromoviridae	Unknown
39		Broad bean wilt	Broad bean wilt virus	Fabavirus	Bromoviridae	Unknown
40		Corky bark	Grapevine virus B	Vitivirus	Betaflexiviridae	Unknown
41		Enation	Undetermined viruslike			Unknown
42		Fanleaf degeneration (infectious degeneration and decline)	Grapevine fanleaf virus	Nepovirus	Secoviridae	Transmitted by nematodes.
43		Grapevine Bulgarian latent	Grapevine Bulgarian latent virus	Tymovirus	Tymoviridae	Unknown
44		Grapevine chrome mosaic	Grapevine chrome mosaic virus	Begomovirus	Geminiviridae	Unknown
45		Grapevine red blotch	Grapevine red blotch-associated virus	Grablovirus	Geminiviridae	Unknown
46		Leafroll	Grapevine leafroll-associated viruses	Ampelovirus	Closteroviridae	Transmitted by ealybugs.
47		Peach rosette mosaic virus decline	Peach rosette mosaic virus nematodes.	Ilarvirus	Bromoviridae	Transmitted by
48		Petunia asteroid mosaic	Petunia asteroid mosaic virus	Potyvirus	Potyviridae	Unknown
49		Raspberry ringspot	Raspberry ringspot virus	Nepovirus	Secoviridae	Unknown
50		Strawberry latent ringspot	Strawberry latent ringspot virus	Nepovirus	Secoviridae	Transmitted by nematodes.

[Table Contd.

Contd. Table]

SN	Crop	Disease	Virus	Genus	Family	Vector
51		Tobacco mosaic	Tobacco mosaic virus	Tobamovirus	Virgaviridae	Transmitted by mechanical contact and possibly by seed.
52		Tobacco necrosis	Tobacco necrosis virus	Ilarvirus	Bromoviridae	Unknown
53		Tobacco ringspot virus decline	Tobacco ringspot virus	Nepovirus	Secoviridae	Unknown
54		Tomato black ring	Tomato black ring virus	Nepovirus	Secoviridae	Unknown
55		Tomato ringspot virus decline	Tomato ringspot virus	Nepovirus	Secoviridae	Transmitted by dagger nematodes.
56	Chilli	Alfalfa mosaic	Alfalfa mosaic virus (AMV)	Alfamovirus	Bromoviridae	Transmitted by aphids in a non-persistent manner.
57		Beet curly top	Beet curly top virus (BCTV)	Curtovirus	Geminiviridae	Transmitted by leafhoppers.
58		Chili leaf curl	Chili leaf curl virus (ChiLCV); Possibly strain of Tobacco leaf curl virus (TLCV)	Begomovirus	Geminiviridae	Transmitted by whiteflies.
59		Chilli veinal mottle	Chilli veinal mottle virus (ChiVMV)	Potyvirus	Potyviridae	Unknown
60		Chino del tomate	Chino del tomate virus (CdTV)	Begomovirus	Geminiviridae	Unknown
61		Cucumber mosaic	Cucumber mosaic virus (CMV)	Cucumovirus	Bromoviridae	Transmitted by aphids in a non-persistent manner.
62		Pepper golden mosaic complex (previously Texas Pepper Serrano Golden Mosaic and Pepper Mild Tigre Viruses)	Serrano golden mosaic virus (SGMV) Pepper mild tigre virus (PMTV) and others	Begomovirus	Geminiviridae	Transmitted by whiteflies.
63		Pepper huasteco	Pepper huasteco virus (PHV)	Begomovirus	Geminiviridae	Unknown

[Table Contd.

Contd. Table]

SN	Crop	Disease	Virus	Genus	Family	Vector
64		Pepper leaf curl	Pepper leaf curl virus (PepLCV)	Begomovirus	Geminiviridae	Transmitted by whiteflies.
65		Pepper mild mottle	Pepper mild mottle virus (PMMV)	Tobamovirus	Virgaviridae	Unknown
66		Pepper mottle	Pepper mottle virus (PepMoV)	Potyvirus	Potyviridae	Transmitted by thrips.
67		Pepper veinal mottle	Pepper veinal mottle virus (PVMV)	Potyvirus	Potyviridae	Unknown
68		Potato virus Y	Potato virus Y (PVY)	Potyvirus	Potyviridae	Transmitted by aphids.
69		Sinaloa tomato leaf curl	Sinaloa tomato leaf curl virus (TCLV)	Begomovirus	Geminiviridae	Transmitted by whiteflies.
70		Tobacco etch	Tobacco etch virus (TEV)	Potyvirus	Potyviridae	Transmitted by aphids in a non-persistent manner.
71		Tobacco mosaic and Tomato mosaic	Tobacco mosaic virus (TMV) and Tomato mosaic virus (ToMV)	Tobamovirus	Virgaviridae	Transmitted by mechanical contact and possibly by seed.
72		Tomato spotted wilt	Tomato spotted wilt virus (TSWV)	Tospovirus	Peribunyaviridae	Transmitted by thrips.
73	Tomato	Common mosaic of tomato (internal browning of fruit)	Tobacco mosaic virus (TMV)	Tobamovirus	Virgaviridae	Transmitted by mechanical contact and possibly by seed.
74		Curly top	Curtovirus	Curtovirus	Geminiviridae	Transmitted by beet leafhopper.
75		Potato virus Y	Potato virus Y	Potyvirus	Potyviridae	Transmitted by aphids.
76		Pseudo curly top	Tomato pseudo-curly top virus	Begomovirus	Geminiviridae	Unknown
77		Tomato bushy stunt	Tomato bushy stunt virus	Tombusvirus	Tombusviridae	Transmitted by whiteflies.
78		Tomato etch	Tobacco etch virus	Potyvirus	Potyviridae	Transmitted by aphids in a non-persistent manner.

[Table Contd.

Contd. Table]

SN	Crop	Disease	Virus	Genus	Family	Vector
79		Tomato fern leaf	Cucumber mosaic virus	Cucumovirus	Bromoviridae	Transmitted by aphids in a non-persistent manner.
80		Tomato mosaic	Tomato mosaic virus (ToMV)	Tobamovirus	Virgaviridae	Transmitted by mechanical contact and possibly by seed.
81		Tomato mottle	Tomato mottle geminivirus	Begomovirus	Geminiviridae	Unknown
82		Tomato necrosis	Alfalfa mosaic virus	Alfamovirus	Bromoviridae	Transmitted by aphids in a non-persistent manner.
83		Tomato spotted wilt	Tomato spotted wilt virus	Tospovirus	Peribunyaviridae	Transmitted by thrips.
84		Tomato yellow leaf curl	Tomato yellow leaf curl virus	Begomovirus	Geminiviridae	Transmitted by whiteflies.
85		Tomato yellow top	Tomato yellow top virus			Unknown
86	Bottle Gourd	Mosaic	Cucumber mosaic virus (CMV)	Cucumovirus	Bromoviridae	Transmitted by aphids in a non-persistent manner.
87		Mosaic	Watermelon mosaic virus (WMV)	Potyvirus	Potyviridae	Transmitted by aphids.
88		Mosaic	Zucchini yellow mosaic virus (ZYMV)		Potyvirus	Potyviridae Transmitted by aphids.
89		Mosaic	Squash mosaic virus (SqMV)	Comovirus	Secoviridae	Transmitted by aphids.
90	Cucumber	Mosaic	Cucumber mosaic virus (CMV)	Cucumovirus	Bromoviridae	Transmitted by aphids in a non-persistent manner.
91		Mottle Mosaic	Cucumber green mottle mosaic virus (CGMMV)		Tobamovirus	Virgaviridae Unknown
92		Mosaic	Cucumber vein yellowing virus (CVYV)		Ipomovirus	Potyviridae Unknown
93		Mosaic	Zucchini yellow mosaic virus (ZYMV)		Potyvirus	Potyviridae Transmitted by aphids.
94		Mosaic	Watermelon mosaic virus (WMV)	Potyvirus	Potyviridae	Transmitted by aphids.

[Table Contd.

Contd. Table]

SN	Crop	Disease	Virus	Genus	Family	Vector
95	Watermelon	Mosaic	Watermelon mosaic virus (WMV)	Potyvirus	Potyviridae	Transmitted by aphids.
96		Mosaic	Cucumber mosaic virus (CMV)	Cucumovirus	Bromoviridae	Transmitted by aphids in a non-persistent manner.
97		Yellow Mosaic	Zucchini yellow mosaic virus (ZYMV)		Potyvirus	Potyviridae Transmitted by aphids.
98		Mosaic	Squash mosaic virus (SqMV)	Comovirus	Secoviridae	Transmitted by aphids.
99		Ringspot	Papaya ringspot virus (PRSV)	Potyvirus	Potyviridae	Transmitted by aphids.
100	Potato	Alfalfa mosaic virus	Alfalfa mosaic virus (AMV)	Alfamovirus	Bromoviridae	Transmitted by aphids in a non-persistent manner.
101		Andean potato mottle virus	Andean potato mottle virus (APMV)	Comovirus	Secoviridae	Transmitted by aphids.
102		Beet curly top virus	Beet curly top virus (BCTV)	Curtovirus	Geminiviridae	Transmitted by leafhoppers.
103		Cucumber mosaic virus	Cucumber mosaic virus (CMV)	Cucumovirus	Bromoviridae	Transmitted by aphids in a non-persistent manner.
104		Potato aucuba mosaic virus	Potato aucuba mosaic virus (PAMV)	Potexvirus	Alphaflexiviridae	Unknown
105		Potato black ringspot virus	Potato black ringspot virus (PBRSV)	Nepovirus	Secoviridae	Unknown
106		Potato deforming mosaic virus	Potato deforming mosaic virus subgroup III (PDMV)	Carlavirus	Betaflexiviridae	Unknown
107		Potato latent virus	Potato latent virus (PLV)	Carlavirus	Betaflexiviridae	Unknown
108		Potato leafroll virus	Potato leafroll virus (PLRV)	Polerovirus	Solemoviridae	Unknown
109		Potato mop-top virus (spraing of tubers)	Potato mop-top virus (PMTV)	Furovirus	Virgaviridae	Unknown

[Table Contd.

Contd. Table]

SN	Crop	Disease	Virus	Genus	Family	Vector
110		Potato rugose mosaic	Potato virus Y (PVY strains O N and C)	Potyvirus	Potyviridae	Transmitted by aphids.
111		Potato stem mottle (spraing of tubers)	Tobacco rattle virus (TRV)	Tobravirus	Virgaviridae	Transmitted by nematodes.
112		Potato yellow dwarf virus	Potato yellow dwarf virus (PYDV)	Nucleorhabdovirus	Rhabdoviridae	Unknown
113		Potato yellow mosaic virus	Potato yellow mosaic virus (PYMV)	Begomovirus	Geminiviridae	Transmitted by aphids.
114		Potato yellow vein virus	Potato yellow vein virus (PYVV)	Crinivirus	Closteroviridae	Unknown
115		Potato yellowing virus	Potato yellowing virus (PYV)	Alfamovirus	Bromoviridae	Unknown
116		Potato virus A	Potato virus A (PVA)	Potyvirus	Potyviridae	Unknown
117		Potato virus M	Potato virus M (PVM)	Carlavirus	Betaflexiviridae	Unknown
118		Potato virus S	Potato virus S (PVS)	Carlavirus	Betaflexiviridae	Unknown
119		Potato virus H	Potato virus H (PVH)	Carlavirus	Betaflexiviridae	Unknown
120		Potato virus T	Potato virus T	Trichovirus	Betaflexiviridae	Unknown
121		Potato virus U	Potato virus U (PVU)	Nepovirus	Secoviridae	Unknown
122		Potato virus V	Potato virus V (PVV)	Potyvirus	Potyviridae	Unknown
123		Potato virus X	Potato virus X (PVX)	Potexvirus	Alphaflexiviridae	Transmitted by aphids.
124		Potato virus Y	Potato virus Y (PVY)	Potyvirus	Potyviridae	Transmitted by aphids.
125		Solanum apical leaf curling virus	Solanum apical leaf curling virus (SALCV) subgroup III	Begomovirus	Geminiviridae	Unknown
126		Tobacco mosaic virus	Tobacco mosaic virus (TMV)	Tobamovirus	Virgaviridae	Transmitted by mechanical contact.

[Table Contd.

Contd. Table]

SN	Crop	Disease	Virus	Genus	Family	Vector
127		Tobacco necrosis virus	Tobacco necrosis virus (TNV)	Necrovirus	Virgaviridae	Unknown
128		Tobacco rattle virus	Tobacco rattle virus (TRV)	Tobravirus	Virgaviridae	Transmitted by nematodes.
129		Tobacco streak virus	Tobacco streak virus (TSV)	Ilarvirus	Bromoviridae	Transmitted by leafhoppers.
130		Tomato black ring virus	Tomato black ring virus (ToBRV)	Nepovirus	Secoviridae	Unknown
131		Tomato mosaic virus	Tomato mosaic virus (ToMV)	Tobamovirus	Virgaviridae	Transmitted by mechanical contact and possibly by seed.
132		Tomato spotted wilt virus	Tomato spotted wilt virus (TSWV)	Tospovirus	Peribunyaviridae	Transmitted by thrips.
133		Tomato yellow mosaic virus	Tomato yellow mosaic virus (ToYMV)	Tobamovirus	Virgaviridae	Transmitted by whiteflies.
134		Wild potato mosaic virus	Wild potato mosaic virus (WPMV)	Potyvirus	Potyviridae	Transmitted by aphids.

Transmission of *Virgaviridae* viruses primarily occurs through mechanical means, such as contaminated tools or hands during cultivation practices. These viruses can also be transmitted through infected seeds, which serve as an important source of virus inoculum in agricultural systems. In some cases, *Virgaviridae* viruses may also be transmitted by insect vectors, although this mode of transmission is less common compared to other plant virus families (Martelli *et al.*, 2002).

Management of the Virus diseases: Developing a management strategy to effectively manage virus diseases in Horticultural crops presents several challenges due to various factors. Firstly, both the insect vectors responsible for transmitting the viruses and the viruses themselves often have a wide range of host plants, making it challenging to contain their spread (Jones, 2019). Additionally, insect vectors tend to develop resistance to control measures relatively quickly, complicating efforts to manage their populations (Smith *et al.*, 2020). Moreover, viruses infecting vegetable crops have a tendency to evolve rapidly, potentially rendering previously effective control methods ineffective (Brown & Johnson, 2018). Furthermore, there is a limited availability of plant varieties with resistance to these viruses, further complicating management efforts (Roberts, 2017). To address these challenges, integrated strategies are recommended, which combine multiple approaches such as host plant resistance, induced resistance, biological control, cultural practices, and judicious use of chemical pesticides (Gomez *et al.*, 2021). By integrating these approaches, it becomes possible to develop more effective and sustainable management strategies for viral diseases in vegetable crops, ultimately reducing the reliance on pesticides and minimizing their environmental impact (Zhang & Wang, 2020).

1. **Host Plant Resistance:** Host plant resistance against viruses in horticultural crops involves developing cultivars that possess genetic traits conferring resistance or tolerance to specific viral pathogens. For instance, tomato (*Solanum lycopersicum*) cultivars like 'Tygress', 'Typhoon', 'Tyking', and 'Tycoon' exhibit varying degrees of resistance to *Tomato yellow leaf curl virus* (TYLCV) (McGrath, 2016; Moriones& Navas-Castillo, 2020), while resistant potato (*Solanum tuberosum*) cultivars such as 'Russet Burbank' and 'Ranger Russet' combat *Potato virus Y* (PVY) (Whitworth *et al.*, 2009; Gray *et al.*, 2010). Pepper (*Capsicum annuum*) cultivars like 'CM334' are resistant to *Pepper mild mottle virus* (PMMoV) (Thompson *et al.*, 2007), and melon (*Cucumis melo*) cultivars such as 'PI 313970' provide resistance against *Cucumber mosaic virus* (CMV) (Provvidenti *et al.*, 1979). Squash (*Cucurbita* spp.) cultivars like 'Elite', 'Ezra', and 'Rustler' are resistant to *Zucchini yellow mosaic virus* (ZYMV) (Provvidenti& Gonsalves, 1981; Thomas *et al.*, 1984).

Table: List of resistant varieties identified against the viral diseases of horticultural crops

Crop	Viral Disease	Resistant Varieties	Source/Reference
Tomato	Tomato Yellow Leaf Curl Virus (TYLCV)	Pusa Rohini, Pusa Sadabahar	Indian Institute of Vegetable Research (IIVR)
Tomato	Tomato Leaf Curl Virus (ToLCV)	Arka Rakshak, Arka Abha	Indian Institute of Horticultural Research (IIHR)
Potato	Potato Virus Y (PVY)	Kufri Swarna, Kufri Jyoti	Indian Institute of Potato Research (IIPR)
Brinjal	Brinjal Leaf Curl Virus (BLCV)	Manik, Pusa Uttam	Indian Institute of Vegetable Research (IIVR)
Cucumber	Cucumber Mosaic Virus (CMV)	Poinsett, Ashley	Indian Agricultural Research Institute (IARI)
Chilli	Chilli Leaf Curl Virus (CLCuV)	Pusa Jwala, Pusa Sadabahar	Indian Institute of Vegetable Research (IIVR)
Pumpkin	Cucumber Mosaic Virus (CMV)	Arka Suryamukhi, Pusa Vishwas	Indian Institute of Vegetable Research (IIVR)
Cauliflower	Cauliflower Mosaic Virus (CaMV)	Snowball Y Improved, Pusa Snowball K-1	Indian Institute of Vegetable Research (IIVR)
Carrot	Carrot Motley Dwarf Virus (CMDV)	Nantes, Kuroda	Indian Agricultural Research Institute (IARI)
Capsicum	Pepper Veinal Mottle Virus (PVMV)	California Wonder, Indra	Indian Institute of Vegetable Research (IIVR)
Muskmelon	Cucumber Mosaic Virus (CMV)	Arka Jyoti, Arka Manikara	Indian Institute of Horticultural Research (IIHR)
Watermelon	Watermelon Mosaic Virus (WMV)	Charleston Gray, Sugar Baby	Indian Institute of Vegetable Research (IIVR)
Papaya	Papaya Ringspot Virus (PRSV)	Coorg Honey Dew, Pusa Delicious	Indian Council of Agricultural Research (ICAR)
Banana	Banana bunchy top virus (BBTV)	Grand Naine, Robusta	National Research Centre for Banana (NRCB)
Peach	Peach Mosaic Virus (PMV)	Flordagold, Suncrest	National Horticultural Research and Development Foundation (NHRDF)
Guava	Guava Leaf Roll Virus (GLRV)	Allahabad Safeda, Sardar	Indian Agricultural Research Institute (IARI)

2. **Induced Resistance:** Induced resistance against viruses in horticultural crops involves activating the plant's defense mechanisms through various external stimuli, such as chemical elicitors or beneficial microorganisms, to enhance the plant's resistance to viral infections. For instance, salicylic acid treatment has been shown to induce systemic acquired resistance (SAR) against viruses like *Cucumber mosaic virus* (CMV) in cucumber (*Cucumis sativus*) (Kanwar *et al.*, 2015) and *Tomato mosaic virus* (ToMV) in tomato (*Solanum lycopersicum*) (Oliveira *et al.*, 2015). Beneficial microorganisms such as *Trichoderma harzianum* have been found to induce resistance against *Papaya ringspot virus* (PRSV) in papaya (*Carica papaya*) (Kamala *et al.*, 2016), while *Bacillus amyloliquefaciens* treatment has shown effectiveness against *Watermelon mosaic virus* (WMV) in watermelon (*Citrullus lanatus*) (Liu *et al.*,2020). Additionally, plant growth-promoting rhizobacteria (PGPR) have been demonstrated to elicit resistance against *Pepper leaf curl virus* (PepLCV) in chili pepper (*Capsicum annuum*) (Bharti *et al.*, 2016) and reduce symptoms of *Zucchini yellow mosaic virus* (ZYMV) infection in squash (*Cucurbita pepo*) (Raupach & Kloepper, 2000). Use of *Pseudomonas fluorescens* for seed treatment @ 300 CFU/ml for 24 h, seedling dip @ 300CFU/ml for 2 h, soil drenching with 25 ml @ 300CFU/ml per plant and foliar spray@300CFU/ml is found to induce systemic resistance on tomato plants against the Tomato spotted wilt virus (Nagendran *et al.*, 2017).
3. **Biological Control:** Biological control against viruses in horticultural crops involves utilizing natural enemies or beneficial organisms to suppress viral populations or vector populations, thereby reducing disease incidence. Predatory insects such as ladybugs (*Coccinellidae*) and lacewings (*Chrysopidae*) can help control populations of virus-transmitting insect vectors like aphids, thrips, and whiteflies, indirectly limiting the spread of viral diseases in horticultural crops (Koch *et al.*, 2006). Release of predatory mites such as *Amblyseiusswirski*, *A. cucumeris* and *Stratiolaelapsscimitus* @ 50-100 nos./m2 are found necessary in managing the thrips population (Nagendran *et al.*, 2017). Certain entomopathogenic fungi, such as *Beauveria bassiana, Verticillium lecani* and *Metarhizium anisopliae,* @ 1litre/ha are effective in infecting and kill insect vectors of plant viruses, making them effective biocontrol agents for suppressing vector populations and reducing virus transmission in crops (Lacey *et al.*, 2015). RNA interference (RNAi) is another promising approach where genetically engineered crops expressing double-stranded RNA (dsRNA) targeting viral genes can induce RNAi-mediated degradation of viral RNA in insect vectors, thereby reducing virus transmission (Tenllado & Díaz-Ruíz, 2001). Additionally, antiviral plant

endophytes, such as certain fungi and bacteria residing within plant tissues, can trigger systemic resistance mechanisms, leading to reduced symptom severity and viral replication in horticultural crops (Marquez *et al.*, 2007).

4. **Cultural Practices:** Cultural practices aim to create conditions that are unfavorable for virus transmission and multiplication. Practices such as crop rotation, intercropping, and sanitation (removing and destroying infected plant debris) can help in reductionof virus inoculum. In banana cultivation, adopting strict sanitation practices by removing and destroying infected plant parts can effectively manage *Banana bunchy top virus* (BBTV) and *Banana streak virus* (BSV) (Kumar *et al.*, 2018). Cultural practices influence vector population and virus incidence, playing a significant role in managing viral diseases by affecting the population dynamics of insect vectors and the spread of viruses (Smith *et al.*, 2020). Maintaining a weed-free nursery away from the cropping field and using insect-proof netting in the nursery are effective measures to prevent virus transmission (Jones & Wang, 2019). Creating a buffer zone around the field free of weeds and destroying weeds before flowering are additional strategies to minimize the risk of virus transmission (Brown & Johnson, 2018). Furthermore, avoiding overlapping susceptible crops, sowing crops to minimize vector invasion during sensitive stages, and rouging infected plants soon after transplanting are essential cultural practices for virus management (Roberts, 2017). Intercropping with border crops and selecting transplants produced away from production fields also contribute in reducing virus incidence (Gomez *et al.*, 2021; Thomas & Smith, 2016). Moreover, preventing the movement of virus-infected plants through experimental sources is crucial to limit virus spread (White & Black, 2020).

5. **Vector Management:** Vector management against viruses in horticultural crops encompasses various strategies aimed at managing insect vectors responsible for virus transmission. This includes the application of insecticides targeting vector populations, such as neonicotinoids and pyrethroids, to reduce aphids, thrips, and whiteflies (Snyder *et al.*, 2008). Biological control agents like parasitoids and predators, such as *Aphidiuscolemani* and *Aphelinusabdominalis*, are utilized to manage aphid populations and consequently lower virus transmission rates (Naranjo, 2009). Cultural practices like crop sanitation, weed management, and crop rotation play a vital role in reducing vector populations by removing weed hosts and alternate host plants (Smith *et al.*, 2020). Physical barriers such as insect-proof netting or screens are effective in preventing insect vectors from accessing crops, particularly in greenhouse or high-tunnel production systems (Shipp & Wang, 2006). Additionally, breeding for host plant resistance against insect vectors can help

in vector management by developing crop varieties that are less attractive or less susceptible to feeding by vectors (Kennedy *et al.*, 2013).

6. **Rogueing and Removal of old crops, weeds, and volunteer plants that act as viruses and vectors reservoirs:** Rogueing involves promptly removing and destroying plants showing symptoms of viral infection early in the crop cycle to prevent further spread within the field. Prompt rogueing of infected plants helps reduce virus inoculum and minimizes secondary spread by vectors. It is essential to carefully inspect crops regularly for symptoms such as leaf mosaic, chlorosis, necrosis, or stunting, and remove infected plants immediately to prevent virus dissemination (Smith *et al.*, 2020). In grapevine cultivation, rogueing infected vines showing symptoms of *Grapevine leafroll-associated virus* (GLRaV) can help prevent the spread of the virus to healthy vines within the vineyard (Martelli *et al.*, 2016). Cucurbit crops, weeds, and volunteer plants can serve as reservoirs for viruses and their insect vectors, contributing to the persistence and spread of viral diseases. To prevent virus buildup, it is important to promptly destroy and remove residues of old cucurbit crops after harvest. Additionally, controlling weeds and eliminating volunteer plants in and around fields helps reduce virus reservoirs and minimize virus transmission (Brown & Johnson, 2018).
7. **Use of Certified virus-free seed and planting material:** The utilization of certified virus-free seeds and planting materials is paramount in preventing viral diseases in horticultural crops. Certification programs, such as those implemented in various countries, ensure rigorous testing and inspection to guarantee the absence of viruses in seeds and plants before distribution to growers (Hill *et al.*, 2017). Tissue culture techniques serve as a prominent method for producing virus-free planting materials, particularly in crops like bananas, where viruses such as *Banana bunchy top virus* (BBTV) and *Banana streak virus* (BSV) are prevalent (Vu *et al.*, 2021). Additionally, heat treatment has been effective in eliminating viruses from planting materials, exemplified in crops like potatoes to control *Potato virus Y* (PVY) and *Potato leafroll virus* (PLRV) (Clark *et al.*, 2012). Indexing programs also play a crucial role, ensuring periodic testing of planting materials to maintain the virus-free status and health of horticultural crops (Nguyen *et al.*, 2018). For instance, in grapevine production, certified virus-free planting materials are indispensable to prevent the spread of *Grapevine leafroll-associated virus* (GLRaV) and other grapevine viruses, with tissue culture techniques commonly used for producing virus-free grapevine plantlets (Martelli *et al.*, 2016).
8. **Cross-Protection:** Cross protection is a method used to manage viral diseases in horticultural crops by infecting plants with mild strains of the virus to

confer resistance against more severe strains. This technique relies on the phenomenon known as "superinfection exclusion," where a plant already infected with a mild strain of a virus is less susceptible to infection by a more severe strain. One example of cross protection is the control of *Citrus tristeza virus* (CTV) in citrus orchards. Mild strains of CTV, such as T36, are used to inoculate citrus trees, providing protection against more virulent strains that cause severe symptoms and yield losses (Hilf *et al.*, 2005). Similarly, in tomato cultivation, mild strains of *Tomato mosaic virus* (ToMV) can be used to protect against more damaging strains of the virus (Oliveira *et al.*, 2015). Cross protection offers an environmentally friendly and sustainable approach to managing viral diseases in horticultural crops, reducing the reliance on chemical pesticides.

9. **Genetic Engineering:** Genetic engineering offers a powerful tool for developing virus-resistant horticultural crops by introducing genes that confer resistance to specific viruses. One common approach involves incorporating genes encoding viral coat proteins or other viral proteins into the plant genome. Upon viral infection, these proteins trigger the plant's defense mechanisms, resulting in resistance against the invading virus. For example, transgenic papaya expressing the coat protein gene of *Papaya ringspot virus* (PRSV) has been successfully commercialized to control PRSV infection in papaya orchards (Gonsalves, 2010). Similarly, transgenic squash expressing the coat protein gene of *Zucchini yellow mosaic virus* (ZYMV) has demonstrated resistance to ZYMV infection (Tian *et al.*, 2016). Additionally, RNA interference (RNAi) technology has been employed to develop virus-resistant crops by silencing specific viral genes upon infection. For instance, RNAi-mediated resistance against *Cucumber mosaic virus* (CMV) has been achieved in cucumber plants (Kanwar *et al.*, 2015). These examples highlight the potential of genetic engineering to provide durable and environmentally sustainable solutions for managing viral diseases in horticultural crops.

10. **Mechanical Practices:** Mechanical practices play a crucial role in managing viral diseases in horticultural crops, particularly in monitoring and trapping insect pests that transmit viruses. Regular monitoring using sticky traps, pheromone traps, or yellow sticky cards aids in early detection of vector insects such as aphids, whiteflies, and thrips, enabling timely intervention measures like insecticide applications or biological control (Smith *et al.*, 2020). Mulching with black silver mulch serves as an effective strategy, utilizing its reflective properties to confuse and repel insect vectors while promoting early plant growth and regulating soil moisture, ultimately reducing viral infections and weed growth. Studies have demonstrated its efficacy in reducing virus incidence and insect vector populations in various vegetable crops like

tomatoes, peppers, cucumbers, and squash (Nguyen *et al.*, 2018). Additionally, the use of blue sticky traps, specifically targeting thrips, aids in monitoring, mass trapping, and disrupting mating behavior, significantly reducing thrips populations and virus transmission in vegetable crops under field conditions. The strategic placement of traps throughout the field ensures adequate coverage and effective management of vector populations (Smith *et al.*, 2020; Nguyen *et al.*, 2018).

SUMMARY

Various families of plant viruses pose significant threats to horticultural crops worldwide, each with their own characteristics and impacts on agriculture. Caulimoviridae, Geminiviridae, Alphaflexiviridae, Bromoviridae, Comoviridae, Tombusviridae, Tymoviridae, and Virgaviridae are among these families. They share common traits such as broad host ranges, diverse symptoms, and potential for economic losses. Understanding their structures, genome organizations, host interactions, symptoms, and transmission mechanisms is crucial for developing effective management strategies.For instance, Caulimoviridae viruses, with their diverse symptoms and potential economic losses, require comprehensive management strategies. Similarly, Geminiviridae viruses pose significant threats due to their efficient modes of transmission and broad host ranges. Effective management strategies are also vital for Alphaflexiviridae, Bromoviridae, Comoviridae, Tombusviridae, Tymoviridae, and Virgaviridae, given their impacts on horticultural crops.Specifically, Nanoviridae, characterized by small, non-enveloped particles and a multipartite genome structure, infects a wide range of horticultural crops, primarily transmitted through insect vectors like aphids. Closteroviridae, on the other hand, possesses flexuous filamentous particles and single-stranded, positive-sense RNA genomes, causing symptoms like leaf chlorosis and fruit disorders, primarily transmitted through propagation materials and insect vectors.Luteoviridae, with its non-enveloped, quasi-spherical particles and single-stranded RNA genomes, primarily spreads through insect vectors like aphids, causing symptoms such as leaf yellowing and stunting. Potyviridae, known for its wide host range and efficient transmission, necessitates comprehensive understanding and management to mitigate its impact on horticultural crops. Secoviridae, characterized by non-enveloped, spherical particles and single-stranded RNA genomes, primarily spreads through mechanical means or insect vectors, causing symptoms like leaf mottling and reduced fruit quality. Tymoviridae and Virgaviridae also require effective management strategies to mitigate their impacts on horticultural crops.

Managing virus diseases in horticultural crops presents multifaceted challenges due to factors such as wide host ranges, rapid evolution of viruses, and development of insect resistance to control measures. To address these challenges, integrated management strategies are recommended, encompassing various approaches including host plant resistance, induced resistance, biological control, cultural practices, and vector management.Host plant resistance involves developing cultivars with genetic traits conferring resistance to specific viral pathogens. For instance, various cultivars of tomato, potato, pepper, melon, and squash exhibit resistance to viruses such as Tomato yellow leaf curl virus, Potato virus Y, Pepper mild mottle virus, Cucumber mosaic virus, and Zucchini yellow mosaic virus, respectively.

Induced resistance activates the plant's defense mechanisms through chemical elicitors or beneficial microorganisms, enhancing resistance to viral infections. Treatments with salicylic acid, Trichoderma harzianum, Bacillus amyloliquefaciens, and plant growth-promoting rhizobacteria have shown effectiveness against viruses in crops like cucumber, tomato, papaya, watermelon, chili pepper, and squash.

Biological control utilizes natural enemies or beneficial organisms to suppress viral or vector populations. Predatory insects, entomopathogenic fungi, RNA interference technology, and antiviral plant endophytes are examples of biological control agents effective against insect vectors of plant viruses.Cultural practices aim to create unfavorable conditions for virus transmission and multiplication. Practices such as crop rotation, intercropping, sanitation, and use of insect-proof netting help reduce virus inoculum and vector populations.Vector management involves strategies such as insecticide applications, biological control, cultural practices, and breeding for host plant resistance to manage insect vectors responsible for virus transmission.Rogueing and removal of infected plants, old crops, weeds, and volunteer plants prevent further spread of viruses within fields and reduce virus reservoirs.Utilization of certified virus-free seeds and planting materials, tissue culture techniques, heat treatment, and indexing programs help prevent viral diseases in horticultural crops.Cross-protection involves infecting plants with mild strains of viruses to confer resistance against more severe strains, reducing the reliance on chemical pesticides.Genetic engineering offers durable and sustainable solutions by introducing virus-resistant genes or employing RNA interference technology to develop virus-resistant crops.Mechanical practices such as regular monitoring, trapping of insect pests, mulching, and strategic placement of traps aid in early detection and management of vector populations, ultimately reducing virus transmission in horticultural crops.

REFERENCES

Adams MJ, Adkins S, Bragard C, Gilmer D, Li D, MacFarlane SA, Yoshikawa, N (2017) Potyviridae. In: King, A. M. Q., Adams, M. J., Carstens, E. B., and Lefkowitz, E. J. (Eds.), Virus taxonomy: Classification and nomenclature of viruses (Ninth Report of the International Committee on Taxonomy of Viruses) (pp. 1051-1060) Academic Press.

Adams MJ, Antoniw JF and Bar-Joseph M (2004) Carlaviruses: emerging pathogens in the family Betaflexiviridae. *Molecular Plant Pathology*, 5(1): 75-85.

Adams MJ, Antoniw JF and Fauquet CM (2012) Molecular criteria for genus and species discrimination within the family Potyviridae. *Archives of Virology*, 157(8): 1689-1698.

Adams MJ, Antoniw JF and Kreuze, J (2009) Virgaviridae: a new family of rod-shaped plant viruses. *Archives of Virology*, 154(12): 1967-1972.

Adams MJ, King AMQ, Carstens E Band Lefkowitz EJ (2017) Ratification vote on taxonomic proposals to the International Committee on Taxonomy of Viruses (2016) *Archives of Virology*, 162(8): 2505-2533.

Bar-Joseph M, Dawson WO and Garnsey SM (1989) Citrus tristeza virus: Making an ally from an enemy. *Annual Review of Phytopathology*, 27(1): 291-316.

Bharti N, Pandey SS and Barnawal D (2016) Induction of systemic resistance in chili (Capsicum annuum L.) against Pepper leaf curl virus (PepLCV) by plant growth promoting rhizobacteria. *Microbiological Research*, 193, 103-113.

Brault V and Uzest M (2014) Virus-like particle arrays associated with luteovirus infection. *Advances in virus research*, 90, 205-242.

Brown A and Johnson C (2018) Virus evolution in vegetable crops: challenges and opportunities. *Journal of Plant Pathology*, 100(2): 205-217.

Chung BYW, Miller WA, Atkins JF and Firth AE (2008) An overlapping essential gene in the Potyviridae. *Proceedings of the National Academy of Sciences*, 105(15): 5897–5902. https://doi.org/10.1073/pnas.0800468105

Clark MF, *et al.* (2012) Elimination of Potato leafroll virus and Potato virus Y from seed potatoes using heat therapy. *Plant Disease*, 96(5): 650-660.

Closteroviridae. In The Springer Index of Viruses (pp. 319-324) Springer, New York, pp. 319-324.

Dietzgen RG, Mann KS, Johnson KN and Mitter N (2003) Nepovirus proteins. In Advances in Virus Research. *Academic Press*. 62: 249-323

Gomez R, *et al.* (2021) Integrated strategies for managing virus diseases in horticultural crops. *Horticultural Science*, 48(3): 345-359.

Gonsalves D (2010) Transgenic papaya: A case for managing risks of papaya ringspot virus in Hawaii. *Phytopathology*, 100(7): 696-701.

Gray SM, De Boer SH, Lorenzen JH and Karasev AV (2010) Potato virus Y: a century of evolution. *Journal of Plant Pathology*, 92(4): 458-479.

Harrison BD (2013) Virus variation in relation to resistance-breaking in plants. *Euphytica*, 187(1): 19-32.

Heyraud-Nitschke F, Schumacher S, Laufs J, Schaefer S, Schell J and Gronenborn, B (2013) Determination of the origin cleavage and joining domain of geminivirus Rep proteins. *Nucleic Acids Research*, 31(17): 5034-5044.

Hilf ME, *et al.* (2005) Evidence for common ancestry among Tobamovirus gene 2 proteins and the coat proteins of existing Tobamoviruses. *Virology*, 332(2): 305-317.

Hill JH, *et al.* (2017) The role of certified seed in preventing the spread of plant viruses and viroids. *Plant Disease*, 101(8): 1372-1383.

Hohn T, Richert-Pöggeler KR and Staginnus C (2008) Rapid evolution of viral DNA replication systems. *Trends in Biochemical Sciences*, 33(11): 578-582.

Hyodo K, Okuno T and Taneda A (2019) A virus vector-based toolbox for functional plant virology: From techniques to targeted applications. *Frontiers in Microbiology*, 10, 1277.

Jones L (2019) Challenges in controlling virus diseases in horticultural crops. *Crop Protection*, 123, 105-113.

Jones L and Wang J (2019) Cultural practices for managing virus diseases in vegetable crops. *Crop Protection*, 123, 105-113.

Kamala S, Devaki K and Sreenivasulu P (2016) Induction of systemic resistance in papaya (*Carica papaya* L.) against Papaya ringspot virus by salicylic acid and Trichoderma harzianum. *Archives of Phytopathology and Plant Protection*, 49(17-18): 293-305.

Kanwar MK, Bhardwaj R and Arora P (2015) Effects of salicylic acid on induction of systemic acquired resistance and its potential to reduce Cucumber mosaic virus titer in *Cucumis sativus* L. *Journal of Plant Interactions*, 10(1): 109-117.

Karasev AV and Bar-Joseph M (2012) An overview of the family

Kennedy GG, *et al.* (2013) Selection of crops and cultivars for use in insect exclusion experiments that test plant resistance to insects. *Journal of Economic Entomology*, 106(1): 366-374.

Koch RL, Burkness EC and Hutchinson WD (2006) Phytophagous insect-microbial interactions: Importance in sustainable agricultural systems. In Phytophagous Insects: Host-Seeking Behavior, *Physiology and Mechanisms of Adaptation* (pp. 21-44) CRC Press.

Kumar A, *et al.* (2020) Horticulture in India: Current scenario and future prospects. *International Journal of Current Microbiology and Applied Sciences*, 9(1): 3211-3222.

Liu Y, Li G, Zhang L, Sun J, Zhang S and Sun G (2020) Bacillus amyloliquefaciens L-S60 improves watermelon quality and reduces the transmission of watermelon mosaic virus. *Biological Control*, 141, 104139.

Marquez LM, Redman RS, Rodriguez RJ and Roossinck MJ (2007) A virus in a fungus in a plant: three-way symbiosis required for thermal tolerance. *Science*, 315(5811): 513-515.

Martelli GP, *et al.* (2016) Control of grapevine virus diseases: A review. *Sustainable Agriculture Reviews*, 18, 55-127.

McGrath MT (2016) Host plant resistance to Tomato yellow leaf curl virus: A review. *Hort Science*, 51(10): 1219-1226.

Miller WA, Rasochová L, Barley J and Dreher TW (1997) Initiation and termination of viral RNA translation. *Journal of General Virology*, 78(10): 2607-2616.

Moriones E and Navas-Castillo J (2020) Tomato yellow leaf curl virus, an emerging virus complex causing epidemics worldwide. *Virus Research*, 281, 197961.

Nagendran K, Pandey KK, Rai AB and Singh B (2017) Viruses of Vegetables Crops Symptomatology, Diagnostics and Management. IIVR Technical Bulletin No.75 IIVR, Varanasi, pp.48

Nguyen HD, *et al.* (2018) Indexing programs for virus detection in horticultural crops: A review. *Plant Disease*, 102(9): 1686-1701.

Oliveira MG, de Lima JO, Fachinello JC and Bender RJ (2015) Application of salicylic acid on the control of Tomato mosaic virus in tomato plants. *Revista Brasileira de Fruticultura*, 37(2): 305-312.

Provvidenti R, Gonsalves D and Humaydan HS (1981) Zucchini yellow mosaic virus resistance in Cucurbita pepo. *Hort Science*, 16(1): 99-100.

Raupach GS and Kloepper JW (2000) Biocontrol of cucumber diseases in the field by plant growth-promoting rhizobacteria with and without methyl bromide fumigation. *Plant Disease*, 84(10): 1073-1075.

Roberts M (2017) Limited availability of virus-resistant plant varieties: implications for management strategies. *Plant Disease Management*, 86(4): 301-312.

Shipp JL and Wang K (2006) Evaluation of insect-exclusion screens for reducing aphid vectors of Potato leafroll virus in potato fields. *Crop Protection*, 25(5): 478-488.

Singh A, *et al.* (2021) Status and challenges of horticulture in India: An overview. *Journal of Horticulture, Forestry and Biotechnology*, 25(2): 93-106.

Smith P, *et al.* (2020) Insect vector resistance to control measures: a global perspective. *Pest Management Science*, 75(9): 2403-2415.

Snyder WE, *et al.* (2008) Should we expect population reductions of invasive species using conventional pesticides to be more effective than those using biopesticides? *PLoS ONE*, 3(7): e2944.

Tenllado F and Díaz-Ruíz JR (2001) Double strand ed RNA mediated interference with plant virus infection. *Journal of Virology*, 75(24): 12288-12297.

Thomas CE, Provvidenti R and Smith FF (1984) Inheritance of resistance in Cucurbita pepo to zucchini yellow mosaic virus. *Hort Science*, 19(5): 666-667.

Thomas C and Smith F (2016) Cultural practices for managing virus diseases in horticultural crops. *Journal of Horticultural Science & Biotechnology*, 91(2): 127-139.

Thompson JR, Gudmestad NC and Gonsalves D (2007) Sources of resistance in Capsicum chinense germ plasm to Pepper mild mottle virus. *Hort Science*, 42(2): 363-366.

Tian T, *et al.* (2016) Engineering resistance against Zucchini yellow mosaic virus by the CRISPR/Cas9 system in cucumber. *Plant Cell Reports*, 35(11): 2455-2462.

Vu HTT, *et al.* (2021) Application of tissue culture techniques for virus elimination in banana: A review. *Frontiers in Plant Science*, 12, 654925.

White G and Black R (2020) Biosecurity measures to prevent the spread of virus diseases in experimental trials. *Plant Disease Management*, 104(5): 812-820.

Whitworth JL, Novy RG, Hall DG and Crosslin JM (2009) Potato virus Y management in potato. *Plant Disease*, 93(11): 1184-1197.

Zerbini FM, Briddon RW, Idris A, Martin DP, Moriones E, Navas-Castillo, J, Varsani, A (2017). ICTV virus taxonomy profile: Geminiviridae. *Journal of General Virology*, 98(2): 131-133.

Sustainable Agriculture and Biodiversity: Innovations and Insights, Pages: 228–243
Edited by: Bibhuti Bhusan Sahoo, Harvinder Kumar Singh, Niranjan Chinara and Amitabh Panda

CHAPTER - 9

ORGANIC AND NATURAL FARMING: CURRENT STATUS OF FOOD AND LIVELIHOOD THROUGH SUSTAINABLE VEGETABLE FARMING

Samarth R. Patel[1], Nilima Karmakar[2]* and Kamlesh G. Patel[2]

[1]*Central Instrumentation Laboratory, Department of Soil Science and Agricultural Chemistry, N. M. College of Agriculture, Navsari Agricultural University, Navsari-396450, Gujarat*

[2]*Department of Soil Science and Agricultural Chemistry, N. M. College of Agriculture, Navsari Agricultural University, Navsari-396450, Gujarat*

**Corresponding email: nilimanau13@gmail.com*

INTRODUCTIONS

Due to excessive utilization of agrochemicals like Fertilizers, Pesticides, Insecticides, and Fungicides are used globally at huge levels, there will be approximately 3 billion people worldwide affected (Mishra *et al.*, 2023). According to one research website, the global market for agrochemicals reached $243.1 billion in 2019 and will reach $299.8 billion by 2024, at a compound annual growth rate (CAGR) of 4.3% for the period of 2019-2024. The global market for crop protection chemicals is expected to grow from $87.6 billion in 2023 to $114.7 billion by the end of 2028, at a compound annual growth rate (CAGR) of 5.5% from 2023 to 2028 (Retrieved from www.bccresearch.com). World Health Organization stated that thousands of pesticides are used to prevent crop damage from pests. Many of the older, less costly (off-patent) pesticides, such as dichlorodiphenyltrichloroethane (DDT) and lindane, can remain for years in soil and water. The toxicity of an individual pesticide always depends on its function and several factors. For instance, herbicides are less toxic than insecticides and also, dose-dependent toxicity effects always remain variable from person to person. Moreover, it can always happen by ways of exposure like swallowing, inhaling, or contact with the skin. Often contact with pesticides by humans in large quantities causes consequences of acute poisoning or long-term ill effects that may trigger uncontrolled cell cycle regulations (cancer) and also on the human reproductive system (World Health

Organization, 2021; www.who.int). The problem of such a health threat is widely spreading in developing nations rapidly. Several factors like unawareness and poor education, limited financial support of agriculture schemes by governments, and limitless uses of resources, are usually some key reasons behind the severity of health impacts (Anaduaka *et al.,* 2023). Thus, there is a need to set the limit of the traces and agrochemical residues with the help of high throughput technologies like machine learning and highly sophisticated instruments (Feng *et al.*, 2021). Also, the threat of health issues can be catalyzed by reverse farming methods like "Natural" and Organic farming that help to enhance the ease of living lives of humans (Reganold and Wachter, 2016; Smith *et al.*, 2020 and Bharucha *et al.*, 2020). Here we proposed the term "Regenerative Sustainable Farming" which means "Back to Root Culture" as per our ancient farming system like Natural and organic.

AGROCHEMICALS AND THEIR CONSEQUENCES ON THE ENVIRONMENT

There are several examples of the effects of synthetic agrochemicals on different areas of the environment as described in **Fig 1.**

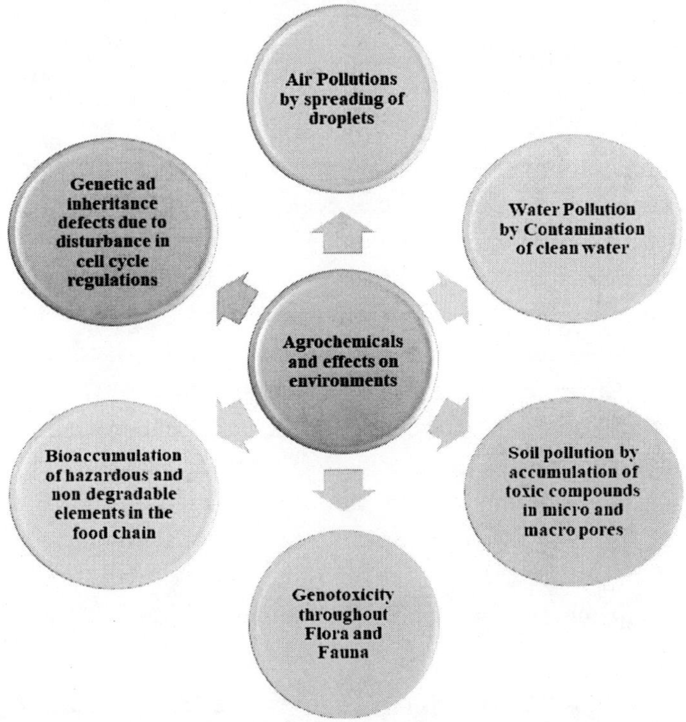

Fig. 1. Different effects on environment and living entities by Synthetic Agrochemicals

Kariukia *et al.* (2023) stated that due to agricultural activity air pollution by agrochemicals increases with the method of spraying. More minute droplets were in the air disseminated with drone sprayer rather than knapsack sprayer-like techniques. Thus, scientists concluded the several factors responsible for air pollution during the dispersion of agrochemicals which were studied through air by the κ-turbulence model. They affirmed that due to air potential movement, tiny droplets can move through all components of the hydrologic cycle.

Along with air pollution, the second most important topic is soil and groundwater contamination by agrochemicals. Because of the excessive addition of phosphorous, nitrogen, and persistent pesticides, the quality of Soil layers are deteriorating very rapidly. It can cause severe health issues as it is the root source of clean consumable water to people (Srivastav, 2020). Surprisingly, Very high and carcinogenic levels of agrochemical concentrations have been detected in global water resources. This is above their threshold limits and can be assumed true because over 4 million tons of pesticide are used on the entire earth yearly (Rad, *et al.*, 2022). Moreover, excessive agrochemicals can block soil layers' micro and macro pore structures. It facilitates heterotrophic respirations decreasing the availability of organic carbon that can alternatively lead to degradation of microbial activity of soil microbiota (Hasnain *et al.*, 2023).

Various pesticides like fungicides, acaricides, insecticides, etc. have been associated with an increase in the incidence of genotoxicity. Scientists provided proof of genotoxic effects triggered by these compounds such as chromosomal aberrations, unnatural nucleic acid damage, sister chromatid abnormal exchange, micronuclei rupturing, uncontrolled cell cycle regulations, disturbance in signaling cascade in gene expression and biosynthesis pathways, etc. (de Castro and Poli, 2020). Over penetrance of pesticides is a common phenomenon in aquatic and terrestrial ecosystems. Through the food chain, bioaccumulation and biomagnification of toxic pesticides have led to sublethal to lethal impacts on every living creature on this earth (Gupta and Gupta, 2020). In humans and animals, there are reported disturbances of the endocrine systems differently due to continuous exposure to agrochemicals. This is due to mimicking receptors as agrochemicals in endocrine systems and disturbing the original system. It can exert toxicity through a non-nuclear receptor-mediated mechanism of action (Warner *et al.*, 2020; Ren *et al.*, 2020). It can lead to epimutation as transgenerational inheritance changes in humans and animals (Montjean *et al.*, 2022).

Thus, because of these destructive causes of excessive use of agrochemicals humans need to come back to natural farming techniques as per ancestors used in ancient Indian civilizations. Thus, we need to adopt Regenerative sustainable farming. Moreover, in "olericulture" uncontrolled agrochemicals are used nowadays and hence we will focus on the Major vegetable crops.

Natural and Organic Farming as a Sustainable Approach

At present time, more than half of the Indian population which are accounting ten percent of the global population will be prone to food deficiencies that might be very severe by the middle of this century. Therefore, with good quality food and low-cost inputs, there is a need for a natural farming system that can improve soil health and the physical condition of farms by reducing hazardous inputs (Smith *et al.*, 2020). In the era of agrochemical revolutions in the fourth industrial revolution, a 'quality of life' is the most significant element and broad concept of sustainable agriculture (Harwood, 1995). Low-budget Natural Farming (LBNF) is an important and recent term used in Indian agriculture. With the heterodox natural farming agronomies, this method is a more significant emerging in India. It can counter sustainable agriculture developments with its technoscientific or organic guises (Munster, 2017).

Table 1: Comparison of Natural / Organic and Morden Farming (Source: Fukuoka, 1985)

Natural / Organic Farming	Morden Farming
i. Non- Action	i. Action
ii. Return to Nature	ii. Away from Nature with the expansion of Human wants and Desire
iii. Centrifugal and Convergent	iii. Centripetal and Divergent
iv. Economically viable	iv. Economically Unviable
v. Low labour inputs required	v. Higher labour inputs required
vi. No machineries required	vi. Morden types of machinery require
vii. Less to almost zero pollution	vii. More pollution of air, water, and soil
viii.Increasing or maintaining the quality of food	viii. Decline quality of food
ix. Natural energy required *i.e.* Bullock for soil preparation, Water elevation for natural flow, rain, manures from waste leaves, and biodegradable organic matters like dung and urines of Cattles	ix. Synthetic inputs are required i.e. tractors, cultivators, fertilizers, pesticides like agrochemicals, irrigation pumps, Petroleum for energy fulfillment etc.

Durán-Lara *et al.* (2020) represent the concepts of biopesticides, biofertilizers, and byproducts of natural farming for organic cultivation of crops. For instance, composting, vermicomposting, or biochar are sorts of biofertilizers that can reduce the toxicity effects of heavy metals which also might be present in the urine and dung of animals. Also, crop rotation, green mulching, and proper tilling with biopesticide blending (i.e. essential oils-based, fungi and strains) can help reduce the hazardous synthetic materials in the food chain and bioaccumulation in living tissues. That also benefits the low-cost farming budget and the environment to protect the natural integrity of flora and fauna. Organic certifications can benefit more factors responsible for organic tilling like crop diversification, intercropping, seminatural habitat, etc. (Tscharntke *et al.*, 2021; Akanmu *et al.*, 2023). Organic agricultural practices enhance the positive effects of the environment namely nutrient cycle, soil erosion, carbon sequestration, and many other ecological patterns (Gamage *et al.*, 2023). It is also able to facilitate the restoration process of the Soil Organic Carbon (SOC) which can be enhanced by the embedding of the legume cropping system in organic farming (Babu *et al.*, 2023).

As food safety and quality gain more importance in the market hence, there is a natural way of cultivation to support such a vision. This can be achieved by organic and natural farming for India-like nations where population is one of the major concern. It can be used as a war footstep and become the biggest sustainable approach for India due to the huge availability of natural resources, major biodiversity, large areas of natural habitats, and various agroclimatic zones (Das *et al.*, 2020; Selvan *et al.*, 2023). Unfortunately, the organic farming development in the entire nation has been facing such issues as continuous production, economic challenges, market aspects, behavioral changes and institutional, microplastic pollutions, social factors, and regulatory elements (Luczka and Kalinowski, 2020; Zhao *et al.*, 2023; Kumar *et al.*, 2023).

The demand for organically produced food is increasing by almost 25% in developed countries due to literacy and education in those countries (Hebbar *et al.*, 2023). Thus, it can be a great opportunity also for the farmers of developing countries like India to earn high income through reverse or ancient farming methods and export those products to highly developed countries, as per statistics India secured the highest rank of organic producers in the year 2021 with sharing 1.5 % of total organic production in the entire world. Also, it is facilitated by the setting up of policies fostering agroecology and organic agriculture in India with the help of the framework of the Indian Natural Development Program (APEDA, 2023; Willer *et al.*, 2024).

In terms of organic vegetables produce, some major recent case studies of those important vegetable crops are elaborated and explained further.

The recent case studies on important vegetable crops:

Potato

Ghoshal and Dutta (2023) used bulky organic manure (where along with FYM @ 10 $t.ha^{-1}$, vermicompost @ 20 $t.ha^{-1}$ was applied twice, one with ½ quantity of vermicompost along with FYM as basal application before 7 days of planting and remaining ½ as a split application at 45 days after planting (DAP)) with Vivek Krishi (where along with FYM @ 10 $t.ha^{-1}$ as basal application 7 days before planting, enriched Sanjeevani @ 10% was applied twice as soil drenching split application) combinations to provide the best conditions for growth, yield, and quality traits in *Solanum tuberosum* L. for the eastern Indian plateau during the rabi seasons of 2019-21. They compare these results with conventional farming (where only FYM @ 10 t ha^{-1} was applied as basal dose 7 days before planting).

Keijzer *et al.* (2022) launched the scientific program Bioimpuls. It aimed to extract a substantial impulse to both the organic and conventional potato breeding sectors by enlarging access to various sources of late blight resistance. The Bioimpuls activities over 11 years range from providing true seed populations for variety selection with five available sources of R-genes against *Phytophthora infestans*, early and advanced introgression breeding with six new R-genes, and education and communication. They stacked the genes to be resistant against blight and genotypes which will be launched in the market for organic cultivation in the next few years.

Imani *et al.* (2021) aimed to explore the effective factors underpinning the willingness of potato farmers (n=200) in the Ardabil Plain to adopt organic farming using an extended model of the "Theory of Planned Behavior." The results of structural equations showed that attitude, subjective norms, perceived behavioral control, moral norms, and environmental concerns accounted for 72% of the variance of potato farmers' intention to engage in organic farming and also environmental concerns captured 43% of this variance.

Eremeev *et al.* (2020) carried out an experiment consisting of "five-field" crop rotation with two different organic (Org 0 as control and Org II with winter cover crops plus added composted manure) and conventional (Conv 0 as control and Conv II with mineral N, 150 kg ha^{-1}) farming systems. The results showed that the hydrolytic activity of soil microbes decreased in every farming system under potato cultivation. Interestingly, after potato cultivation, the lowest and highest decrease in the soil microbial hydrolytic activity was seen in the Org II and Conv II systems, respectively. The highest soil microbial hydrolytic activity was measured in system Org II where incorporation of biomass from winter cover crops and cattle manure was used. Finally, SOC and total N were higher in organic farming systems and there were no significant changes after potato cultivation.

Tomato

Parmar and Patel (2021) observed that Panchagavya @ 3 % at 25 DATP (Days after transplanting) exhibits maximum yield in *Solanum lycopersicum* L. The experiment was conducted during the Rabi season 2017–18 at the Regional Horticultural Research Station, ASPEE College of Horticulture and Forestry, Navsari Agricultural University, Navsari, Gujarat, India. They studied twelve treatment combinations, including vermiwash, novel spray, water spray, Panchagavya, and Moringa extracts.

Vélez-Terreros *et al.* (2021) surprisingly found that conventional and organic tomatoes did not have much difference in terms of the quality contents. Based on fresh weight the respective concentration ranges were 1.00–63.8 mg/100 g and 10.7–40.0 mg/100 g for ascorbic acid, 0.02–337.0 mg/100 g, and 0.44–422 mg/100 g for lycopene, and 0.0058–4.44 mg/100 g and 0.0061–3.90 mg/100 g for β-carotene. For polyphenol and mineral content, the results varied depending on farming technique. However, they concluded that the environment and its integrity are important aspects of organic farming and also it could be able to achieve better market opportunities.

Gao *et al.* (2023) stated that the influence of organic fertilizers on tomato yield and quality is shown to be complex and dependent on soil organic matter, total soil nitrogen, organic fertilizers kinds etc. They also mentioned that various organic fertilizers can enhance the tomato yield by 42.18%. Compared to the control group, soluble solids, soluble sugar, lycopene, vitamin C, and nitrate were raised by 11.86%, 42.18%, 23.95%, 18.97%, and 8.36%, respectively.

Murariu *et al.* (2021) examined the presence of rich antioxidants compounds like lycopene, vit. C and carotene were enhanced by organically under-plastic tunnel farming systems as compared to open field systems. They recommended suitable genotypes for the organic cultivation of tomatoes. In their study, Germplasm G3 2 is highlighted by the highest content of carotene (7.4 mg^{-1}100 g-1 F.W.) and lycopene (8.4 mg^{-1}100 g^{-1} F.W.), and Germplasm G5 by the highest content in vitamin C (16.8 mg^{-1}100 g^{-1} F.W.).

Tripodi *et al.* (2021) used genetic tools genome-wide association mapping studies (GWAS), to explore the framework for the development of markers linked to candidate genes of different tomato germplasms cultivated under organic conditions. These genes can be able to understand regarding genomic-assisted breeding with genotyping and phenotyping under low input organic conditions.

Onion

Rempelos *et al.* (2023) studied the long-term effects of national projects in the European Union. In the year 2004, they investigated the effect of various organic fertilizers and input levels to check pest and disease infestations and yield attributes. They followed the 8-year diverse legume-rich rotation of organic systems with the blending of the cattle manure fertilizer with $(\text{split})^3$ plot design with four blocks and replicates. Onion crops were planted on the 5th, 11th, and 24th of May and harvested by hand on the 6th of October, 25th of September, and 8th of October in the 2005, 2006, and 2007 growing seasons, respectively. Results showed that good quality of the onion bulb was produced which was confirmed with phenolics studies with reverse phase-HPLC.

Abdelrahman *et al.* (2021) carried out the open field experiment and studied the integrated effect of biofertilizer-PGPR and biocontrol agents with compost to improve green onion growth characteristics and yield under the organic farming system. During the experiment, the soil microbial enzyme activity, total NPK and plant uptake, defense enzymes activity, growth characteristics, and vegetative yield were observed. They concluded that onion treated with compost + biofertilizing-PGPR + *Bacillus subtilis* + *Trichoderma viride* (T7) gave the highest values in all estimated parameters except defense enzyme activity. So, it can be recommended as an integrated fertilizing program to facilitate green onion growth, increase crop production, and decrease input values.

Elouattassi *et al.* (2024) explored the effects of organic fertilization on onions intercropped with pepper, fennel, and carrots within an agroecological context. The study was conducted over two seasons using a randomized complete block design with 12 treatments. Notably, organic fertilization yielded results similar to mineral fertilization (NPK). Among intercropping combinations, onion-carrot displayed the highest yield and growth characteristics. Only, Carrot intercropping led to increased plant height, with inorganic fertilization demonstrating stronger effects than organic fertilization. Moreover, Principal component analysis revealed that onion-carrot intercropping with organic fertilizer nearly matched chemical NPK yield levels.

Jangir *et al.* (2022) examined the new integrated nano-organic farming to achieve sustainable food production. They carried out two field trials in consecutive years and reported boosting in yield through root treatment with nano-pyrite (FeS_2) plus soil application of goat dropping (Test) as compared to the use of goat dropping alone (Control). The total biomass (bulb + leaf) weight of the test sample was 4.75 kg (n = 86), while control samples weighed 3.5 kg (n = 86). The total onion bulb weight for the control and test was 2.6 kg and 3.6 kg, respectively. There was a significant increase in the anthocyanin content in test bulbs and flavanol content in test leaves.

Brinjal

Ashokkumar *et al.* (2020) applied various treatments like FYM @2.5 and 5.0 kg and vermicompost @1.25 and 2.5 kg per plant with humic acid @10 and 20 g per plant as soil application along with 0.1 and 0.2 % for foliar spray. The highest yield was found in the vermicompost @2.5 kg per plant with humic acid 0.1 % foliar spray. The application of the same treatments markedly enhanced nutrient uptake and was found best in the Brinjal (*Solanum melongena* L.) plant.

Mahalik (2020) compared the inorganic and organic matter in terms of fertilizers and their effects on *Solanum melongena.* The chemical fertilizer comprised NPK and pesticide Chlorpyrifos and the organic fertilizer used was a mixer of cow dung, vegetable peels, dead plant organic matter, and a combination of organic and inorganic matters. Samples of fruits of *Solanum melongena* were collected in two different fields. Heavy metal analysis was done using X-ray fluorescence spectroscopy. These agrochemicals cause the accumulation of different heavy metals in the soil like zinc and copper, which can also accumulate in the plants out of which some are very much carcinogenic and cause health disorders. They conclude that chemical fertilizers increase yield but also harm the environment and the long-term uses can be catastrophic. So besides the use of chemical products, we can use biological products for good health and a better environment.

Rathore *et al.* (2023) used fermented liquid biofertilizers of different origins and organic manure compositions (panchagavya and jeevamrut) through soil preparation before planting Brinjal. They found that soil microbial biomass of carbon value of 66.1 mg g^{-1} soil. The brinjal yield was maximal under treatment T4 (combined use of organics and fermented liquid organic nutrient formulations) with values of 389.2 q ha^{-1} and 153.7 q ha^{-1} for 2017 and 2018, respectively.

Tripathy (2020) used four non-chemical alternative growing conditions *viz.* E1C1: Amritjal (1%), E1C2: Absolute Control (Inherent Fertility Status of Experimental Soil); E2C1: Amritjal (1%) + Homa Induction (Agnihotra), and E2C2: Only Homa Induction (Agnihotra) by employing five local varieties of Brinjal. They concluded that the Swarna Neelima variety of brinjal is highly suitable for cultivation using alternative growing conditions of E2C1:Amritjal (1%) + Homa Induction (Agnihotra).

Techniques for detecting contaminants in organically grown vegetables:

In the context of detecting contaminants in organically produced vegetables, methods like Gas Chromatography and High-performance liquid chromatography adjoined with various detectors can be used. Moreover, the different nanosensors that can be developed through nanotechnology and sensor detection techniques combined are used for the detection of contaminants at the field level before packaging vegetables (Narenderan *et al.*, 2020). In the era of organic farming, the positive phytomeliorative effect of organic agents on physicochemical and biological indicators of soils was reported (Nasiyev *et al.*, 2021). Nowadays, highly sophisticated instrumentations are used to detect contaminants like metals and toxic elements from organically produced vegetables. They are known as Inductively Coupled Plasma Optical Emission Spectroscopy (ICP-OES), Inductively Coupled Plasma Mass Spectrometry (ICP-MS), Atomic Absorption Spectrometry (AAS) (Popoviæ-Djordjeviæ *et al.*, 2022; Mazarakioti *et al.*, 2022; Sadee, 2022; Iftikhar-Ul-Haq *et al.*, 2021, Nugraha *et al.*, 2024). Organically grown vegetables have high phenolics, high antioxidants, higher soluble solids, higher soluble sugar, and lower nitrate content. These types of examined parameters should be used to confirm the organic certification of vegetables (Rahman *et al.*, 2021).

Moreover, upgraded chromatography techniques like Liquid Chromatography with tandem mass spectrometry (LC-MS/MS), Gas Chromatography with Electron Capture Detector/ Nitrogen-Phosphorus Detector (GC-ECD/NPD), High-resolution mass spectrometry (HR-MS) with advanced metabolomic studies can able to detect trace amount of contamination of pesticide residues in organically grown vegetables (Kazimierczak *et al.*, 2022; Mihailova *et al.*, 2021; del Mar Gómez-Ramos *et al.*, 2020). Recently, consumers have accepted organically produced vegetables like food based on the "Omics" technologies certification even with high pay in the 21st century. However, there is a need for awareness like advanced certification and technologies that can collect the consumers' trust in the market (Castellini *et al.*, 2022; Bashir *et al.*, 2023).

Conclusion

In the era of high use of chemicals in the cultivation of crops, the world of today needs a safe, natural, and sustainable approach. These can be facilitated by degnerative sustainable farming like organic and natural farming. This farming method can be utilized with priorities of the food safety and quality in terms of natural values. Here in this chapter, we have provided the latest and current case studies about different vegetable crops like Potatoes, Tomato, Onion, and Brinjal. These can be used for the commercial production of natural and agrochemical contaminant-free food products in the market. Because of the highest agrochemicals like pesticides and fertilizers used in vegetable farming, this current status and their different reviewed materials and methods can be safe alternatives in quality farming in India.

REFERENCES

Abdelrahman H, Zaghloul RA, Abou-Aly HA, Ragab AA and Elmaghraby MM K (2021) Application of some organic farming methods to enhancement the growth and production of green onion. *Journal of Agricultural Chemistry and Biotechnology*, 12(4): 79-89.

Akanmu AO, Olowe M, Phiri AT, Nirere D, Odebode AJ, Karemera Umuhoza NJ and Babalola OO (2023) Bioresources in organic farming: implications for sustainable agricultural systems. *Horticulturae* 9(6): 659.

Anaduaka EG, Uchendu NO, Asomadu RO, Ezugwu AL, Okeke ES and Ezeorba TPC (2023) Widespread use of toxic agrochemicals and pesticides for agricultural products storage in Africa and developing countries: Possible panacea for ecotoxicology and health implications. *Heliyon*. e15173.

APEDA (2023) https://www.apeda.gov.in/apedawebsite/organic/Organic_Products.htm

Ashokkumar N, Muruganantham M and Dharani S (2020) Effect of organic amendments in Brinjal (*Solanum melongena* L.) *Int. J. Adv. Res. Biol. Sci*, 7(2): 143-148.

Babu S, Singh R, Avasthe R, Kumar S, Rathore SS, Singh VK and Petrosillo, I (2023) Soil carbon dynamics under organic farming: Impact of tillage and cropping diversity. *Ecological Indicators*, 147, 109940.

Bashir T, Ul Haq SA, Masoom S, Ibdah M and Husaini AM (2023) Quality trait improvement in horticultural crops: OMICS and modern biotechnological approaches. *Molecular Biology Reports*, 50(10): 8729-8742.

Bharucha ZP, Mitjans SB and Pretty, J (2020) Towards redesign at scale through zero budgets natural farming in Andhra Pradesh, India. *International Journal of Agricultural Sustainability*, 18(1): 1-20.

Castellini G, Sesini G, Iannello P, Lombi L, Lozza E, Lucini L and Graffigna G (2022) "Omics" technologies for the certification of organic vegetables: Consumers' orientation in Italy and the main determinants of their acceptance. *Food Control*, 141, 109209.

Das S, Chatterjee A and Pal TK (2020) Organic farming in India: a vision towards a healthy nation. *Food Quality and Safety*, 4(2): 69-76.

de Castro VLS and Poli, P (2020) Pollution: Genotoxicity of agrotoxic compounds. In Managing Global Resources and Universal Processes CRC Press. pp. 81-97

del Mar Gómez-Ramos M, Nannou C, Bueno MJM, Goday A, Murcia-Morales M, Ferrer C, and Fernández-Alba AR (2020) Pesticide residues evaluation of organic crops. A critical appraisal. *Food chemistry*: X, 5, 100079.

Durán-Lara EF, Valderrama A, Marican A (2020) Natural organic compounds for application in organic farming. *Agriculture*, 10(2): 41.

Elouattassi Y, Ferioun M, El Ghachtouli N, Derraz, K and Rachidi F (2024) Enhancing onion growth and yield through agroecological practices: Organic fertilization and intercropping. *Ecological Frontiers*, 44(3): 547-557.

Eremeev V, Talgre L, Kuht J, Mäeorg E, Esmaeilzadeh-Salestani K, Alaru M, Luik A (2020) The soil microbial hydrolytic activity, content of nitrogen and organic carbon were enhanced by organic farming management using cover crops and composts in potato cultivation. *Acta Agriculturae Scandinavica, Section B—Soil & Plant Science*, 70(1): 87-94.

Feng C, Xu Q, Qiu X, Ji J, Lin Y, Le S, She J, Lu D and Wang G (2021) Evaluation and application of machine learning-based retention time prediction for suspect screening of pesticides and pesticide transformation products in LC-HRMS. *Chemosphere*, 271, 129447.

Fukuoka M (1985) The natural way of farming. Tokyo: Japan Publications, 17.

Gamage A, Gangahagedara R, Gamage J, Jayasinghe N, Kodikara N, Suraweera P and Merah O (2023) Role of organic farming for achieving sustainability in agriculture. *Farming System*, 1(1): 100005.

Gao F, Li H, Mu X, Gao H, Zhang Y, Li R and Ye L (2023) Effects of organic fertilizer application on tomato yield and quality: a meta-analysis. *Applied Sciences*, 13(4): 2184.

Ghoshal S and Dutta AK (2023) Expression of Quantitative and Qualitative Traits in Potato (*Solanum tuberosum* L.) Grown through Organic Production Intervention in Eastern Indian Plateau. *Current Agriculture Research Journal*, 11(2)

Global Crop Protection Chemicals Market (CHM188A) BCC Publishing. Jun 2024. https://www.bccresearch.com/market-research/chemicals/agrochemicals-global-markets-report.html

Gupta S and Gupta K (2020) Bioaccumulation of pesticides and its impact on biological systems. Pesticides in Crop Production: Physiological and Biochemical Action, 55-67.

Harwood RR (1995) Natural Farming Systems: Meeting the Goals of Sustainable Agriculture. Michigan, USA. Michigan State University.

Hasnain M, Munir N, Abideen Z, Zulfiqar F, Koyro HW, El-Naggar A and Yong JWH (2023) Biochar-plant interaction and detoxification strategies under abiotic stresses for achieving agricultural resilience: A critical review. *Ecotoxicology and Environmental Safety* 249, 114408.

Hebbar SS, Nair AK, Kumar MS and Prabhakar M (2023) Organic Farming in Vegetable Crops. In Organic Crop Production Management Apple Academic Press. pp. 201-220

Iftikhar-Ul-Haq I, Ahmed E, Sharif A, Ahmed M, and Ahmad W (2021) Optimization of ultrasound-assisted extraction of essential and non-essential/ toxic trace metals in vegetables and their determination by FAAS and ICP-OES: An evaluation of human health risk. *Food Analytical Methods*, 14(11): 2262-2275.

Imani B, Allahyari MS, Bondori A, Emami N and El Bilali H (2021) Adoption of organic potato production in Ardabil Plain, Iran: an application of the extended theory of planned behaviour. *Potato Research*, 64, 177-195.

Jangir H, Kaler B, Srivastava G and Das M (2022) Nano pyrite root treatment in conjunction with soil application of goat droppings boost onion yield and anthocyanin and flavonoids content: A nano-organic farming model towards sustainability.

Kariukia KD, Ngugib KM and Mwitic KME (2023) Investigation of air pollution due to agrochemicals over large farms by k-õ turbulence model. International *Journal of Advances in Applied Mathematics and Mechanics*, 1(2): 53 – 64.

Kazimierczak R, Œrednicka-Tober D, Golba J, Nowacka A, Ho³odyñska-Kulas A, Kopczyñska K and Gnusowski B (2022) Evaluation of pesticide residues occurrence in random samples of organic fruits and vegetables marketed in Poland. *Foods*, 11(13): 1963.

Keijzer P, Van Bueren EL, Engelen, CJM and Hutten RCB (2022) Breeding late blight resistant potatoes for organic farming—a collaborative model of participatory plant breeding: The Bioimpuls Project. *Potato Research*, 65(2): 349-377.

Kumar J, Rana S, Rani V and Ahuja A (2023) What affects organic farming adoption in emerging economies? A missing link in the Indian agriculture sector. International Journal of Emerging Markets.

Luczka W and Kalinowski S (2020) Barriers to the development of organic farming: A polish case study. *Agriculture*, 10(11): 536.

Mahalik, G. (2020) Comparative Study of the Effect of Organic Fertilizer and Chemical Fertilizer on Brinjal Plant (*Solanum melongena* L.) *Indian Journal of Natural Sciences*. 10(60): 976-997.

Mazarakioti EC, Zotos A, Thomatou AA, Kontogeorgos A, Patakas A and Ladavos A (2022) Inductively coupled plasma-mass spectrometry (ICP-MS): a useful tool in the authenticity of agricultural products' and foods' origin. *Foods*, 11(22): 3705.

Mihailova A, Kelly SD, Chevallier OP, Elliott CT, Maestroni BM and Cannavan, A (2021) High-resolution mass spectrometry-based metabolomics for the discrimination between organic and conventional crops: A review. *Trends in Food Science & Technology*, 110, 142-154.

Mishra MK, Mishra SK, Panday L and Pandey R (2023) Agrochemicals and Their Impact on Environment. *Recent Research and Innovation in Plant Protection*. 47-56. https://www.researchgate.net/publication/375485756_Agrochemicals_and_Their_ Impact_on_Environment

Montjean D, Neyroud AS, Yefimova MG, Benkhalifa M, Cabry R and Ravel C (2022) Impact of endocrine disruptors upon non-genetic inheritance. *International Journal of Molecular Sciences*, 23(6): 3350.

Münster D (2017) Zero budget natural farming and bovine entanglements in South India. *RCC Perspectives*, (1): 25-32.

Murariu OC, Brezeanu C, Jităreanu CD, Robu T, Irimia LM, Trofin AE and Brezeanu PM (2021) The functional quality of improved tomato genotypes grown in open fields and in plastic tunnels under organic farming. *Agriculture*, 11(7): 609.

Narenderan ST, Meyyanathan SN and Babu BJFRI (2020) Review of pesticide residue analysis in fruits and vegetables. Pre-treatment, extraction and detection techniques. *Food Research International*, 133, 109141.

Nasiyev B, Vassilina T, Zhylkybay A, Shibaikin V and Salykova A (2021) Physicochemical and biological indicators of soils in an organic farming system. *The Scientific World Journal*, 2021(1): 9970957.

Nugraha F, Oktaviani SM and Luliana S (2024) Analysis of Heavy Metal Cadmium (Cd) and Lead (Pb) in Eggplant (*Solanum melongena* L.) in Pontianak City by Using Atomic Absorption Spectrophotometry (AAS) *Biology, Medicine, & Natural Product Chemistry*, 13(1): 83-90.

Parmar MN and Patel SY (2021) Effect of organic spray on yield and its attributes of tomato (*Solanum lycopersicum* L.) cv. GT 2 under south Gujarat condition.

Popoviæ-Djordjeviæ JB, Kostiæ A•, Rajkoviæ MB, Miljkoviæ I, Krstiæ Đ, Caruso G and Brèeski I (2022) Organically vs. conventionally grown vegetables: Multi-elemental analysis and nutritional evaluation. *Biological Trace Element Research*, 200, 426-436.

Rad SM, Ray AK and Barghi S (2022) Water pollution and agriculture pesticide. *Clean Technologies*, 4(4): 1088-1102.

Rahman SME, Mele MA, Lee YT and Islam MZ (2021) Consumer preference, quality and safety of organic and conventional fresh fruits, vegetables and cereals. *Foods*, 10(1): 105.

Rathore G, Kaushal R, Sharma V, Sharma G, Chaudhary S, Dhaliwal SS and Hossain A (2023) Evaluation of the usefulness of fermented liquid organic formulations and manures for improving the soil fertility and productivity of brinjal (*Solanum melongen*a L.) *Agriculture*, 13(2): 417.

Reganold JP and Wachter JM (2016) Organic agriculture in the twenty-first century. *Nature plants*, 2(2): 1-8.

Rempelos L, Barañski M, Sufar EK, Gilroy J, Shotton P, Leifert H and Leifert C (2023) Effect of climatic conditions and agronomic practices used in organic and conventional crop production on yield and nutritional composition parameters in potato, cabbage, lettuce and onion; results from the long-term NFSC-trials. *Agronomy*, 13(5): 1225.

Ren XM, Kuo Y and Blumberg B (2020) Agrochemicals and obesity. *Molecular and cellular endocrinology*, 515, 110926.

Sadee BA (2022) Determination of trace metals in vegetables using ICP-MS. *Zanco Journal of Pure and Applied Sciences*, 34(3): 73-83.

Selvan T, Panmei L, Murasing KK, Guleria V, Ramesh KR, Bhardwaj DR and Deshmukh HK (2023) Circular economy in agriculture: Unleashing the potential of integrated organic farming for food security and sustainable development. *Frontiers in Sustainable Food Systems*, 7, 1170380.

Smith J, Yeluripati J, Smith P and Nayak DR (2020) Potential yield challenges to scale-up of zero budget natural farming. *Nature sustainability*, 3(3): 247-252.

Srivastav AL (2020) Chemical fertilizers and pesticides: role in groundwater contamination. In Agrochemicals detection, treatment and remediation Butterworth-Heinemann. pp. 143-159

Tripathy S (2020) Homa Farming Of Brinjal (Effect of Agnihotra and Amritjal on Growth, Yield and Quality of Brinjal (*Solanum melongena* L.) Scholar's Press ISBN: 978-631-8-93349-6. http://dx.doi.org/10.2139/ssrn.3851237

Tripodi P, Soler S, Campanelli G, Díez MJ, Esposito S, Sestili S and Cardi T (2021) Genome-wide association mapping for agronomic, fruit quality and root architectural traits in tomato under organic farming conditions. *BMC Plant Biology*, 21, 1-22.

Tscharntke T, Grass I, Wanger TC, Westphal C and Batáry P (2021) Beyond organic farming–harnessing biodiversity-friendly landscapes. *Trends in ecology & evolution* 36(10): 919-930.

Vélez-Terreros PY, Romero-Estévez D, Yánez-Jácome GS, Simbaña-Farinango K and Navarrete H (2021) Comparison of major nutrients and minerals between organic and conventional tomatoes. A review. *Journal of Food Composition and Analysis*, 100, 103922.

Warner GR, Mourikes VE, Neff AM, Brehm E and Flaws JA (2020) Mechanisms of action of agrochemicals acting as endocrine disrupting chemicals. *Molecular and cellular endocrinology* 502, 110680.

WHO (2021) https://www.who.int/news-room/fact-sheets/detail/pesticide-residues-in-food

Willer H, Trávnièek J and Schlatter S (2024) The World of Organic Agriculture. *Statistics and Emerging Trends* 2024.

World Health Organization (2021) Report on the Environmental and Health Impacts of Pesticides and Fertilizers and Ways to Minimize Them-Draft Synthesis Report.

Zhao H, Zhou Y, Lu Z, Ren X, Barcelo D, Zhang Z and Wang Q (2023) Microplastic pollution in organic farming development cannot be ignored in China: Perspective of commercial organic fertilizer. *Journal of hazardous materials*, 460, 132478.

Sustainable Agriculture and Biodiversity: Innovations and Insights, Pages: 244–258
Edited by: Bibhuti Bhusan Sahoo, Harvinder Kumar Singh, Niranjan Chinara and Amitabh Panda

CHAPTER - 10

THE IMPACT OF GLOBAL WARMING AND INSECTICIDE RESISTANCE ON AGRICULTURE

Uttam Kumar Behera[1*], Pravasini Behera[2], Chittaranjan Sahoo[1], Manoj Kumar Rout[1] and Sushant Swain[3]

[1]AICRP on Seed Crop OUAT, Bhubaneshwar-751003, Odisha
[2]AICRP on Honeybees & Pollinators,
Department of Entomology, OUAT, Bhubaneshwar-751003, Odisha
[3]Director of Research, OUAT, Bhubaneshwar-751003, Odisha
**Corresponding email: uttambehera@ouat.ac.in*

INTRODUCTION

Global warming, a primary component of climate change, is the long-term increase in Earth's average surface temperature caused by the excessive accumulation of greenhouse gases, such as carbon dioxide (CO_2), methane (CH_4), and nitrous oxide (N_2O), in the atmosphere. This phenomenon is predominantly driven by human activities, including the burning of fossil fuels, deforestation, and industrial processes.

According to the Intergovernmental Panel on Climate Change (IPCC), a leading international body for assessing the science related to climate change, global temperatures have risen by approximately 1.1°C above pre-industrial levels. Unless significant mitigation efforts are implemented, this warming trend is projected to continue, leading to severe consequences for ecosystems, societies, and economies worldwide.

The impact of global warming extends far beyond rising temperatures. It significantly influences various aspects of the environment, including weather patterns, sea levels, and ecosystems. Agricultural systems in particular are increasingly vulnerable to the effects of global warming, which can alter growing conditions, pest dynamics, and crop yields. Changes in temperature and precipitation patterns can disrupt traditional farming practices, necessitating the adaptation of agricultural strategies to ensure food security and sustainability.

SIGNIFICANCE OF INSECTICIDE RESISTANCE IN THE CONTEXT OF PEST MANAGEMENT

Insecticide resistance occurs when pest populations evolve mechanisms that enable them to survive exposure to insecticides that were once effective in controlling them. This resistance develops through genetic mutations that provide an advantage to pests in environments where insecticides are used. Resistance mechanisms, such as increased activity of detoxifying enzymes or changes in feeding preferences or avoidance of treated areas, can be biochemical or behavioural (WHO, 2019).

The significance of insecticide resistance is profound. It challenges the effectiveness of pest control programs, leading to increased pest populations and greater economic losses for farmers. Resistant pests can cause more damage to crops, requiring more frequent and higher doses of insecticides, which can have further environmental and health impacts. The development of resistance also increases the risk of pest outbreaks, complicating management efforts and reducing the sustainability of pest control strategies (WHO, 2019).

RELATIONSHIP BETWEEN GLOBAL WARMING AND INSECTICIDE RESISTANCE IN INDIAN AGRICULTURE

The connection of global warming and insecticide resistance poses a significant challenge to Indian agriculture that contributes highly towards the nation's economy and food security. This relationship has far-reaching implications for crop yields, pest management strategies, and the overall sustainability of agricultural practices. Considering India's diverse agricultural landscape and dependence on agriculture for economic stability, the challenges posed by global warming and insecticide resistance are matter of concern.

1. **Agricultural productivity:** India's agricultural sector is highly vulnerable to climate variations. Global warming can alter pest life cycles, resulting in more frequent and severe pest outbreaks. Understanding how these changes interact with existing resistance patterns is essential for developing effective pest management strategies and maintaining crop productivity (Subedi *et al.*, 2023). Increased pest pressure and diminished pesticide efficacy can significantly reduce crop yields, threatening food security and farmers' livelihoods.
2. **Ecosystem health:** Climate change can alter pest behaviour and resistance, disrupting ecosystems where insect and other pests play a role in natural processes. Such disruptions may affect biodiversity and the ecosystem services on which humans depend. A healthy ecosystem can survive disruptions,

recover from stress, and provide essential services. Ecosystem health assessment involves monitoring of indicators like species diversity, habitat quality, water quality, soil health, and ecological processes, which are important to support human and planetary well-being (IPBES Secretariat, 2019).

3. **Increased pest pressure:** Climate change can intensify pest pressure in agricultural and natural ecosystems by influencing pest populations, behaviour, and distribution. Warmer temperatures, changes in precipitation patterns, and extreme weather events can create favourable conditions for pests to thrive and expand their range, often leading to more frequent and severe infestations. This increased pest pressure can result in higher crop losses, reduced yields, and increased reliance on pesticides, potentially affecting ecosystem health and agricultural sustainability. Effective pest management strategies need to adapt to these changes, integrating climate-resilient practices to maintain productivity and reduce environmental impact (IPBES Secretariat, 2019).
4. **Evolution of resistance:** Pests evolve resistance to pesticides and other control measures over time, a process accelerated by repeated and widespread use of chemical controls (Tabashnik and Carrière, 2017). This evolution poses significant challenges in agriculture, as pests become less susceptible to standard treatments, leading to reduced effectiveness of pest control, increased crop losses, and higher management costs (Gould *et al.* 2018). Climate change may further complicate this issue by altering pest biology and hastening resistance development due to more favourable conditions for pest reproduction and survival. Managing resistance requires an integrated approach, combining biological, cultural, and chemical methods, and encouraging practices like crop rotation, bio-pesticide use, and reduced chemical reliance to slow resistance development (Dara, 2019).
5. **Reduced pesticide efficacy:** The effectiveness of pesticides is decreasing as pest populations evolve resistance over time due to frequent and repeated chemical use (Bass *et al.*, 2014). This resistance reduces the effectiveness of conventional pest control, increasing crop losses, production costs, and reliance on higher pesticide doses, which can further harm ecosystems and human health (Sparks and Nauen, 2015). Climate change comprehends this issue by altering pest behaviour, physiology, and range, which may speed up the development of resistance and intensify pest pressures in new areas (Gressel, 2015). In response, integrated pest management (IPM) strategies that incorporate biological, cultural, and chemical approaches are essential to maintain effective pest control, minimize resistance, and ensure agricultural sustainability (Casida and Durkin, 2017).
6. **Economic implications:** The development of pesticide resistance increases production costs due to more frequent pesticide applications and reduced

crop yields. Farmers face higher management costs and crop losses, which can impact overall agricultural profitability. These challenges also raise the cost of food production, affecting both local economies and global markets (Sparks & Nauen, 2015).

7. **Sustainable agriculture:** Sustainable agriculture involves practices that maintain soil health, conserve water, and reduce pesticide use while ensuring long-term crop productivity. Integrated pest management (IPM) and crop diversification are key strategies to reduce reliance on chemicals and promote ecological balance, enhancing resilience to climate change and ensuring food security (Gressel, 2015).

The challenges like pesticide resistance and reduced efficacy can be defied by shifting a conventional approach to sustainable agricultural practices. By integrating approaches like IPM, crop diversification, and the judicious use of pesticides, it can enhance pest control, protect biodiversity, and ensure long-term food security. Collaboration across the agricultural sector, climate adaptation, and on-going research should emphasize to develop more effective, eco-friendly pest management solutions. This universal and holistic approach is essential for harmonising agricultural productivity with environmental conservation.

IMPACT OF GLOBAL WARMING ON PEST POPULATIONS

1. **Effect of increasing temperature on pest biology and behaviour:** The rising temperature due to the consequence of global warming causes a direct and profound impact on pest biology and behaviour. Temperature critically influences the development, survival, and reproduction of insects. As global temperatures increase, several key changes in pest populations have been observed:
   a. **Quicker development:** Higher temperatures significantly speed up the developmental rates of insects, reducing the time required for pests to develop from eggs to adults. This leads to more generations within a single growing season, resulting in rapid population growth and frequent infestations. For example, warmer temperatures have been shown to reduce the developmental time of the cotton bollworm (*Helicoverpa armigera*), leading to increased population density and greater pest pressures on crops (Subedi *et al.*, 2023).
   b. **Extended geographic range:** Warmer temperatures enable many pest species to expand their geographic range into previously cooler regions. As these areas become suitable for pest survival and reproduction, pests can migrate into new territories, overwhelming existing pest management strategies. For instance, the pine beetle (*Dendroctonus ponderosae*)

has migrated to higher latitudes and elevations due to increasing temperatures (Raza *et al.*, 2021). Similarly, the European corn borer (*Ostrinia nubilalis*) has moved northward into cooler regions, affecting crops in areas previously less impacted by this pest (Kumari *et al.*, 2020).

c. **Increased feeding rates:** Higher temperatures enhance the metabolic rates of pests, resulting in increased feeding activity. This leads to greater crop damage as pests consume more plant material. For example, aphids show increased reproduction rates and higher population densities under warmer conditions, causing more severe infestations and potential crop damage (Gupta & Pathak, 2016).

d. **Behavioural changes:** Climate change and pesticide exposure can trigger behavioural shifts in pests, such as altered feeding patterns, migration, and resistance development. These changes complicate pest management and require adaptive strategies to effectively control pest populations and mitigate crop damage (Bass *et al.*, 2014; Gressel, 2015).

e. **Increased reproduction rates:** Warmer temperatures can enhance reproductive rates in many pest species. Higher temperatures often lead to increased fecundity (egg production) and more frequent generations, resulting in larger and more frequent pest outbreaks. For example, the rice weevil (*Sitophilus oryzae*) shows increased reproductive rates with rising temperatures, exacerbating pest problems in stored grains (Raza *et al.*, 2021).

2. **Effect of changed rainfall patterns on pest life cycles and habitat:** Changes in rainfall patterns due to global warming have profound impacts on pest populations, influencing their life cycles and habitat suitability:

   a. **Altered life cycles**: Variations in rainfall can modify the timing and duration of pest life stages. For instance, increased rainfall creates favourable conditions for pests that thrive in wet environments, such as mosquitoes, leading to extended breeding seasons and higher population densities (Reiter, 2001; IPCC, 2021).

   b. **Habitat suitability**: Alterations in rainfall impact habitat availability and quality for pests. Drought conditions can confine pests into smaller areas, intensifying their potential to cause damage. On the contrary, excessive rainfall can create environments conducive to pests breeding in standing water or waterlogged soils (Bentz *et al.*, 2010).

   c. **Plant-pest interactions**: Rainfall variations can affect plant health and susceptibility to pests. Water-stressed plants are often more vulnerable to infestations, while excessive moisture can promote fungal growth, which interacts with pests and exacerbates crop damage (Mittler & Blumwald, 2010; Rosenzweig *et al.*, 2014).

d. **Pest migration**: Changed rainfall patterns influence pest migration. Moisture variations affect pest movement between agricultural fields and natural habitats, potentially increasing pest pressures in certain areas (Bebber *et al.*, 2013).

3. **Influence of CO_2 concentration on pest reproduction and plant-pest interactions:** Rising atmospheric CO_2 concentrations, consequent upon by global warming, impact pest populations, especially through changed plant-pest interactions and pest reproduction:

   a. **Enhanced plant growth**: Elevated CO_2 levels stimulate plant growth, resulting in increased biomass and potentially higher yields. However, these changes often lead to shifts in plant nutritional quality. Plants grown under elevated CO_2 typically have higher carbohydrate levels and reduced essential nutrients, affecting pest feeding and reproduction (Zavala *et al.*, 2013).

   b. **Increased pest reproduction**: Elevated CO_2 can enhance the reproductive success of certain pests. For example, studies have shown that increased CO_2 levels lead to higher fecundity and faster development rates in pests like the cabbage white butterfly (*Pieris rapae*), resulting in larger and more damaging populations (Dermody *et al.*, 2008).

   c. **Altered plant-pest interaction**: Changes in CO_2 levels affect plant chemistry and pest interactions. These shifts can reduce the effectiveness of plant defenses, making them more attractive or susceptible to herbivores. Elevated CO_2 can decrease the production of defensive compounds in plants, increasing their vulnerability to pests (Hamilton *et al.*, 2005).

   d. **Shifts in pest communities**: Changed plant quality and availability under elevated CO_2 conditions can lead to shifts in pest communities. Some pests become more prevalent due to their adaptability to the changed plant conditions, while others may decline, altering the structure and composition of pest communities (DeLucia *et al.*, 2012).

MECHANISMS OF INSECTICIDE RESISTANCE

Insecticide resistance poses a noteworthy challenge in pest management, triggered by various mechanisms that allow pests to survive and reproduce despite exposure to chemical control agents. These mechanisms need to be understood for developing effective strategies to manage and mitigate resistance. The primary mechanisms of insecticide resistance include genetic mutations, biochemical adaptations, and behavioural changes.

1. **Genetic basis of mutations and adaptation:** Genetic mutations and adaptations have consistent effect on development of insecticide resistance. These changes occur in individual insects and can result in populations that survive pesticide exposure which would typically be lethal. Mutations affecting detoxification enzymes or target site modifications enable pests to adapt to chemical controls. Selective pressure from repeated pesticide use drives these genetic changes, which spread across populations over generations, leading to resistance (Plapp, 1984 and Roush & McKenzie, 1987). Considering these genetic mechanisms the development of effective pest management strategies is very essential to mitigate resistance.
   a. **Point mutations**: Point mutations are a well-documented genetic mechanism that can lead to insecticide resistance in pest populations. These mutations involve a single nucleotide change within a gene, which alters the amino acid sequence of the encoded protein. Such changes can significantly impact the protein's structure and function, especially when the mutation occurs in genes encoding target proteins of insecticides. For instance, mutations in the **ace1** gene, which codes for the acetylcholinesterase enzyme—a target for organophosphates and carbamates—can decrease the binding affinity of these insecticides, making them less effective. Similarly, mutations in the **kdr** (knockdown resistance) gene, which encodes voltage-gated sodium channels targeted by pyrethroids, result in reduced sensitivity to these insecticides, contributing to resistance (Dong *et al.*, 2014).
   b. **Gene amplification**: Resistance can also be associated with the amplification of genes encoding detoxification enzymes. For instance, the overexpression of cytochrome P450 monooxygenases due to gene amplification has been linked to resistance in pests like the tobacco budworm (*Chloridea virescens*). This gene amplification increases enzyme levels, allowing more efficient metaboliation and detoxification of insecticides (Li *et al.*, 2007).
   c. **Genetic recombination**: Recombination events can contribute to resistance by creating new gene variants that confer a survival advantage. Recombination between alleles of resistance genes can result in novel resistance traits in pest populations, complicating resistance management (Bass & Field, 2011).
2. **Biochemical mechanisms: enzymatic changes and detoxification processes:** Biochemical mechanisms of insecticide resistance involve alterations in metabolic processes, primarily through the action of detoxification enzymes. These adaptations enable pests to neutralize or secrete insecticides before they can exert toxic effects.

a. **Detoxification enzymes**: A major biochemical mechanism involves increased activity of detoxification enzymes such as esterases, glutathione S-transferases (GSTs), and cytochrome P450 monooxygenases. For example, esterases can hydrolyse insecticides like organophosphates, rendering them inactive. Elevated esterase activity has been linked to resistance in the house fly (*Musca domestica*) (Devonshire & Field, 1991).

b. **Cytochrome P450 monooxygenases**: Cytochrome P450 enzymes are key for the oxidative metabolism of various insecticides. Increased expression of P450 genes can enhance the breakdown of insecticides such as pyrethroids and neonicotinoids. This mechanism has been observed in pests like the cotton bollworm (*Helicoverpa armigera*) and the cabbage aphid (*Brevicoryne brassicae*) (Shi *et al.*, 2021 and Ahmad and Akhtar (2013).

c. **Glutathione S-Transferases (GSTs)**: GSTs are involved in conjugating insecticides with glutathione, facilitating their excretion. Overexpression of GSTs has been associated with resistance to insecticides like organophosphates and carbamates. In the aphid *Myzus persicae*, increased GST activity has been linked to resistance to multiple insecticides (David *et al.*, 2005).

3. **Epigenetics, climate change and insect resistance:** Epigenetics, which involves heritable changes in gene expression without altering the DNA sequence, is pivotal in understanding how insects adapt to climate change and develop resistance to various stresses, including insecticides.

   a. **Epigenetic modifications and stress response:** Arthropod pests can swiftly adapt to new environmental stresses, including insecticides and climate change, due to their ability to rapidly adjust both genetically and epigenetically. Epigenetic regulation, through processes such as DNA methylation and histone acetylation, significantly influences insecticide resistance and climatic adaptation. (Mogilicherla and Roy (2023).

   b. **Adaptive mechanisms to climate change:** Insects may use epigenetic mechanisms to adapt rapidly to climate change by modulating genes related to thermotolerance, reproductive timing, and developmental rates. The fruit fly (*Drosophila melanogaster*) demonstrates how temperature-induced epigenetic changes can affect developmental timing and stress tolerance (Rutherford & Lindquist, 1998).

   c. **Insecticide resistance and epigenetics:** Insecticides and climate change affect insect development, physiology, and inheritance, with recent research revealing their impact on epigenetic regulation at the molecular level. Epigenetic mechanisms modify gene expression in response to environmental stresses without changing the genetic code. Notably, epigenetic modifications significantly influence insecticide resistance, as they can regulate gene expression and metabolic pathways involved in detoxification, such as those mediated by cytochrome P450 enzymes in response to pesticide exposure. (Mogilicherla and Roy (2023).

d. **Trans-generational effects:** Epigenetic changes can be inherited across generations, potentially impacting long-term resistance patterns and population dynamics. For example, epigenetic modifications related to insecticide resistance can persist, affecting future generations and resistance development (Jablonka & Lamb, 2005).

e. **Implications for pest management:** Integrating epigenetic insights into pest management strategies can enhance resistance monitoring and management. Understanding epigenetic mechanisms can lead to more effective and sustainable pest control practices under changing environmental conditions (Bossdorf *et al.*, 2008).

STRATEGIES FOR RESISTANCE MANAGEMENT

The convergence of global warming and insecticide resistance poses a significant challenge to sustainable agriculture and public health. To address this complex issue, a multifaceted approach is necessary.

1. **Adapting IPM practices to mitigate resistance and climate change:** Integrated Pest Management (IPM) practices are the key for managing both pesticide resistance and the challenges posed by climate change. Adapting IPM includes strategies such as rotating chemical classes, using bio-pesticides, and promoting cultural practices like crop diversification and pest-resistant varieties. These practices reduce dependency on chemical pesticides, thus slowing the evolution of resistance. Additionally, climate change can alter pest dynamics, making it essential to regularly update IPM strategies based on climatic conditions and pest behaviour. By incorporating climate-smart IPM, farmers can better manage resistance while maintaining crop productivity and ecosystem health (Gould *et al.*, 2018; Dara, 2019).
2. **Monitoring and early detection**: Regular pest monitoring through techniques like pheromone traps and pest scouting helps detect resistance early, enabling timely interventions. Early detection allows farmers to apply targeted control measures, reducing pesticide use and minimizing environmental impact. Monitoring pest under IPM practices enhances pest management by taking early decisions and reducing the risk of resistance development (Gould *et al.*, 2018; Dara, 2019).

3. **Cultural practices:** Cultural practices are integral part in IPM, offering non-chemical methods to reduce pest populations and mitigate resistance. Techniques such as crop rotation, intercropping, and adjusting planting time disrupt pest life cycles and habitats. Removal of crop residues and maintaining field hygiene also reduce pest breeding grounds. These methods enhance natural pest control, minimize reliance on pesticides, and delay resistance development. These methods can sustainably manage pest populations by reducing environmental impacts and preserving agricultural ecosystems (Klumper & Qaim, 2014).
4. **Chemical control to mitigate resistance due to global warming:** Global warming intensifies pest challenges by altering their life cycles and expanding their habitats, increasing the risk of resistance. Chemical control an unavoidable tool but must be employed judiciously. Strategies like rotating pesticides with different modes of action, using reduced-risk green pesticides, and integrating chemical control with other IPM practices is the need of hour. Precision application technologies and adherence to recommended dosages help minimize selection pressure on pests. These measures along with continuous resistance monitoring, ensure effective chemical control to mitigate resistance development in a warm climate (Gould *et al.*, 2018; Sparks & Nauen, 2015).
5. **Role of biological control in global warming:** Biological control plays a major role in mitigating resistance exacerbated by global warming by providing sustainable alternatives to chemical pesticides. Climate change impacts pest and biocontrol agent dynamics, necessitating the selection of resilient natural enemies like parasitoids, predators, and pathogens. Enhancing biodiversity through habitat management fosters ecosystem resilience, while combining biocontrol with cultural and mechanical methods reduces reliance on chemicals. Effective biocontrol strategies, such as augmentative releases and conservation of natural enemies, delay pest resistance at the same time minimizing environmental impact in a changing climate (Dara, 2019; Van Lenteren *et al.*, 2018).
6. **Educational programs, training, and awareness in sustainable agriculture amid changing climate:** Educational programs and training in promoting sustainable agriculture in the face of climate change is the basics of sustainability. Farmers need updated knowledge about climate-resilient practices, pest resistance management, and integrated pest management (IPM) techniques. Awareness campaigns empower communities to adopt eco-friendly solutions like biocontrol agents, crop diversification, and water-efficient practices. Extension services and workshops bridge the knowledge gap, ensuring the adoption of technologies and strategies tailored to local conditions. These initiatives build capacity, improve decision-making, and enhance

resilience, helping farmers sustainably manage resources and mitigate challenges posed by climate change (Pretty, 2008; Rickards & Howden, 2012).

CONCLUSION

Global warming poses significant challenges to Indian agriculture, intensifying pest pressures and fostering insecticide resistance. Rising temperatures and erratic rainfall patterns are expected to expand pest habitats and accelerate life cycles, increasing crop vulnerabilities. To combat these challenges, the scientists and the stakeholders must prioritize sustainable agricultural practices, including integrated pest management (IPM), climate-resilient crop varieties, and precision farming technologies. Strengthening research on biological control and fostering farmer education can reduce reliance on chemical pesticides. Collaborative efforts among policymakers, researchers, and farmers will be dynamic to promoting eco-friendly practices, ensuring food security, and sustaining agricultural productivity in the face of climate change.

REFERENCES

Ahmad M and Akhtar S (2013) Development of insecticide resistance in field populations of *Brevicoryne brassicae* (Hemiptera: Aphididae) in Pakistan. *Journal of Economic Entomology, 106*(2): 954-958. https://doi.org/10.1603/ec12233

Bass C and Field LM (2011) Gene amplification and insecticide resistance. *Pest Management Science, 67*(8): 886-890. https://doi.org/10.1002/ps.2189

Bass C, Puinean AM, Zimmer CT, Denholm I, Field LM, Foster SP, Gutbrod O, Nauen R, Slater R and Williamson MS (2014) The evolution of insecticide resistance in the peach potato aphid, *Myzus persicae*. *Insect Biochemistry and Molecular Biology, 51*, 41-51. DOI: 10.1016/j.ibmb.2014.05.003

Bebber DP, Ramotowski MA and Gurr SJ (2013) Crop pests and pathogens move polewards in a warming world. *Nature Climate Change, 3*(11): 985-988. https://doi.org/10.1038/nclimate1990

Bentz BJ, Régnière J, Fettig CJ, Hansen EM, Hayes JL, Hicke JA, Kelsey RG, Negrón JF and Seybold SJ (2010) Climate change and bark beetles of the western United States and Canada: Direct and indirect effects. *Bioscience, 60*(8): 602-613. https://doi.org/10.1525/bio.2010.60.8.6

Bossdorf O, Richards CL and Pigliucci M (2008) Epigenetics for ecologists. *Ecology Letters, 11*(11): 1066-1077. DOI:.org/10.1111/j.1461-0248.2007.01130.x

Casida JE and Durkin KA (2017) Pesticide chemical research in toxicology: Lessons from nature. *Chemical Research in Toxicology*, 30(1): 94-104. DOI: 10.1021/acs.chemrestox.6b00361

Dara SK (2019) The new integrated pest management paradigm for the modern age. *Journal of Integrated Pest Management*, 10(1): 12. DOI: 10.1093/jipm/pmz010

David JP, Strode C, Vontas J, Nikou D, Vaughan A, Pignatelli PM, Louis C, Hemingway J and Ranson H (2005) The *Anopheles gambiae* detoxification chip: A highly specific microarray to study metabolic-based insecticide resistance in malaria vectors. *Proceedings of the National Academy of Sciences, 102*(11): 4080-4084. DOI:.org/10.1073/pnas.0409348102

DeLucia EH, Nabity PD, Zavala JA and Berenbaum MR (2012) Climate change: Resetting plant-insect interactions. *Plant Physiology, 160*(4): 1677-1685. https://doi.org/10.1104/pp.112.204750

Dermody O, O'Neill, BF, Zangerl AR, Berenbaum MR and DeLucia EH (2008) Effects of elevated CO_2 and O_3 on leaf damage and insect abundance in a soybean agroecosystem. *Arthropod-Plant Interactions, 2*(3): 125-135. https://doi.org/10.1007/s11829-008-9043.

Devonshire AL and Field LM (1991) Gene amplification and insecticide resistance. *Annual Review of Entomology, 36*(1): 1-21. https://doi.org/10.1146/annurev.en.36.010191.000245

Dong K, Du Y, Rinkevich F, Nomura Y, Xu P, Wang L, Silver K and Zhorov B S (2014) Molecular biology of insect sodium channels and pyrethroid resistance. *Insect Biochemistry and Molecular Biology, 50*, 1-17. https://doi.org/10.1016/j.ibmb.2014.03.012

Gould F, Brown ZS and Kuzma J (2018) *Wicked evolution: Can we address the sociobiological dilemma of pesticide resistance? Science, 360*(6390): 728-732. DOI: 10.1126/science.aar3780

Gressel J (2015). *Herbicide resistance in plants and weeds: A challenge for the 21st century. Pest Management Science*, 71(3): 408-415. DOI: 10.1002/ps.3938

Gupta A and Pathak H (2016) *Climate Change and Agriculture in India: A Thematic Report*. National Mission on Strategic Knowledge for Climate Change (NMSKCC) under the National Action Plan on Climate Change (NAPCC). Department of Science and Technology, Government of India,

New Delhi. https://dst.gov.in/sites/default/files/Report_DST_CC_ Agriculture.pdf

Hamilton JG, Dermody O, Aldea M, Zangerl AR, Rogers A, Berenbaum MR and DeLucia EH (2005) Anthropogenic changes in tropospheric composition increase susceptibility of soybean to insect herbivory. *Environmental Entomology, 34*(2): 479-485. https://doi.org/10.1603/0046-225X-34.2.479

Intergovernmental Panel on Climate Change (IPCC):https://www.ipcc.ch/

IPBES Secretariat (2019) *Global Assessment Report on Biodiversity and Ecosystem Services of the Intergovernmental Science-Policy Platform on Biodiversity and Ecosystem Services*. IPBES. DOI: 10.5281/zenodo.3831673

IPCC (2021) Climate Change 2021: The Physical Science Basis. Contribution of Working Group I to the Sixth Assessment Report of the Intergovernmental Panel on Climate Change. https://doi.org/10.1017/9781009157896

Jablonka E and Lamb MJ (2005) Evolution in Four Dimensions: Genetic, Epigenetic, Behavioral and Symbolic Variation in the History of Life. *MIT Press*.

Klumper W and Qaim M (2014) A meta-analysis of the impacts of genetically modified crops. PLoS ONE, 9(11): e111629. DOI: 10.1371/journal.pone.0111629.

Kumari S, George SG, Meshram M and Kumar P (2020) A review on climate change and its impact on agriculture in India. *Current Journal of Applied Science and Technology, 39*(44), Article 31152. https://doi.org/10.9734/cjast/2020/v39i4431152

Li X, Schuler MA and Berenbaum MR (2007) Molecular mechanisms of metabolic resistance to synthetic and natural xenobiotics. *Annual Review of Entomology, 52*(1): 231-253. https://doi.org/10.1146/annurev.ento.51.110104. 151104

Mittler R and Blumwald E (2010) Genetic engineering for modern agriculture: Challenges and perspectives. *Annual Review of Plant Biology, 61*(1): 443-462. https://DOI: 10.1146/annurev-arplant-042809-112116

Mogilicherla K and Roy A (2023). Epigenetic regulations as drivers of insecticide resistance and resilience to climate change in arthropod pests. *Frontiers in Genetics*, 13, Article 1044980. https://doi.org/10.3389/fgene.2022.1044980

Plapp FW Jr (1984) The genetic basis of insecticide resistance in the house fly: Evidence that a single locus plays a major role in metabolic resistance to insecticides. *Pesticide Biochemistry and Physiology, 22*(2): 194-201. DOI: 10.1016/0048-3575(84)90089-0

Pretty J (2008) Agricultural sustainability: Concepts, principles and evidence. Philosophical Transactions of the Royal Society B: *Biological Sciences*, 363(1491): 447-465. DOI: 10.1098/rstb.2007.2163.

Raza H, Bilal H, Raza A, Shoukat MB, Tariq R, Rasheed A, Shahid M, Sharif MU, Maqsood M and Ramzan M (2021) Impact of climate change on agriculture and food security: A review. *Indian Journal of Pure and Applied Biosciences, 9*(3): 110-118. http://dx.doi.org/10.18782/2582-2845.8699

Reiter P (2001) Climate change and mosquito-borne disease. *Environmental Health Perspectives,* 109(Suppl 1): 141-161. https://doi.org/10.1289/ehp.01109s1141

Rickards L and Howden SM (2012) Transformational adaptation: Agriculture and climate change. *Crop and Pasture Science, 63*(3): 240-250. DOI: 10.1071/CP11172.

Rosenzweig C, Iglesias A, Yang XB, Epstein PR and Chivian E (2014) Climate change and extreme weather events: Implications for food production, plant diseases and pests. *Global Change and Human Health, 2*(2): 90-104. https://doi.org/10.1023/A:1015086831467

Roush RT and McKenzie JA (1987) Ecology of resistance to insecticides: A laboratory perspective. *Annual Review of Entomology, 32*, 361-380. DOI: 10.1146/annurev.en.32.010187.002045

Rutherford SL and Lindquist S (1998) Hsp90 as a capacitor for morphological evolution. *Nature, 396*(6709): 336-342. https://doi.org/10.1038/24550

Shi Y, Jiang Q, Yang Y, Feyereisen R and Wu Y (2021) Pyrethroid metabolism by eleven *Helicoverpa armigera* P450s from the CYP6B and CYP9A subfamilies. *Insect Biochemistry and Molecular Biology, 132*, 103597. https://doi.org/10.1016/j.ibmb.2021.103597

Sparks TC and Nauen R (2015) *IRAC:* Mode of action classification and insecticide resistance management. Pesticide Biochemistry and Physiology, 121, 122-128. DOI: 10.1016/j.pestbp.2014.11.014

Subedi B, Poudel A and Aryal S (2023) The impact of climate change on insect pest biology and ecology: Implications for pest management strategies, crop production and food security. *Journal of Agriculture and Food Research, 14,* 100733. https://doi.org/10.1016/j.jafr.2023.100733

Tabashnik BE and Carrière Y (2017) Surge in insect resistance to transgenic crops and prospects for sustainability. *Nature Biotechnology*, 35(10): 926-935. DOI: 10.1038/nbt.3974

Van Lenteren JC, Bolckmans K, Köhl J, Ravensberg WJ and Urbaneja A (2018) *Biological control using invertebrates and microorganisms: Plenty of new opportunities. Bio Control, 63*(1): 39-59. DOI: 10.1007/s10526-017-9801-4.

World Health Organization (WHO) (2019) Global vector control response 2017-2030. World Health Organization, https://www.who.int/publications/i/item/9789241512978

Zavala JA, Nabity PD and DeLucia EH (2013) An emerging understanding of mechanisms governing insect herbivory under elevated CO_2. *Annual Review of Entomology, 58*, 79-97. https://doi.org/10.1146/annurev-ento-120811-153544

Sustainable Agriculture and Biodiversity: Innovations and Insights, Pages: 259–278
Edited by: Bibhuti Bhusan Sahoo, Harvinder Kumar Singh, Niranjan Chinara and Amitabh Panda

CHAPTER - 11

MUSHROOM PRODUCTION: A STEP TOWARDS NONGREEN REVOLUTION

Niranjan Chinara[1*] and Kailash Behari Mohapatra[2]

[1]All India Coordinated Research Project on Fruits,
Odisha University of Agriculture and Technology, Bhubaneswar-751003, Odisha
[2]Department of Plant Pathology,
Odisha University of Agriculture and Technology, Bhubaneswar-751003, Odisha
**Corresponding email: niranjanchinara@gmail.com*

INTRODUCTION

Mushrooms are the part of the fungal diversity that has been known around 300 million years. Mushrooms are the macro-fungi with specialized fruiting bodies, which may be epigeous or hypogeous and are large enough to be seen by naked eye and picked by hand. They do not contain chlorophyll that present in case of green plants. They grow on dead and decaying organic matters or mycorrhizal association with specific plants. Mushrooms are of four groups: (i) Those that are fleshy and edible fall into the edible mushroom category i.e., *Agaricus bisporus*; (ii) Mushroom that are considered to have medicinal applications are referred to as medicinal mushrooms i.e., *Ganoderma lucidum*; (iii) Those that are proved to be or poisonous or suspected of being poisonous are named poisonous mushrooms i.e., *Amanita phalloides*; (iv) those in a miscellaneous category, which include a large number of mushrooms whose properties remains less defined.

Since ancient times humans have used mushrooms as food. Ancient peoples such as the Greeks, Egyptians, Romans, Chinese and Mexicans all considered mushrooms to be delicious, recognized their medicinal properties and often used them in religious ceremonies. The Greeks referred mushrooms as a powerful weapon for the military, the Romans as the "medicine of the God" and the Chinese as the "elixir of life". Archaeological and other evidence suggests that ancient people used mushrooms in religious ceremonies. The high esteem in which mushrooms were held in different civilizations indicated the restrictions on the use of many species to their rulers.

Currently 14,000 mushroom species are known to exist. Out of which about 10,000 species are considered to posses varying degrees of edibility and almost 3000 species are regarded as prime edible mushrooms. To date only 200 of them are experimentally grown, 100 of them economically cultivated, approximately 60 commercially cultivated and about 10 have reached to industrial scale production in many countries. Moreover, about 2000 are considered as medicinal mushrooms with varieties of health attributes. The number of poisonous mushroom is usually reported to be relatively small, but there is an estimate of about 10 % may have poisonous attributes while 30 species are considered to be lethal.

IMPORTANCE

Agriculture continues to be the main strength of Indian economy. With the variety of agricultural crops grown today the country has achieved food security by producing about 260 million tonnes of food grains. However, the struggle to achieve nutritional security is still on. In future, the ever increasing population, depleting agricultural land, climate changes, water shortage and need for quality food products at competitive rates are going to be the vital issues. Hence, it is imperative to diversify our agricultural activities in areas like horticulture to meet these challenges and to provide food and nutritional security to our people. Mushrooms are one such component that not only use vertical space but also help in addressing the issues of quality food, health and environmental sustainability. There is need to promote both mushroom production as well as consumption for meeting the changing needs of food items. Fortunately, mushroom trade has gained importance in the recent years possibly for the global shift towards vegetarian food and recognition of mushroom as a functional food. Mushroom cultivation also offers scope to recycle agro-wastes into good quality protein, otherwise wasted in the field. This unique horticultural venture has tremendous scope to meet the food shortage without undue pressure on land.

INTEGRATED FARMING SYSTEM AND MUSHROOM CULTIVATION

Integrated farming is a farming system where high quality food, feed, fibre and renewable energy are produced by using resources such as soil, water, air and nature as well as regulating factors to farm sustainably and with as little polluting inputs as possible. Being bound to sustainable development, the underlying three dimensions; 'economic development', 'social development' and 'environmental protection' are thoroughly considered in the practical implementation of integrated farming. However, the need for profitability is a decisive pre-requisite. To be sustainable, the system must be profitable, as profits generate the possibility to support all activities outlined in the integrated farming network. Mushrooms are one such component that not only impart diversification but also help in addressing the problems of quality food, health and environment related issues. One of the major areas that contribute towards goal of conservation of natural resources as well as increased productivity is recycling of agro-wastes. Such kind of activity can enhance income and impart higher level of sustainability. Unlike most of the development efforts, in mushroom cultivation, the emphasis has been on economic development with improvement in quality of life of the rural poor through an integrated effort. Labour intensive nature of cultivation, significant value addition, dependence on locally available, abundant and cheap raw material, low initial investment, simple technology and limited space requirement make it a worthwhile proposition. Mushroom, being an indoor crop, happens to be women friendly in nature. This brings about removal of poverty and social development in harmony and therefore, fits well in any farming system. Integrating mushroom cultivation in existing farming systems will supplement the income of rural poor, provide gainful employment and will lead to inclusive growth.

MUSHROOM: A FUNCTIONAL FOOD

It is well recognized that mushroom contains all essential components of a balance food (Table 1 and 2). In fact, being rich in highly digestible lysine rich protein, vitamins and minerals, mushroom lacks fats and are low in carbohydrate (low calorie food). They are rich in folic acid, phosphorus, potassium, calcium, copper, iron and vitamin-B complex. In place of starch, mushroom contains sorbitol and linolenic acid. They are excellent sources of thiamine, riboflavin, niacin, pantothenic acid, biotin, folic acid and vitamin B-12 (Table 3). Further, mushrooms are also reported to have high medicinal attributes. A large number of mushrooms contain biological active polysaccharide protein complex having anti-tumour, immuno modulating and antioxidant properties. *Ganoderma lucidum* showed hypo-glycemic, hypo-lipidemic, endurance enhancing, immune- modulatory, anti-hypertensive, antiplatelet aggregation, hepato-protective, anti-HIV and anti-atherosclerotic effects. Many medicinal mushrooms also showed hematological, antiviral, anti-tumour, antioxidant, cardiovascular and renal effects. The interesting fact is that the K: Na ratio is very high and it is suitable for people who suffer from hypertension. Besides its nutritional and medicinal values, mushrooms are relished as a source of food due to the pleasant aroma, taste and its fleshy nature.

Table 1: Proximate composition (per unit fresh weight) of the cultivated mushrooms.

Mushrooms/ vegetables	Moisture	Protein	Fat	Carbohydrate	Fibre	Ash	Calorie
Button mushroom (*Agaricus bisporus*)	90.1	2.9	0.3	5.0	0.9	0.8	36
Oyster mushroom (*Pleurotus sajor-caju*)	90.2	2.5	0.2	5.2	1.3	0.6	35
Paddy straw mushroom (*Volvariella volvacea*)	90.1	2.1	1.0	4.7	1.1	1.0	36
Cabbage	91.9	1.8	0.1	4.6	1.0	0.6	27
Cauliflower	90.8	2.6	0.4	4.0	1.2	1.0	30
Potato	74.7	1.6	0.1	22.6	0.4	0.6	97

(**Source:** Rai and Sohi, 1988)

Table 2: Protein content in food stuffs

Food stuff	Protein in edible portion (per unit fresh weight)
Button mushroom (*Agaricus bisporus*)	3.7
Paddy straw mushroom (*Volvariella volvacea*)	3.9
Peas	2.6
Cabbagc	1.5
Cauliflower	2.7
Potato	2.1
Apple	0.3
Banana	1.1
Chicken	20.01
Fish	10.0
Egg	12.8
Milk (cow)	3.2

(**Source:** Kaul, 1983)

Table 3. Essential amino acids (per cent crude protein) in edible mushrooms

Amino acid	Button mushroom (*Agaricus bisporus*)	Oyster mushroom (*Pleurotus sajor-caju*)	Paddy straw mushroom (*Volvariella volvacea*)
Leucine	7.5	7.0	4.5
Isoleucine	4.5	4.4	3.4
Valine	2.5	5.3	5.4
Tryptophan	2.0	1.2	1.5

[Table Contd.

Contd. Table]

Amino acid	Button mushroom (*Agaricus bisporus*)	Oyster mushroom (*Pleurotus sajor-caju*)	Paddy straw mushroom (*Volvariella volvacea*)
Lysine	9.1	5.7	7.1
Threonine	5.5	5.0	3.5
Phenyl alanine	4.2	5.0	2.6
Methionine	0.9	1.8	1.1
Histidine	2.7	2.2	3.8

(**Source:** Bano and Rajanathnam, 1982; Li and Chang, 1982)

BACKGROUND INFORMATION

It is estimated that around 200 billion tonnes of organic matter are generated annually through the process of photosynthesis (Zhang, 2008). A chunk of this organic matter and many agro-industrial wastes furnish large volumes of solid wastes, residues and byproducts which pose serious environmental pollution (Lal, 2005). In fact, the physico-chemical properties of such lingo-cellulosic materials offer great scope for their enormous biotechnological importance. Global efforts are being made involving innovative technologies and disposal methods for the utilization of these lingo-cellulosic materials into more profitable materials by solid state fermentation (Chang, 2006). Among various methods, mushroom cultivation is the most economical and relatively short biological process for the biotransformation of such materials into protein rich food (Chiu and Moore, 2001). The production of tropical mushrooms like oyster (*Pleurotus* spp.), paddy straw mushroom (*Volvariella volvacea*) and milky mushroom (*Calocybe indica*) utilizing locally available substrates such as paddy straw, wheat, soybean, cotton wastes, coffee waste, water hyacinth, saw dust, sugar cane bagasse, wild grasses and various categories of refuse and ligno-cellulosic wastes have great potential to exploit and transform it into a highly nutritious food within short period of time (Singh *et al.*, 2011). Cotton wastes and sunflower stalks are examples of agro-industrial wastes which are of limited/no economic use. In recent years, cotton wastes have become popular as substrates for straw mushroom (Chang, 1982) and oyster mushroom (Hamlyn, 1989) cultivation. Oyster mushroom *Pleurotus ostreatus*) is the third mushroom of the world produced by China (Li, 2012). Its production in India is also becoming popular in almost all the states having temperate, sub-tropical and tropical climate. Paddy straw mushroom (*Volvariella volvacea*) is another mushroom dominating whole of Odisha and available in every nook and corners of the state. It is the sixth mushroom of the world in production (Ahlawat and Tewari, 2007). Paddy straw mushroom is a fast growing fungus with a crop duration of 14-15 days. Outdoor cultivation of paddy straw mushroom under trees like coconut, casurina, cashewnut etc. is very much successful during summer (Thakur and Mohapatra, 2013). Odisha is the only state where paddy straw mushroom is grown commercially for eight months a year involving common people. The paddy straw mushroom is successfully cultivated in the plains of Kerala throughout the year where the temperature ranges between 28-32°C (Prakasam, 2012). In India, four mushroom varieties such as *Agaricus bisporus*, *Pleurotus* spp., *Volvariella* spp. and *Calocybe indica* have been recommended for the year round cultivation. The Indian subcontinent is known worldwide for its varied agro-climatic zones with a variety of habitats that favour rich mushroom biodiversity (Verma, 2013). Among tropical mushrooms, oyster, paddy straw and milky mushrooms are very much in cultivation in India. Besides, cold loving white button mushroom is also in cultivation in suitable agro-climatic situations of the country.

GLOBAL SCENARIO OF MUSHROOM PRODUCTION

China is the largest producer, consumer and exporter of mushrooms in the world followed by Japan and Poland (Table 4). China grows over 60 mushroom species in a small to commercial scale. Ten most widely cultivated mushrooms in China include *Pleurotus ostreatus, Lentinus edodes, Auricularia auricula, Agaricus bisporus, Flammulina velutipes, A. polytricha, P. cornucopiae, Coprinus comatus, Agrocybe chaxinggu* and *Volvariella volvacea* (Li, 2012). The world production of mushroom in China in 2010 was 22.020 million tonnes, of which the major share goes to oyster mushroom. Oyster mushroom (*Pleurotus ostreatus*) is the third largest mushroom of the world produced by China with a production of 4.929 million tonnes (Chinese Edible Fungi Association, 2011). The three major mushroom producing countries as per FAO data namely China, Japan and Poland account for more than 60 % of the world production.

Table 4: Production trends of different countries with respect to fresh mushroom production (million tons).

Sl. No.	Country	2000	2010	2021
1	China	6.644	22.020	41.127
2	Japan	0.424	0.470	0.469
3	Poland	0.109	0.230	0.379
4	USA	0.384	0.359	0.344
5	Netherland	0.265	0.266	0.260

[Table Contd.

Contd. Table]

Sl. No.	Country	2000	2010	2021
6	India	0.024	0.041	0.243
7	Spain	0.063	0.133	0.164
8	Canada	0.080	0.115	0.138
9	Russia	0.006	0.005	0.111
10	France	0.204	0.119	0.099

Source: FAOSTAT, 2023

NATIONAL SCENARIO

Mushroom production in the country started in the 70s but growth rate, both in terms of productivity as well as production has been phenomenal. In seventies and eighties button mushroom was grown as a seasonal crop in hills, but with the development of technologies for environmental controls and increased understanding of the cropping systems, mushroom production shot up from mere 5,000 tonnes in 1990 to 1,29,782 tonnes in 2016 (Sharma *et al.*2017). Today, commercially grown species are button and oyster mushrooms, followed by other tropical mushrooms like paddy straw mushroom, milky mushroom etc. Two to three crops of button mushroom are grown seasonally in temperate regions with minor adjustments of temperature in the growing rooms; while one crop of button mushroom is raised in North Western plains of India seasonally. Oyster, paddy straw and milky mushrooms are grown seasonally in the tropical/ subtropical areas from March to October. The areas where these mushrooms are popularly grown are Odisha, Maharashtra, Tamil Nadu, Kerala, Andhra Pradesh, Karnataka and North Eastern region of India. Some commercial units are already in operation located in different regions of our country and producing the quality mushrooms for export. The present production of white button mushroom is about 80 % of the total production of mushrooms in the country. The state wise production of mushroom in India is presented in Table 5.

Table 5. State wise mushroom production in India (tonnes) (Sharma *et al.*2017)

Sl.No.	State	Button	Oyster	Milky	Others	Total production
1	Panjab	16000	2000	0	0	18000
2	Odisha	126	6310	0	9550	15986
3	Haryana	15000	50	50	0	15100
4	Maharastra	10000	2000	50	0	12050

[Table Contd.

Contd. Table]

Sl.No.	State	Button	Oyster	Milky	Others	Total production
5	Gujrat	10000	1200	0	0	11200
6	Uttarakhand	8189	1228	819	0	10236
7	Tamil Nadu	6500	2000	1500	0	10000
8	Himachal Pradesh	9000	110	30	10	9150
9	Uttar Pradesh	7000	100	0	0	7100
10	Goa	4200	20	0	0	4220
11	Other states	8661	6254	735	1090	16740
	Total	**94676**	**21172**	**3184**	**10650**	**129782**

STATUS OF MUSHROOM PRODUCTION IN ODISHA

The research on edible mushroom in Odisha made its humble beginning in the Department of Plant Pathology, College of Agriculture, Orissa University of Agriculture and Technology, Bhubaneswar in 1972 with a view to generate profitable and sustainable production technology. Having achieved success in developing mushroom cultivation and spawn production technology, research efforts were further strengthened and transfer of technology was initiated with the establishment of Centre of Tropical Mushroom Research and Training in the University with the financial support of Government of Odisha in the year, 1991-92. This research organization paved the way for initiation of commercial mushroom cultivation in the state within two years of its establishment.

In depth study of production of spawn and mushroom cultivation particularly paddy straw mushroom, oyster mushroom, milky mushroom and button mushroom were undertaken. Farmer training programmes and demonstrations on spawn production and mushroom cultivation were then extended to all over the state. Technical assistance was provided for development of individual/ group/ private sectors in establishing spawn production units and mushroom cultivation centres. The overall activities of the centre further gained momentum with the establishment of All India Coordinated Research Project on Mushroom in the University during 2009-10 with the financial assistance from Indian Council of Agricultural Research, New Delhi. At present the total mushroom production of the state has reached upto 32,050 tonnes/annum contributing to over 10 per cent of the country's production (Table 6).

Table 6: Mushroom production in Odisha (tonnes)

Sl.No.	Mushroom	2009	2022
1.	Paddy straw	5,000	21,500
2	Oyster	810	9,500
3	Milky	0	0
4	Button	36	1,050
	Total	**5,846**	**32,050**

Paddy straw mushroom has been a popular vegetarian diet of the people of Odisha since long. People in rural areas are in the habit of collecting this mushroom grown naturally during the rainy season on straw piles. However, it could not find a regular place in the diet due to its non-availability in other seasons. Now the scenario has changed altogether and straw mushroom is being cultivated 8-10 months a year (Fig. 1). Besides, methods for off-season cultivation of paddy straw mushroom in the coastal ecosystem of the state are in the offing. The conditions in most of the agro-climatic situations of Odisha are favourable for taking up cultivation of a number of edible mushrooms. Diverse agro-wastes suitable for mushroom production are also available in abundance. Paddy straw, the main input for mushroom cultivation is plenty available at cheaper rates. Besides, cultivation is comparatively easier, as it involves minimum investment, labour and space. Straw mushroom is cultivated largely as an intercrop in the coconut plantations in the coastal agro-ecological situation. However, in the inland districts it is cultivated under thatched roof. Large numbers of farmers rely on the cultivation of *V. volvacea* as a secondary source of income, making use of their waste paddy straw. The enterprise has assumed the proportion of a cottage industry among the rice farmers in the hot, steamy climate in the East and South-Eastern coastal plain zone of Odisha (Mohapatra *et al.*, 2013).

Fig. 1. Paddy straw mushroom

Odisha produces about 3.05 million tonnes of paddy straw per annum and a major part of it is left out to decompose naturally or burnt *in situ*. Therefore, farmers utilize paddy straw for straw mushroom cultivation owing to its availability at cheaper rates throughout the state. In most cases, cultivation is done in non-pasteurized substrate with incorporation of pulse powder or wheat bran. The acidic medium in hot and humid climate is more prone to competitor moulds and diseases. Therefore, the biological efficiency hovers at 10 per cent which is low. However, there is scope for improvement in biological efficiency by using more productive strains of the species, producing good quality spawn and effective substrate management. Further, higher and more stable yields (30-45 %) could be obtained through adoption of indoor method of farming (Ahlawat and Tewari, 2007).

Besides, pasteurized paddy straw substrate supplemented with cotton seed hulls could give better productivity.

OUTDOOR CULTIVATION OF PADDY STRAW MUSHROOM

The outdoor cultivation method was introduced in the middle of the 1980's. This method is ideal in rural areas where paddy straw is abundant after each paddy harvest. Low cost house for growing paddy straw mushroom has been developed. Besides, outdoor cultivation under trees like coconut, cashew nut, casurina etc. is very much successful during summer (Fig. 2 and 3). However, outdoor cultivation leads to uncertainty in production with irregular and low yields due to difficulties in controlling temperature, relative humidity and insect pest problems. Further, yield deteriorates owing to site contamination out of moulds and mushroom diseases. The traditional outdoor method uses the bed type approach and utilizes a number of agricultural wastes like dried paddy straw, rice stubbles, water hyacinth, banana leaves and stalks (Reyes and Abella, 1997).

Fig. 2. Paddy straw mushroom in coconut plantation

Fig. 2. Paddy straw mushroom in bamboo plantation

Good quality paddy straw bundles are collected, made to 45 cm length by trimming both the ends and soaked in 2 % solution of calcium carbonate for six hours. Bundles are taken out, excess water drained off to 65 % substrate moisture and beds are raised having dimension of 1.5ft x 1.5ft x 1.5ft (length x breadth x height). Requirement of spawn and organic additive has been standardized at 3 % each of the dry weight of the substrate. Three layered raised beds are prepared for spawning and organic supplementation at 1:1:2 proportion. Beds are topped with a thin layer of straw followed by covering with polythene sheet that maintains appropriate temperature for mycelial ramification (30-35°C) and fruit body formation (28-30°C). Polythene sheet is removed after the emergence of pin heads, watering done as and when required and the fruits of the first flush are harvested at 14-15 days followed by providing the polythene cover once again. The second harvest from the second flush is harvested at 21-22 days. One bed thus prepared, may yield approximately one kilogram of fruits with a biological efficiency of 15 % with appropriate aftercare.

MILKY MUSHROOM

Fig. 4. Milky mushroom

About two decades ago, milky mushroom (*Calocybe indica*) was identified as a wild edible mushroom in India (Fig. 4). However, complete commercial production techniques were developed for the first time in Tamil Nadu (Krishnamoorthy, 1995). Milky mushroom is a potentially new species to the world mushroom portfolio. It is a robust, fleshy, milky white, umbrella like mushroom which resembles button mushroom (Pani, 2012). The species is suitable for hot humid climate and can be cultivated indoor in high temperature and high humidity areas. It grows well in the temperature range of 25-35°C and relative humidity at more than 80 %. The cultivation technology is very simple, involves less cost and no special compost is needed for the cultivation.

The cultivation process resembles that of oyster mushroom with the additional process of casing. The mushroom can be harvested at 40 days after spawning with the total crop cycle of 60 days. The milk mushroom has an extended shelf life of 3-5 days compared to other cultivated species, making it more amenable to handling, transportation and storage. Its robust size, sustainable yield, attractive colour, delicacy, long shelf life and lucrative market value attracted the attention of both mushroom consumers and prospective growers. It is a low fat mushroom with good amount of essential amino acids (Ruhul *et al.,* 2010).

OYSTER MUSHROOM

Fig. 5. Oyster mushroom

Oyster mushroom (*Pleurotus* spp.) has species suitable for both temperate and sub-tropical regions (Fig. 5). It is the third largest cultivated mushroom of the world. In Odisha, cultivation is limited to winter months (November-February) and the production stands at 4095 tonnes/annum contributing to 33 per cent of total mushroom production of the state. The mushroom can be cultivated in a wide range of temperature (22-28°C) employing most cellulosic farm wastes. Oyster mushroom can be grown using a variety of containers such as polythene bags, nylon nets, baskets, shelves, trays etc. The spawn run is rapid and the first crop of mushroom can be harvested in about three weeks time. Gray oyster mushroom (*Pleurotus sajor-caju*), white oyster mushroom (*P. florida*), pink oyster mushroom (*P. eous*) and blue oyster mushroom (*Hypsizygous ulmarius*) are the ruling species of the state. Pink oyster is gaining popularity owing to its attractive colour along with good taste and flavour for small scale semi-urban and urban units. The biological efficiency of oyster mushroom is very high (100 per cent) and the shelf life is better (24 hours) than straw mushroom. Production cost is low with little longer cropping cycle (45 days). Further, it is suitable for post-harvest processing. However, the consumer demand is limited in the state.

BUTTON MUSHROOM

Fig. 6. Button mushroom

The button mushroom (*Agaricus bisporus*) is most popular variety of mushroom in the country (Fig. 6). At global level, it ranks first in terms of production. Punjab is the leading state contributing to 60 per cent of the total production of the country. Being a temperate mushroom, production can be taken up year round in controlled environment or seasonally during winter months. Odisha has just entered into commercial production with 110 tonnes/annum and it is likely to grow further in near future.

The commercial cultivation of button mushroom is done on wheat or rice straw which may be supplemented with wheat or rice bran, calcium carbonate (chalk powder), calcium sulphate (gypsum) and chemical fertilizers. All these things are mixed thoroughly and then aerobically fermented. This process is called as composting.

Compost is suitable for the growth of mushroom mycelium and restrict the growth of other microorganisms. After composting, nutrients of the raw material used for making compost becomes readily available for the growth and development of the mushroom mycelium. On the other hand, it does not favour the growth of other microorganisms. Essential requirements for the cultivation of button mushroom are mushroom house, composting yard, ingredients and raw material for compost, mushroom spawn, casing, equipment, accessories and pesticides.

Compost can be prepared employing long method or short method. Fully prepared compost is dark brown in colour, has no trace of ammonia, no unpleasant odour but smells like fresh hay. The desired quantity of spawn 1-1.5 % of the dry straw) is mixed thoroughly with the required quantity of compost which is then filled in racks, trays or bags. The trays or bags are stacked in the growing room at 22-26°C temperature and 80-85 % RH for mycelial growth. Under favourable environmental conditions spawn run is completed in 15-20 days. Spawn run compost is covered with 4-5 cm thick layer of casing soil for induction of fruit bodies. Within 4-5 days mycelium spreads in the casing soil. Thereafter, temperature of the mushroom house is lowered down to 18°C and maintained between 14-18°C during rest of the fruiting period. In ideal conditions, fruits are observed after 15-20 days of casing. Harvesting is done when the cap size is 3-4.5 cm in diameter. The yield in bag cultivation method is estimated to be 15-20 kg/100 kg compost in about 8 weeks time.

ECONOMICS OF MUSHROOM CULTIVATION

Mushroom cultivation is a profitable enterprise. The cost involved in raising one bed of straw mushroom of 1.5ft x 1.5ft x 1.5ft size comes to Rs.100 /- with a production of one kilogram mushroom within a crop cycle of 21 days. The net return is Rs.100/- per bed assuming the market rate at Rs.200/- per kilogram. Likewise, the cost for raising one bag of oyster mushroom is Rs.50/- per bag with a production of 1.5 kilogram mushroom within a crop cycle of 45 days. The net return is Rs.70/- per bag assuming the market rate at Rs.80/- per kilogram. A model small mushroom production unit (300 sq.ft.) with the investment of Rs.30,000/- accommodating 120 beds of paddy straw mushroom per month during summer and rainy season and 225 bags of oyster mushroom per two months during winter season gives an estimated net income of Rs.10,000/- per month.

PROTECTED (INDOOR) CULTIVATION OF MUSHROOM

It is imperative to say that mushrooms can not be grown year after year with full commercial access, unless proper growing conditions are provided and adequate facilities are available for the control of competitor moulds, diseases and insect pests. Possibly, such conditions can be fulfilled in shelf growing, by the construction of properly insulated and ventilated mushroom houses accommodating store room, spawn running room, cropping room as well as packing and preservation room.

Fig. 7. Paddy straw mushroom in attach house

Fig. 8. Paddy straw mushroom in asbestos house

Fig. 9. Paddy straw mushroom in polly house

In Odisha, raising of simplified and low cost thatched mushroom houses are being encouraged for seasonal cultivation of mushrooms round the year with greater precision. The houses are appropriately designed to maintain required temperature and humidity inside, besides having access to ventilation. House of variable sizes may be raised as per the requirement of a grower. However, bigger houses are often difficult to maintain in terms of mushroom crop requirement and therefore, are not desirable. The vertical space in the mushroom house can be utilized effectively by raising three-tiered structures (shelves), mandatory for indoor cultivation. The thatched roof can be supported by a wooden or bamboo frame with insect proof net protecting the sides. During the hot summer months, humidity of mushroom house can be raised by spreading sand on the floor up to a thickness of 4-6 inches, wetting if sufficiently. Experiments have shown that these low cost houses perform better than outdoor cultivation in terms of productivity. Further, infestation of beds out of competitor moulds and incidence of diseases and insect pests are found to be low in low cost houses. Various modifications of the thatched houses are being designed now-a-days in order to make it more permanent and mushroom friendly. Shade net houses and houses having asbestos roof are therefore viable alternatives to the thatched sheds. In order to maintain appropriate temperature and humidity, foggers or micro-sprinklers can be fitted inside so as to make the summer cultivation of paddy straw mushroom more easier and simpler. The houses can be disinfected at regular intervals by applying germicidal sprays for raising healthy crops. By and large, higher productivity is obtained from these growing rooms in comparison to the outdoor units (Fig. 7, Fig. 8 and Fig. 9).

In view of the higher sale price of the produce, off season cultivation or winter cultivation of paddy straw mushroom is gaining popularity now-a-days in the state. Hence, poly house cultivation of paddy straw mushroom is being popularized during winter season wherever growers are interested. It is possible to maintain required temperature (34-35°C) and humidity (80 %) in the polyhouse through provision of foggers or microsprinklers and therefore, satisfactory yields can be obtained from the tropical mushrooms even during winter months. Low cost polyhouses with 200 micron ultraviolet proof polythene sheets erected even on bamboo frame can be raised by the growers for the purpose. Length wise, on the rear wall, a vent with an exhaust fan has to be provided for evacuation of hot air. Poly mushroom growing houses can be of variable dimensions. However, the appropriate dimension for maintaining suitable microclimate and obtaining good yields could be 20ft x 10ft x 9 (length x breadth x height).

INDOOR CULTIVATION USING PARTIALLY COMPOSTED SUBSTRATE

Fig. 10. Paddy straw mushroom with partially composed subatrate

Indoor cultivation of paddy straw mushroom with partially composted substrate can be a boon for mushroom growers in view of its higher productivity (Quimio, 1993) (Fig. 10). This gives 30-45 % yield in comparison to 10 % obtained from outdoor farming. Semi-industrialization of paddy straw mushroom cultivation can be made possible utilizing this technology. The substrate with paddy straw and cotton waste in equal proportions each partially composted for a period of six days with addition of calcium carbonate and rice bran in appropriate amounts. The compost is spread on shelves in the growing room and steam pasteurized at 60°C for six hours. After natural cooling, spawning is done and left covered with polythene sheet. After the spawn run is over, light and intermittent fresh air are introduced into the growing room for induction of fruit bodies. Suitable temperature and humidity are maintained inside the growing room. A quantity of 30-45 kg of fresh mushrooms can be harvested in two flushes within the crop cycle of 21 days per each 100 kg of compost. Hence, this system is far more productive than the outdoor/indoor cultivation. This system has successfully been initiated in Odisha and efforts are on to popularize it among the farmers.

MUSHROOM SPAWN PRODUCTION

In nature, mushroom spores help in survival of a species from one generation to the next. These spores are extremely small and therefore, difficult to handle as seed. Further, spores need time and specific conditions to germinate and during this time the competitor fungi grow faster to exhaust the available substrate. Therefore, a pure culture of the desired mushroom mycelium is first raised on a convenient artificial culture medium and then added to a substrate to give it an advantage. Spawn is a vegetative mycelium from a selected mushroom grown on

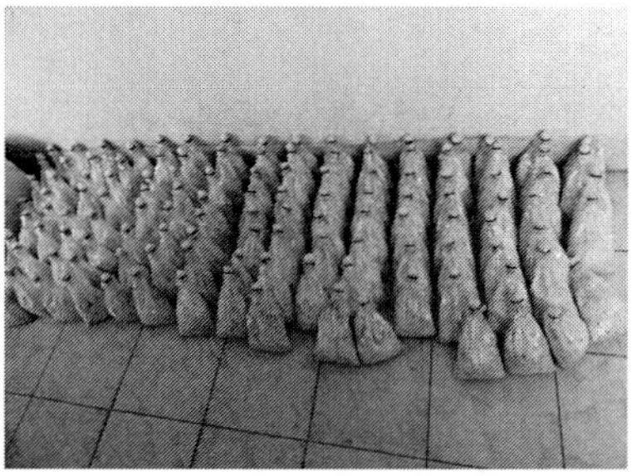

Fig. 11. Mushroom spawn

a convenient medium (Fig. 11). It comprises mycelium of the mushroom and a supporting medium, which provides nutrition to the fungus during its growth. Spawn is used as seed in mushroom cultivation. Appropriate kind and quality of spawn is very important for mushroom cultivation. The isolation, purification and maintenance of mushroom cultures requires technical expertise and aseptic laboratory facilities. It is difficult to maintain pure cultures and/or spawn for small mushroom growers. They have to rely entirely on commercial spawn producers, reliable government and non-government organizations that play a vital role in supplying quality spawn of a desired mushroom strain or variety. At present Odisha is ahead of other states in establishment of commercial spawn production centres. It has been estimated that a quantity of over 7,600 tonnes of spawn is produced per annum from 305 spawn production units present across the state in both private and public sectors in the year of 2021-22. However, mushroom productivity often declines owing to use of poor quality spawn supplied by some spawn producers. In this context, it is imperative to produce quality spawn in order to exploit the yield potential of different mushroom species.

MUSHROOM VALUE ADDITION

Mushrooms are living entities and are governed by a number of factors that lead to post-harvest spoilage and losses. These include infection with micro-organisms, spoilage by enzymes and unfavourable conditions of humidity, temperature, atmosphere during picking/harvesting, transportation, packaging, storage and marketing. Mushrooms contain about 90 % moisture, are highly perishable and cannot be stored for more than 24 hours at the ambient temperature. Once the fruiting body matures and harvested, degradation process starts and it becomes unconsumable after sometime, if not properly handled. Development of brown colour is the first sign of deterioration and is a major factor contributing to quality losses. It is suggested that spoilage might be caused by the action of bacteria on the mushroom tissue and browning of mushroom is due to a combination of auto-enzymatic and microbial action on the tissue. Sound post-harvest practices have since been developed to extend the shelf life of fresh mushrooms.

Harvesting of mushrooms at optimum stage of maturity is of great significance. Harvesting of under matured or over matured fruit bodies results in poor texture, flavor and immediate degradation. White button mushroom are harvested when the cap size reaches 30-45 mm. In general these should be harvested at a stage when the cap diameter is twice the length of stem. Oyster mushrooms are harvested when the fruit body has curled under edges because fully matured fruit body is fragile and difficult to handle. Paddy straw mushrooms should be harvested at button or egg stage. The button stage occurs when the mushroom has begun to differentiate into the cap and stem inside the universal veil which continues until the differentiation is complete and the veil is broken. Due to the fast post-harvest maturity, many changes occur as the caps open and it gets the appearance of flattened umbrella. This is accompanied by loss in weight due to loss of moisture and release of spores. These spores cause pink colouration of fruit bodies and thus causing off-flavour and rendering the produce unmarketable.

Freshly harvested mushrooms are highly perishable and require immediate processing for stable preservation (Fig. 12). Pickling and sun-drying are economically viable methods for preserving mushrooms during the glut period to meet the domestic requirement in India (Fig. 13). However, freezing and freeze-drying gives a product of excellent quality and can be explored commercially for export market.

Fig. 12. Paddy straw mushroom tin cans

Fig. 13. Mushroom pickle

SPENT MUSHROOM SUBSTRATE AS ORGANIC FERTILIZER

Spent mushroom substrate (SMS) is nutrient rich. However, N release from the SMS is very slow. Therefore, plant nutrition may be insufficient. On the other hand, phosphorus, potassium, micronutrient and the growth substances in SMS are present in sufficient quantity. Availability of nutrients from SMS can be enhanced by preparing an organic mineral fertilizer from the spent mushroom substrate.

SMS has the potential of solving several agriculture related problems. However, it has to be suitably processed before using it as a feeding material for vermicomposting, plant disease management, preparation of organic-mineral fertilizer and bio-remediation of contaminated soils besides the purpose of raising horticultural and several other crops.

CONCLUSION

Mushrooms are truly health foods and promising neutraceuticals. Odisha has tremendous potential for mushroom production owing to the availability of agricultural wastes in abundance, manpower and suitable climate. Further, there is increasing demand for quality products both in domestic and export markets. Mushroom being a women friendly crop, could be facilitated well with different Shelf Help Groups (SHGs). To be successful in both domestic and export market, it is essential to produce quality fresh mushrooms and processed products devoid of pesticide residues at competitive rates. It is also important to commercially utilize the spent mushroom substrate left after cultivation for making manure or vermin-compost for additional income and total recycling of agro-wastes. It is worthwhile to mention here that few of our entrepreneurs have got recognition at the national and international levels owing to their excellence in mushroom production and allied activities. This would probably be the appropriate way to search for alternative nutritional sources for our huge population and help achieve non-green revolution.

REFERENCES

Ahlawat OP and Tewari RP (2007) Cultivation technology of paddy straw mushroom (*Volvariella volvacea*). Technical Bulletin, National Research Centre for Mushroom, Chambaghat, Solan (H.P.). p.36.

Ahlawat OP and Vijay B (2004) Effect of casing material fermented with thermophillic fungi on yield of *Agaricus bisporus*. *Indian J. Microbiol.* 44(1): 31-35.

Bano Z and Rajarathnam S (1982) *Pleurotus* mushrooms as a nutritious food. In: S.T. Chang and T.H. Quimio (eds.) Tropical Mushrooms: Their Biological Nature and Cultivation Methods. The Chinese University Press, Hong Kong, 493p.

Beyer M (1996) The impact of the mushroom industry on the environment. *Mushroom News* 44(11): 6-13.

Chang ST (1982) Cultivation of *Volvariella* mushrooms in South East Asia. In: Tropical Mushrooms: Their Biological Nature and Cultivation Methods (Chang and Quimio eds.). The Chinese University Press, Hong Kong, 493p.

Chang ST (2006) The world mushroom industry: Trends and technological development. *Int. J. Med. Mush.* 8(4): 297-314.

Chiu SW and Moore D (2001) Threats to biodiversity caused by traditional mushroom cultivation in China. In: Fungal conservation (Moore, Nauta and Rotheroe eds.), The 21st Century Issue, Cambridge, p 111-119.

Chong C, Cline RA and Rinker DL (1987) Spent mushroom compost and paper mill sludge as soil amendments for containerized nursery crops. Combined proceedings. *International Plant Propagators Society* 37:347-353.

Food and Agriculture Organization (2023) FAOSTAT Statistics database 2023. Food and Agriculture Organization of the United Nations.

Fox R and Chorover J (1999) Seeking a cash crop in spent substrate. http:/aginfo, psu. edu./pea/sgg/mushrooms, html.

Hamlyn PF (1989) Cultivation of edible mushroom as cotton waste, *The Mycologist* 3 (4): 171-173.

Kaul TN (1983) Cultivated Edible Mushrooms. Regional Research Laboratory, Council of Scientific and Industrial Research, Srinagar.

Krishnamoorthy AS (1995) Studies on the cultivation of Milky mushroom, *Calocybe indica* P. and C., Ph.D. thesis, Tamil Nadu Agricultural University, Coimbatore, India, p.124.

Lal R (2005) World crop residue production and implications of its use as a biofuel. *Environment Int.* 31(4): 575-584.

Li GSF and Chang ST (1982) Nutritive value of *Volvariella volvacea.* In: S.T. Chang and T.H. Quimio (eds.) Tropical Mushrooms: Their Biological Nature and Cultivation Methods. The Chinese University Press, Hongkong. 493p.

Li Yu (2012) Present development situation and tendency of edible mushrooms industry in China. *Mushroom Science*, 18:3-9.

Mohapatra KB, Chinara N and Behera B (2013) Low cost cultivation techniques for *Volvariella volvacea* grown in Odisha. Proceedings of National Symposium on 'Mushrooms for Medicinal Value and Nutritional Security under Changing Climatic Conditions', 27-28 December, 2013, UHF, Nuani, Solan (HP), pp.10.

Pani BK (2012) Sporophore production of milky mushroom (*Calocybe indica*) as influenced by depth and time of casing. *Int. J. Advanced Bot. Res. (IJABR)* 2(1): 168-170.

Prakasam V (2012) Mundkur Memorial Lecture: Current Scenario of Mushroom Research in India. *Indian Phytopath.* 65(1): 1-11.

Quimio TH (1993) Indoor cultivation of straw mushroom *Volvariella volvacea. Mushroom Research* 2(2): 87-90.

Rai RD and Sohi HS (1988) How protein rich are mushrooms. *Indian Horticulture*, 33:2-3.

Rajarathnam S, Shashirekha MN, Zakia Bano and Ghosh PK (1997) Renewable lingo-cellulosic wastes, the growth substrates for mushroom production: National strategies. In: Advances in Mushroom Biology and Production (R.D.

Rai, B.L. Dhar and R. N. Verma., eds.), Mushroom Society of India, NRCM, Solan, India, p. 291-304.

Reyes RG and Abella EA (1997) Mycelial and basidiocarp performance of *Pleurotus sajor-caju* on the mushroom spent of *Volvariella volvacea.* Proceedings of International seminar on the "Development of Agri-business and its impact on Agricultural Production in South East Asia, Tokyo NODAI Press, pp.491-497.

Ruhul A, Abul K Nuhu A and Tae SL (2010) Effect of different substrates and casing materials on the growth and yield of *Calocybe indica. Mycobiol.* 38(2): 97-101.

Sharma VP, Annepu SK, Gautam Y, Singh M and Kamal S (2017) Status of mushroom production in India. *Mushroom Research* 26(2):111-120.

Singh, M. (2011) Mushroom production: An agribusiness Activity. In: Mushroom Cultivation, Marketing and Consumption (eds. Singh, M., Vijay, B., Kamal, S. and Wakchaure, G.C.), Directorate of Mushroom Research, ICAR, Solan (HP), pp.1-10.

Szmidt RAK (1994) Recycling of spent mushroom substrates by aerobic composting to produce novel horticultural substrates. *Compost Science and Utilization* 2(3): 63-72.

Thakur MP and Mohapatra KB (2013) Tropical Mushrooms: Present Status, Constraints and Success Story. Proceedings of Indian Mushroom Conference, 16-17 April, 2013. PAU, Ludhiana (Punjab), pp.42-43.

Verma RN (2013) Indian Mushroom Industry-Past and Present. Bulletin 8 of World Society for Mushroom Biology and Mushroom Products. P.16.

Wuest PJ and Fahy, H.K. (1991) Spent mushroom compost traits and uses. *Mushroom News* 39(12):9-15.

Zhang YHP (2008) Reviving the carbohydrate economy via multi-product lignocelluloses bioefficiencies. *J. Ind. Microbial. Biotechnol.* 367-375.

Sustainable Agriculture and Biodiversity: Innovations and Insights, Pages: 279–311
Edited by: Bibhuti Bhusan Sahoo, Harvinder Kumar Singh, Niranjan Chinara and Amitabh Panda

CHAPTER - 12

FISHERIES OF ASIAN BRACKISH WATER LAGOON: STEP TOWARDS SUSTAINABLE DEVELOPMENT DURING PRE AND POST-RESTORATION PHASES

Saibala Parida[1*], Rabindra Nath Samal[2] and Amlan Nayak[2]

[1]*Wetland Research & Training Centre, Barkul-752030, Odisha*
[2]*Chilika Development Authority, Palashpalli, Bhubaneswar-751020, Odisha*
**Corresponding email: saibalparida@yahoo.com*

INTRODUCTION

The fauna of the Asia's largest brackish water lake: Chilika (Mangla, 1989; Dujovny, 2009 and Mohanty & Otta, 2008) was comprehensively studied for the first time by the Zoological Survey of India during 1914-15 which brought out the richness and variety of the fauna and the very remarkable ecological features associated with it. The Importance of lake as a science is well recognized, (Wetzel, 1975) has defined wetlands are among the most productive ecosystems in the world. The functions of wetlands include flood control, aquifer recharge, nutrient absorption, and sediment retention and erosion control. The wetlands provide plant products, animal products, mineral products, energy sources, recreation and water treatment. The attributes include biological diversity, landscape qualities, cultural or spiritual place, evidence of evolution, and wilderness. Immense varieties of species of microbes, plants, insects, amphibians, reptiles, birds, arthro jodda (crusherans) decapode mammals (dolpums) fish and other wildlife depend on wetlands.

Coastal lakes are transitional littoral zones between the sea and mainland brought about by the action of complex physical processes. This is an area of salt and brackish water separated from the adjacent sea by a low-lying sand/shingle barrier. Many large expenses of relatively protected shallow marine waters also differently called as lakes. Following this definition, many water bodies near to the coast and connected with it either at present or in the recent past, come under the category of lagoons/lakes. Because of their hydrodynamic and sediment character, these environments are the unique wetlands. The interplay of energy and sediment operate in a shallow body of water, which is partly enclosed by a barrier spit and communication is made with the sea through one or more inlets in barrier spit (Phleger, 1981). Sedimentary processes like erosion/accretion and other geophysical processes create distinct morphological and sedimentary features.

Location

Chilika is the largest and vibrant lake along the east coast of India. It extends from south west corner of Puri and Khurda districts to adjoining Ganjam district of Odisha state (Fig.1). It is the first Indian Ramsar Site It is designated as an important Ramsar site (No.229) of India on 1.10.1981 and International Importance lying on the east coast of India (Latitude 19°28' and 19°54' North and Longitude 85°05' and 85°38' East). The lake is dynamic assemblage of shallow to very shallow marine, brackish and freshwater habitats / environments (Parida et. al, 2013). The lake's water area fluctuates from minimum of 906 km^2 in summer to maximum of 1165 km^2 in monsoon. The total catchment area of the lake 3560 km^2 (Ghosh & Pattanaik 2005) . The presence of unique salinity gradient enables the eco-system to support a wide range of biodiversity and livelihoods of more

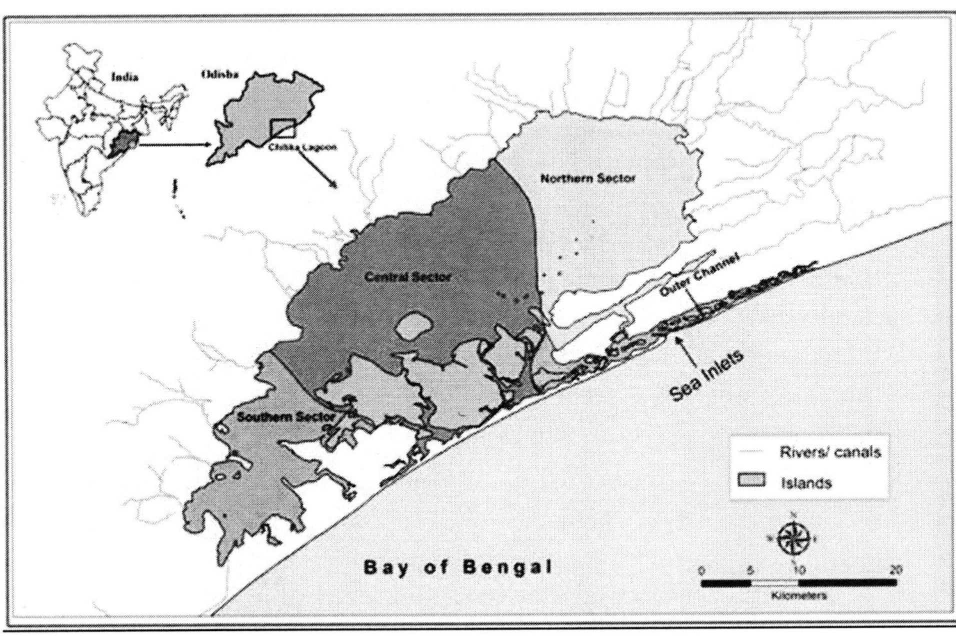

Fig. 1. Location map of Chilika lake

than 0.2 million dependent communities through its ecosystem services. The aesthetic value of lake is unique and the eco-tourism spots added substantially to the total economic value of the ecosystem (Parida & Bhatta, 2023).

It is one of the hot spot of biodiversity in the country, and some rare, vulnerable and endangered species listed in the IUCN Red List of threatened animals inhabit the lake area for at least part of their life cycle. This list includes a number of rare, threatened and endangered species such as Irrawady dolphins (*Orcaella brevirostris*) and a rare limbless skink (*Barakudia insularis*) endemic to Badakuda island of Chilika was described by Nelson & Annandale in 1915 and rediscovered in the wild in 2003. The lake ecologically divided into four sectors, southern sector, central sector, northern sector and outer channel based on the variation of ecological features (Fig.2). Basically, the northern sector is fresh water dominated zone and central sector is a brackish water zone. The southern sector is a higher saline are exhibit with marine habitat. The outer channel is marine in nature with saline water but during monsoon, the water becomes fresh water due to discharge of flood water to the sea through the sector (World Bank, 2005).

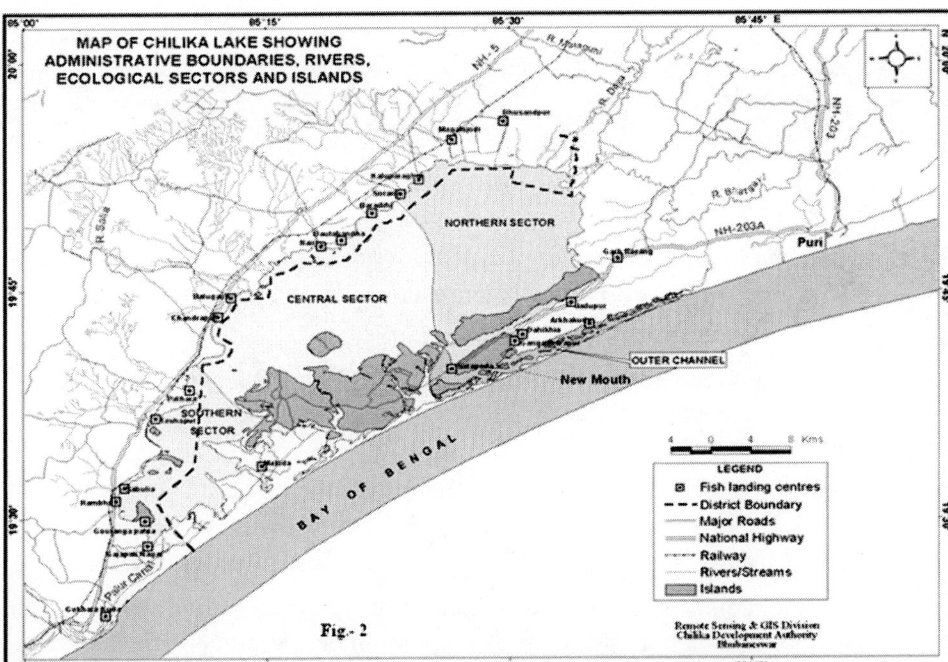

Fig. 2. Sectoral map of Chilika lake

HYDROLOGY AND ENVIRONMENTAL PARAMETERS

Hydrology

i) The freshwater inflows from different sources into Chilika lake are of considerable importance for the lake fisheries. The changes in fish diversity and catch are very sensitive to the salinity and hydrologic dynamics of the lake, which in turn are strongly influenced by freshwater inflows.

ii) Catchment area treatment (restoration of catchment ecology) assumes greater importance with a view to arresting further degradation and sediment inflow into the lake and improving the livelihood of catchment communities.

iii) Preferential operating rule for the new Naraj Barrage is of vital importance to maintain desired hydrologic regime, ecological health and productive fishery of Chilika lake.

Environmental Parameters

i) Salinity, temperature and currents in the lake are influenced by the flood waters of 52 rivers and rivulets in catchments streams during monsoon and post-monsoon seasons.

ii) Northern and central sectors contain almost freshwater in the flood season and so too the outer channel. The salinity of the lake starts rising from January and reaches high values in May or June. The southern sector shows comparatively less fluctuations and appears less affected by floods and seawater influx.

iii) Low transparency value observed in peak summer due to agitation of shallow lake water by wind action and in monsoon due to silt – laden flood water inflow into lake.

The lake water is primarily 'mixohaline' except during peak monsoon period (August – October) when it becomes almost 'limnetic' in northern, central and outer channel sectors.

iv) Lake salinity profile showed cyclical changes in the range 0.34ppt during pre-intervention phase and the same cyclic changes showing unimodal seasonal oscillation in the range 0.10 – 36.9 ppt during post-intervention period.

v) Lake water remained alkaline throughout the year and did not show marked seasonal variations both during pre and post -intervention period. Fluctuations in pH are primarily regulated by heavy freshwater inflows and sea water influxes. Comparatively low pH in rainy season and high values in winter season are observed.

vi) Higher dissolved oxygen values generally recorded in northern sector during high phytoplankton production period in winter and during fresh water inflow period in rains. After opening of the new lake mouth, higher values were recorded during post-monsoon followed by monsoon and summer.

Eco-degradation of the Lake

The unique and fragile ecosystem of Chilika lake with estuarine character and pristine glory of rich fishery resources gradually started losing its ecological characters due to changing coastal process, degraded drainage basin, natural changes in the physiographic and anthropogenic pressure during the last few decades. Being a coastal wetland, Chilika lake is quite sensitive to the changes in the hydrological regimes. During the last few decades, the lake gradually moved towards freshwater ecosystem due to significant decrease in salinity regime. Several natural changes and anthropogenic activities contributing to the eco-degradation of the lake are described as follow. The lake was facing the problems like siltation, shrinkage of area, chocking of the inlet as well as the outer channel connecting the sea, decrease in salinity, weed infestation, decrease in fish productivity, increase in aquaculture and an overall loss of biodiversity before undertaking the hydrological interventions.

Siltation

Siltation of Chilika lake is attributed as one of the principal factor endangering the lake. There are 52 numbers of rivers and rivulets draining into Chilika. High volume of sediment is pumped into the lake by the distributaries of Mahanadi on the northern sector and by other rivers and rivulets on western side. The catchment of the lake (excluding the Mahanadi drainage) is estimated to be 3987.4 square kilometres out of which 37 % is arable land, 32 % is fallow land, 26% is forest land and 5% is wasteland. Chilika Development Authority (CDA) have sponsored a study by the Orissa Remote Sensing Application Centre (ORSA) for preparing a catchment management plan based on the latest satellite data from IRS 1C & 1D, to prioritise the most erosion prone areas, accordingly catchment treatment is being undertaken in a phased manner.

Choking of the Inlet and Outer Channel

Various studies indicate that about 1 million CuM of drift is prevailing in this region of the coast. This leads to shifting of the mouth in north-eastern direction around 500 m per year. The cross section of the outer channel is also significantly reduced due to shoal formation, along the outer channel, which leads to considerable hydraulic head loss and poor flushing action. The recent bathymetry of the Magarmukh (the gate-way between lake and outer channel) indicated that the summer depth is alarmingly low i.e., only 0.3 meters. This is preventing the discharge of fresh water and sediment during monsoon to sea through Outer channel thereby causing flood in the northern sector. Similarly the Nalabana Bird Sanctuary is also remaining submerged during October causing management problems of the Bird Sanctuary. At the same time the tidal ingress into the lake is also reduced. This is also obstructing the prospect of auto-recruitment of the fish and other economic species from sea to the lake thereby reducing the prospect of capture fishery. The breeding migration from sea to lake and vise versa is also adversely affected due to this obstruction. To mitigate this problem, the Central Water and Power Research Station (CWPRS), Pune have recommended to develop a lead channel near Magarmukh to establish proper link between outer channel and the main lake based on their mathematical model studies.

Shrinkage of Water-spread Area

There was shrinkage of 34 sq.km. water spread area of Chilika as computed from Survey of India Topo-sheet (1972-1973) and from the Land Sat-5, data dt. 25.5.1986. The Orissa Remote Sensing Agency was carried out a detail Land-use/Land cover and geo-morphological mapping of Chilika lake catchment by use of IRS-IC, ID Satellite data, sponsored by Chilika Development Authority.

Weed Infestation

Analysis of satellite data revealed that during the years 1973, 1977, 1985 and 1993 the weed covered areas in the lake were 20,60, 200 and 398 sq.kms, respectively. The weeds are emergent floating and submerged type. The predominant weeds encountered are *Potamogeton pectinatus* in the central sector and fringes of southern sector, *Schoenoplectus littoralis* and *Phragmitis karka*, the emergent weed in northern sector. *Gracillaria verrucosa* a red algae is dominant in southern and central sectors of Chilika lake.

Decrease in Fish-landing

The species composition of fish in the lake is complex in the sense that it includes fresh water, brackish, marine, catadramous and anadromous fish species. A large percentage of the commercial fish species are migratory (either catodromous or anadromous) Jhingran (1963) estimated that 63-75% of the annual fish production was contributed by migratory species. This percentage was reduced to 30% due to relative increase in fresh water species. A large percentage of the fish species feed on plant materials and detritus, because of these trophic characteristics, the lake is considered as highly productive from fishery point of view. As per calculation annual fish production from Chilika on an average was about 90 kg/ha during the 1980's (Range-65-122 kg/ha). The poaching of juvenile for seeding the shrimp ponds and firms is also adversely affecting the fishery resource.

The total fish landing of the lake was 6,000 ton an average during last one decade but it is gradually declining, the fish landing of 1997-98 is 1641.5 t. This decline can be attributed to composite factors like over fishing beyond carrying capacity, obstruction of migratory routes affecting auto-recruitment. The Orissa Marine Fisheries Regulation Act (OMFA), prevents operation of off-shore vessels within 5 kms of coast to facilitate the fish recruitment into Chilika as the spawning zone concentrate within 5 kms of restricted fishing zone. The decline is also attributed due to decrease in salinity level of the lake. Due to chocking of the outer channel at the Magarmukh, the migration of *Hilsha* sp. was also adversely affected. Prawn gherry is gradually increasing along the periphery and outer channel of Chilika. The potential environmental impacts that may result from the prawn gherry can be it's effect on land-use, aesthetics, habitat reduction, water quality, construction impact, reduction of wild populations because of their capture for brood stock. The shallow bathymetry makes virtually a large part of the lake suitable for aquaculture. As littoral zone of the lake provide feeding and spawning ground for both resident and migratory fish species, the area when encroached for prawn gheries adversely affecting the productivity.

The Zoological Survey of India (ZSI) after completing the Chilika expedition (1985-1987) opined that the lake ecosystem was tending towards a freshwater ecosystem and warrants urgent restoration measures. Owing to the degraded state of the lake's ecosystem with drastic changes in the ecological character imbalance and overall loss of biodiversity, the Chilika lake was includes in the Montreux Record (threatened list of Ramsar sites) in 1993.

Eco-restoration

From the 1950's to 2000, the lake was in a continual decline, with increase in sediment loads and decreasing salinity. Fishery showed the major decline, invasive weed began to take hold, and the entire lake progressively shrunk in area and volume. The eco-degradation reached a critical stage with loss of ecological characters, change in ecosystem function and overall loss of biodiversity which brought miseries to about 0.2 million local people directly or indirectly depending on the lake's resource (goods) and services. Being a bonafied signatory to Ramsar convention and Chilika being a listed Ramsar site, Governemnt of India took urgent measures to restore Chilika lake through State Government and provided the necessary funds from the tenth and eleventh finance commission. The State Government constituted an authority in the year 1992 called Chilika Development Authority who implemented the eco-restoration programmes in Chilika lake during 2000-2001 and 2001-2002 following ecosystem approach. The restoration programme included five major activity components as under:

i) Hydrological Intervention by opening of new lake mouth.
ii) Dredging of lead channel at Magarmukh, its extension to Daya, Bhargavi and Nuna river confluence area and Magarmukh-Balugaon channel.
iii) Palur canal renovation
iv) Preferential Environmental flow through operation of new Naraj Barrage
v) Catchment treatment

HYDROLOGICAL INTERVENTIONS

For restoration of this unique wetland Chilika Development Authority (CDA) was commissioned the service of the premier institutes of the country for a clear understanding of the root cause of degradation of this complex ecosystem. The National Institute of Oceanography (NIO, Goa) carried out a detail study of the wave climate of the inlet, long shore sediment transport along the coast and the bathymetry of the lead channel for a better understanding of the coastal processes. Chilika Development Authority carried out an extensive hydro-biological study in collaboration with the Wetlands International South Asia (WISA), New Delhi to understand the changes in the hydrological regimes, water quality and biota in reference to the changes in the flow pattern into the lake from the drainage basin. The study revealed that the flow from the western catchments was also changing due to the change of the land use practices resulting in heavy inflow of sediment into the lake.

The root cause of the degradation of the lake ecosystem was identified to be due to the alteration of the hydrological regime changes in the land use pattern in the catchments and unsustainable exploitation of wetland resources. The hydrological alterations leading to the transformation of the lake towards a fresh water ecosystem, was considered as a main threat to the flora and fauna of this unique ecosystem. This was broadly attributed due to the change in the flow pattern from the catchments and the changes in the coastal processes. The Central Water and Power Research Station (CWPRS), Pune carried out the hydrological and three dimensional numerical model studies based on the data generated from the above studies, which served as the boundary condition. The studies concluded that the tidal influx into the lake was considerably reduced because of the shoal formation along the outer channel and continuous shifting of the mouth, which was resulting in significant hydraulic head loss. They further concluded that the salinity flux and tidal flux into the lake would not improve unless the location of the opening of the inlet is brought closer to the lake to an appropriate location Fig. 3.

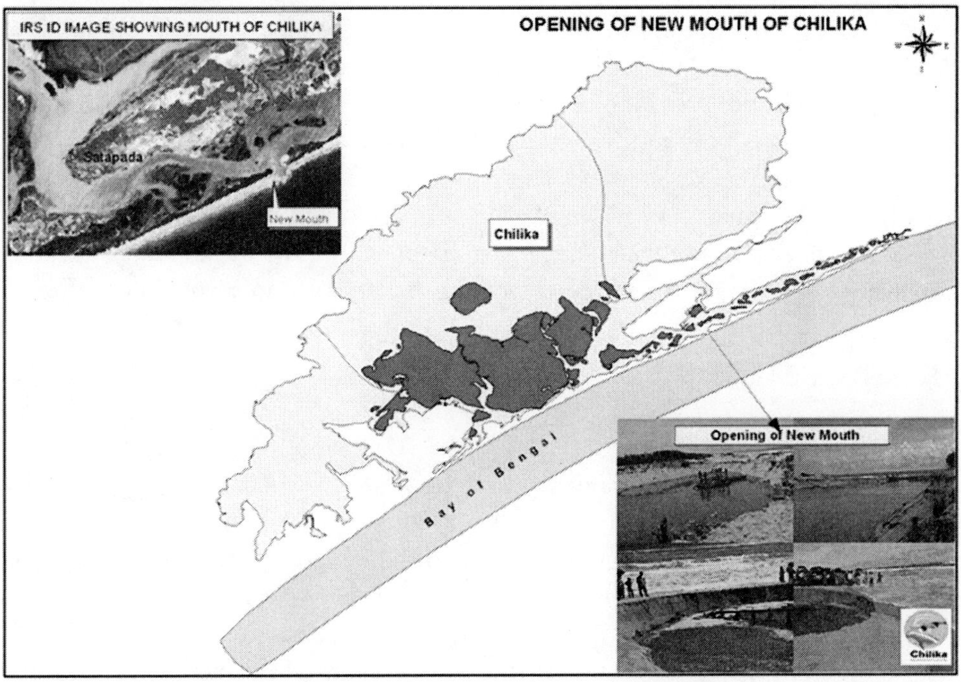

Fig. 3. Opening of New mouth of Chilika

After thorough investigation in the coastal process and river basin functioning by commissioning the services of CWPRS, Pune; IIT, Chennai; NIO, Goa and on the basis of their recommendations, Chilika Development Authority (CDA) successfully carried out the classic hydrological intervention by opening a new mouth at a strategic point in outer channel during September, 2000 based on the three dimensional numerical model studies undertaken by CWPRS, Pune. The mouth was initially 100 m wide 2.5 m below the lowest lake water level.

Dredging of lead channel at Magarmukh, its extension to Daya, Bhargavi and Nuna river confluence area and excavation of Magarmukh-Balugaon channel.

The other hydrological works were excavation of a 3.2 km lead channel at Magarmukh, excavation of 22.6 Kms long lead channel from Magarmukh towards the river confluence points of river Daya, Bhargavi and Nuna and excavation of 3.4 kms long Magarmukh-Balugaon lead channel.

Palur Canal Renovation

Palur canal (16 km) connects the Palur Bay of Chilika lake in the southern sector to the Rushikulya estuary. Due to heavy siltation in Palur canal the exchange of water between the sea and the southern sector of the lake was not taking place. Palur canal plays a vital role in maintaining salinity gradient and the auto recruitment of economic species from the sea to enrich the fishery resources of the southern sector. The canal was renovated by CDA with an objective to enhance the fishery resources of southern sector. It became functional from the year 2003. This renovated canal is providing effective recruitment and migration route for fish and shellfish to southern sector. After renovation of the Palur canal, there has been a significant improvement of the fishery resources of this sector.

FISHING METHODS

Fishing Craft

ICAR-Central Institute of Fisheries Technology (CIFT), Cochin studied extensively the existing fishing craft, gears & fishing methods of Chilika lake and identified the destructive fishing gears with a view to reducing by-catch for conservation of the resources during 2011 (Remesan *et al.,* 2011). Wooden plank-built Chilika type boats called *Danga* or *Naha,* of length 9-15 m, maximum width 1.8-2.1m with 0.45-0.6m depth fitted with 6-9 hp OBM engines are the common fishing crafts in the lake today. Small size flat bottom wooden boats are also used by refugee fishers mainly for hook & line fishing. The commonly used timber for construction of wooden fishing boats. Coaltar is used for preservation of timbers. Most of the boats have front and middle deck made of wood or split bamboo. Two bamboo poles of about 7.5 m each are used for hoisting the polythene sheet made sails of size about 5.5m×4.5m. . The fishing boats in Chilika Lake are used by three categories of stakeholder's namely individual fisher, refugee fisher and commission agents who collect catch from individual fishers and transport to landing centre for disposal. Taking into account all the three categories of fishery boats in Chilika the overall length ranged from 4.0-16.79 m. The minimum, maximum and average size of boats under three categories were as under: (Parida *et al.,* 2019).

Traditional Methods of Fishing

The local fisher folk evolved a complex system of partitioning the fisheries of Chilika lake amongst themselves. The local fishermen large treat the lake waters as common property resource where access rights are managed by themselves both in terms of spatial and temporal contexts (Sekhar N. U., 2004). Each fishing group's excess was determined on the basis of the species they catch. Such regulations among the fishers community had worked well for the long term benefits of the different fishing group. The fishing gears of Chilika lake have been variously described and discussed by many authors (Mitra, 1946; Devasundaram, 1951a & 1954; Jones & Sujansingani, 1952 & 1954; Job & Pantulu, 1953; Mohapatra, 1955; Mitra & Mohapatra, 1957 ; Jhingran & Natarajan, 1966 & 1969 and Parida et al 2019). Fishing in the lake was done through different types of traditional methods such as (i) Bahani (Net fishing), (ii) Janos (large impoundments constructed with split bamboos) in shallow regions, (iii) Dian sources, (iv) Uthapani fisheries and (v) Prawn khati (Trap fishing). Net fishing goes on throughout the year, while 'Jano and Trap' fishing was only seasonal. Trap Fisheries also contribute reasonably to the total fish production of the lake. Since time immemorial, five types of traditional fishery sources were operating in the lake. These are as follows.

Bahani (Net fishing)

Net fishing operations in Chilika are locally called "Bahani" which are carried out in open Chilika, locally known as deeper Chilika. There are different ***Bahani*** fishing grounds known as ***Bahani*** fishery sources (Leased out fisheries). During 1960s the ***Bahani*** fishery contributed about 50-66% of annual landings. Generally, dragnets and gillnets of different mesh specifications and in different local names are operated in the main water body of the lake with the help of plank-built flat bottom boats. 'Bahani' fishing is operated round the year and the fishing grounds are leased out fishery sources. *Noli jal* was more selective in respect of mullets and Asian seabass threadbin, Gizzard shad, Bengal corvine, Grey mullet Thryssa nets were more selective in respect of fish after which they were named. *Khepa jal* was more effective for Balangi, Kabala and Sahala, in addition to other fishes. Among nets *Khadi jal* was very effective for during prawn fishery season. Bahani contributes about 50% to the annual landing. This method of fishing is continuing at present.

Jano (Enclosure fishing)

Jano fisheries of Chilika is a large enclosures of split bamboo screens, which enclose fish in shallow water areas of lake. When the mullet juveniles during recruitment in the lake, groups of fish juveniles moved in large soals for feeding purpose in the lake, fishers enclosed the shallow areas in the night. There are three types of Janos operated in Chilika, one is circular direction in open Chilika and other two are semicircular on the shoreline and U shaped connecting to the land mass. These Jano are used for 5 to 6 months (September to January) and removed after harvesting. ***Jano*** fishery contributed 13-22% of the total lake catches and though wide varieties of fishes are caught from ***Janos,*** mullets particularly *Mugil cephalus* and *Chelon macrolepis* accounted for more than 60% of the ***Jano*** catches. In 1960s a total of 112 ***Janos*** were operating in the lake as seasonal leased out fishery operation (Jhingran & Natarajan, 1969). During sixties more than 112 numbers of Jano (134.76 km^2) were operated in Chilika which was reduced to 83 numbers by 1993. At present it decreased to less than 10 numbers. Most of these Janos were converted to Prawn Gherries in Chilika during the study period. Janos were the main mode of mullet fishing, which are contributing about substial amount to the annual landings. The mullet catch started decline after loss of Jano fishey grounds from 1995 onwards mohanty *et al.* (2004)

Dian Sources

This is a minor traditional fishery operated in shallow land side areas of the margine Jano grounds where fishes from Jano enclosures jump out to the shallow waters outside the split bamboo screen wall. Fishermen catch them in that area which is known as Dian sources. Generally mullets and other fishes are caught in Dian sources. These traditional fisheries are non-existent today being fully occupied by Ghery operators.

Uthapani Fisheries

Fishing is carried out in the shallow areas adjacent to shore lands and around islands during the first rains. Catfishes constitute the bulk amount to Uthapani fishery. These fisheries grounds are mostly converted to illegal Ghery culture activities. Both 'Dian' and 'Uthapani' contributes minor share to the annual landing. During the first rains, when the rainwater from upland flows into the lake, fishes like *Mystus gulio* and other catfishes move landward against water flows in large numbers when they are caught by check bunds with openings. This type of fishing is only seen in island villages of Parikud area. At present most of these type of fishery area converted to prawn Gheries. The fisheries are no more in operation to day and the fishing grounds are mostly, converted to illegal Ghery culture activities. Both 'Dian' and 'Uthapani' fisheries form about 6% of the annual landing.

Prawn Khati (trap fishing)

Traditional 'prawn khatis' or prawn fisheries are leased out fisheries operated at specified areas for a particular period (season). Prawn fishing is carried out by split bamboo traps set with leader line of split bamboo screen having specified length. The bigger traps are known as Dhaudi and the smaller traps are known as Bajas. All penaeid prawns (*Penaeus monodon, Penaeus indicus, Metapenaeus monoceros, Metapenaeus dobsoni, Penaeus semisulcatus* etc.) are generally caught in Dhaudi and smaller prawns (*Metapenaeus sp.* and freshwater prawns) are caught in Baja. Prawn fishery with traps operates during March-August which contributes good amount of prawn to the annual landing. Presently, most of the designated traditional prawn fisheries (fishing grounds) are either converted to 'Prawn gheries' or 'Khanda' fishing which are operated throughout the year. Both Ghery and Khanda fishing are adversely affecting the recruitment routes and the ecology of the lake affecting the sustainable management of fisheries resources in the lake.

PRESENT METHOD OF FISHING

Khanda (Monofilament twine net box)

This is trap net operated intensively throughout the year. Juveniles of fish, prawn and crabs are caught in 'Khandas' since the mesh size of net boxes are also small and are operated throughout the year. Prawn and crab juveniles are caught in large number, which adversely affect their population. Mesh size regulation of 'net boxes' (PUDA) of Khanda, length and spacing of leader lines and their concentration particularly along the recruitment routes i.e, along outer channel, magarmukh, dredge channel and in Palur canal need to be regulated.

Gill Nets and Trammel Nets (Dubi jaal)

Gill nets are passive fishing gears in the lake, mostly made of nylon monophilament (PA) and mainly operated for fin fishes like mullets, seabass, threadfins, catfishes, clupeiforms and sciaenids (Remesan *et al.,* 2011). Surface gill net for sea bass is made of PA multi philament of 210D×8×3 twine specifications with 150mm mesh size. It is operated as surface set net and fishermen drive the fish into the net by disturbing the water. The non-selective gear trammel net (**Dubi jaal**) or a three-walled gill net is operated in the lake to catch fin fishes and prawns. The net is encircled from the boat and the water inside the circle is disturbed using bamboo poles to drive the fish towards the net where they get entangled in the loose netting. This type of non-selective fishing gear needs to be regulated in spatio-temporal scale to reduce threat for biodiversity and killing of brood stock mullets during breeding migration and also to reduce juvenile by-catch (Parida *et al.,* 2019).

Small-meshed Patua Net

The operation of this dragnet is detrimental to the species like *Sahala (Elutheronema tetradactylum)*, *Kundala (Etroplus suratensis) and Boraga(Daysciaena albida)* etc. as the juvenile of above species are caught before they contribute to the lake fishery.

Alimi Net

This giant size boat seine net (500m-800m in length) made of filament nets are operated near the lake new mouth along the outer channel to cover almost the entire width of the channel and catch large shoals of immigrating fishes from the sea into the lake (Hilsa, Threadfins, Mullets etc) as well as the brood stock mullets migrating seaward in large numbers during September-January. During study period 7 units of Alimi jaal were operated by the local villagers near the new mouth along the outer channel round the year. The ***Alimi jaal*** is thus considered as one of the most destructive fishing method in Chilika lake which prevents the immigration of adult fishes into the lake and seaward migrating catadromous fishes. This has become a serious threat to the lake fisheries and rare Irrawady Dolphins of Chilika lake. This net is operated in the outer channel during September/October to December/January before the commencement of south wind.

Boat Seine

As described by Remesan *et al.* (2011), boat seine is used for catching schooling fishes like Patua (*Stolephorous* and *Thryssa* sp.) and Gania (*Strongylura strongylura*). Ocasionally Hilsa and mullets are also caught in this lake. It is large net made of nylon monophilament with 300–500m length, about 1.7m depth 14mm mesh size. The net is operated using motorized boats and the net is released after locating a fish shoal and the boat returns to the same place after encircling the shoal. One end of the net is attached to the boat and the net is hauled by pooling the other end of the net from another motorized boat.

Fishing Gears

The fishing gears of Chilika lake have been described by many authors. Classification of Chilika lake fishing nets given by Jhingaran and Natarajan (1969) has undergone dramatic changes. Cotton and hemp gear materials are completely replaced by synthetic filament twine and monofilaments of different diameters. Many old local names of fishing nets have been changed to new names basing on the diameters of filaments and mesh sizes. Many old traditional nets like Manijal', Dostijal' 'Orajal' etc. are not in operation today. However, all the nets are classified into dragnets, gill nets and cast nets. Other fishing gears are 'Traps' Poluha, Push net, Hook and line, Rod and line, crab lift nets etc.

Fixed Bag Net

The fixed bag net locally known as Behendi jaal was new introduction during 2004-05. This net is operated towards mouth of the lake where the current is strong. The net is made of HDPE twine and it target prawns and small fishes. Mesh size of the net in the mouth region is 30mm-40mm and decreases to 13-10mm in the cod end. The net is fix at 3-4m depth using heavy anchors at both ends of the wings and the upper edge of the mouth is kept open using two bamboo poles. The net is usually set at night and hauled at early morning. Similar type of bag nets used in the tidal creeks of Hooghly-Matlah Estuary in West Bengal, locally called as *Been* or *Bhunti* jaal (Remesan *et al.,* 2011). The operation of this type of fixed bag net during the high tide in the outer channel is harmful as the net captures large quantity of juveniles entering into the lake system: hence the bag net is categorized as a destructive fishing gear in Chilika lake.

Hook and line fishing

Hook and line fishing is much popular in outer channel between Magarmukh and Gabakund. Mostly non-motorized boats with two fishermen are used for line fishing mostly targeting the predatory fishes like thread fins, catfish, seabreem, perches, croakers etc. Nylon monophilament with 0.6mm dia is usually used as line along with Limerick hook of size ranging from 2-5 (Remesan *et al.,* 2011). The Bangladeshi refugee fishers with small non-motorized boats are fishing with hook and lines in Central and Southern sector of the lake. Some refugee fishers catch murrels from Northern sector using hook and line.

Importance of Chilika Fishery

The fish fauna of the lake was described by Chaudhuri (1916a, 1916b, 1917 & 1923) and Hora (1923) and that of decapod crustacea (prawns & crabs) by Kemp (1915). While fish fauna of the lake received early attention, detailed studies on the fisheries and fish catch statistics of the lake were not taken up until 1948 (Jones & Sujansingani, 1954) & (Jhingran & Natarajan, 1969). The necessity of details investigation of the lake fisheries was indicated by Mitra (1946). Efforts towards the studies on the fisheries of the lake were made by Devasundaram (1954) and Jones & Sujansingani (1954), though they were of a rather preliminary nature. In those days the very low local demand for Chilika fish resulted in the export of almost the entire catches in fresh condition to the neighboring states, particularly the Howrah fresh fish market in West Bengal. The existence of only a single channel and route of export (Rail route) facilitated the export of Chilika fish (Jones & Sujansingani, 1954). The average fisheries output from the lake were estimated at 3837 t during 1950s as reported by Jones & Sujansingani, 1954.

Chilika lake is a estuarine ecosystem. Because of it's rich fishery resources, the people who are living in and around the lake are depending upon the lake for their livelihood. The ability of Chilika to support highly productive fishery which is a steady source of livelihood sustenance for the large population of local fishing communities. The lake is considered as one of the most tourist potential place in India. The lake and its environment have attained an international importance because of its fishery potential (Das, 1996). Nearly 86% of fish and shellfish species found in the lake are of migratory nature being mainly marine migrants from the sea and also riverine migrants from the rivers draining into northern sector of the lake. Hence by and large the lake's fish stock is migratory species dependent. The freshwater inflow and spatio-temporal salinity gradient being influenced by semi-diurnal tidal influx from the sea provide migration cues to anadromous and catadromous species to move towards risperidone breeding grounds near river mouth and in the sea respectively and facilitate auto recruitment of the fish and shellfish to the lake. The lake came under the stress of natural changes coupled with anthropogenic pressure during the last 3 decades until 2000 when the first hydrological intervention with opening of a new artificial lake mouth for eco-restoration of the lake was implemented by Chilika Development Authority (CDA) during September, 2000.Villages with fishing communities numbering 131 are located in and around the lake where more than 0.2 million fishers live in more than 25,000 households and the livelihood of these fishing communities is solely dependent on fisheries of the lake (Mohanty *et al.*, 2009).

Fisheries of the lake contribute more than 96% to the total estimated economic value of provisioning services of the lake. Therefore the lake fisheries have tremendous livelihood and social-economic impacts on the communities living in and around Chilika Lake. Enhancement of fish stock in the lake is largely dependent on effective connectivity of the lake with the sea through lake mouth and distributaries of Mahanadi River in the northern sector of the lake. Geologically, ancient ecosystem of Chilika exhibit high diversity of fish and shell fish species, which has been subjected to multi- dimensional stress impacting its overall biodiversity, particularly that of fishes (Mohanty, 2002). Therefore the lake fisheries have tremendous livelihood and social impacts on the communities living in and around Chilika lake. Enhancement of fish stock in the lake is largely dependent on effective connectivity of the lake with the sea through its mouth and distributaries of Mahanadi river in the northern sector of the lake. Geologically, ancient ecosystem of Chilika exhibit high diversity of fish and shell fish species, fish species 336, prawn species 29 and crab species 35, which has been subjected to multi- dimensional stress impacting its overall biodiversity, particularly that of fishes (Suresh *et al.*, 2018).

Large number of fishing villages with large population of fishers depending on the lake fisheries for their livelihoods have been encountering multiple problems and difficulties in accessing to the fishing grounds, lack of infrastructure facilities on the lake shore for landing of daily catches, maintaining post-harvest fish quality, landing and berthing facilities for fishing crafts, hygienic handling, packing and marketing of catches etc. These infrastructure facilities are of paramount importance to achieve the goal of responsible fisheries which will provide wholesome fish to consumers and enhance the economic returns and socio-economics of fishers. For livelihood and socio-economic development of fishing communities of Chilika lake, adequate provision of various infrastructures with landing, storage, preservation, transportation, hygienic fish handling, clean environment of fish landing centres including waste management facilities are imperative in promoting responsible fisheries in general and community development of Chilika, the largest coastal wetland in the Indian subcontinent in particular. Fish and shellfish from the lake were harvested by Chilika fishers using five traditional fishing methods namely, ***Bahani*** (Net fishing), ***Jano*** (Entrapment fishing), ***Prawn Khati*** (Prawn fishing by traditional split bamboo traps), ***Dian*** (Shallow water fishing adjacent to Jano) & ***Uthapani***. The annual average fish landing for the period 1957-65 was estimated at 3653t (Jhingran & Natarajan, 1969), showing more or less the similar trend of fish yield as recorded previously. Quite a number of new records of fish fauna were reported during the period (Rajan *et al.*, 1968). Thus there was dramatic change in fish and fisheries of Chilika lake during this period. Local consumption fresh fish increased while export to neighboring states gradually decreased. The split bamboo trap fishing for prawns was gradually decreased and the barrier nets with HDPE netting box traps (***Khanda*** fishing) was introduced and continued to increase. The century old traditional ***Jano*** fisheries were also gradually abolished and those productive areas were gradually encroached by vested interest groups for expansion ***Gheries.*** The entire spectrum of fishing practices was, thus, shifted from traditional fishing methods to barrier (***Khanda*** fishing), large size pen culture (unauthorized ***Gheries***), trammel net fishing (Three walled gill net with different mesh size), hook and line fishing etc during post-restoration period.

The methodology for collection and estimation of fish catch statistics from Chilika lake as developed by Chilika Development Authority was ratified by ICAR-Central Inland Fisheries Research Institute (CIFRI), Barrackpore, Kolkata and described by Mohapatra *et al* (2007) was followed during the post-restoration period (2001-02 to 2013-14). The number of fish landing centres gradually increased after 2003-04 and at present there are 35 established fish landing centres, two nos. village fish markets and one fish godown cluster (Fig 4). Nine landing centres are located in southern sector followed by northern sector with 8 landing centres. Outer channel and central sectors cover 6 and 5 landing centres

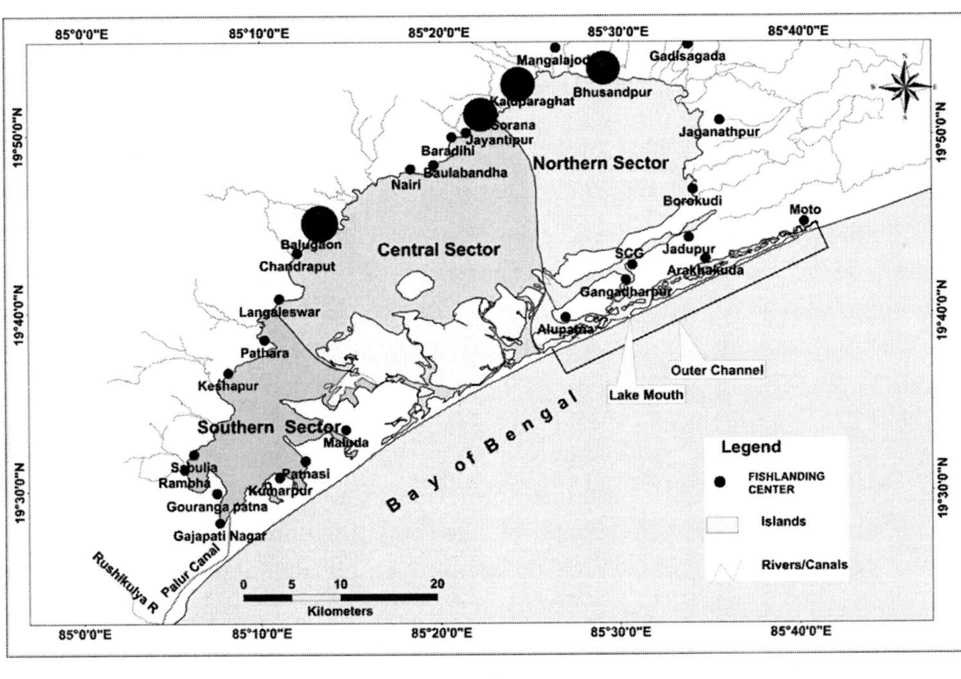

Fig. 4. Chilika map showing fish landing centres

respectively. The monthly fish catch statistics and other related information's are collected from four sectors for six sampling days in a month. Among all landing centres, four centres namely, Bhusandapur, Kalupada, Sorana & Balugaon are considered as major centres which together receive more than 75% of the total catch. The two village markets (Malud & Jadupur) and fish collection godown cluster located between Brahmagiri & Dahikhia Bridge receive/collect shrimp and fish directly from the fishers.

SCENARIO OF BRACKISH WATER LAKE (CHILIKA) FISH LANDINGS DURING THE PRE- AND POST-INTERVENTIONS PERIOD

The Chilika fish landings data was collected since 1929-30 by the office of the erstwhile Assistant Director of Fisheries and presently District Fisheries Officer (Biology & Technology), Balugaon. Due to some environmental as well as anthropogenic pressure such as siltation, shrinkage in water spread area, shifting of lake inlet in North-East direction, ingress of less saline water to the lake, luxuriant growth of fresh water weed in northern sector, use of destructive fishing practices by fishers like zero nets, Alimi nets, Khanda nets and illegal prawn gherries in Chilika shore line, etc. The fish landings of Chilika lake decreased by year to year.

Chilika Development Authority (CDA) has taken up some classic hydrological interventions like opening of a mouth along the outer channel, excavation of inner channels like Magarmukh, Daya, Bhargavi and Luna river confluence channel, Magarmukh-Balugaon channel and renovation of Palur canal in Chilika after consulting premier Scientific and Engineering Institutes of India. After the above hydrological interventions, some positive impacts have been seen on the fish landings of Chilika lake. They are as follows;

(i) The distance between the lake mouths to the main lake decreased. The decrease in distance facilitated effective recruitment of juveniles of fish, prawn and crab.

(ii) It has also helpful for better migration of fish, prawn, crab and their juveniles from sea to the lake and vice versa for feeding and breeding purposes.

(iii) Increase in salinity gradients of the lake help in increasing the population of fish, prawn and crab species in the lake.

(iv) Due to higher salinity in the northern sector of the lake, the freshwater weeds die and decomposed giving more nourishing ground for the fish and prawn.

(v) The Magarmukh-River confluence lead channel helps in flushing of silt laden flood water coming from Mahanadi river system to the lake en-route to sea via this channel. During summer season, more saline water reached to the river mouth through this channel.

(vi) Renovation of Palur canal which connects with Rambha Bay of Chilika to Bay of Bengal Via Rushikulya river mouth acted as the migration route of fish, prawn and crab juveniles to the southern sector of the lake from sea. The mouth was opened during September, 2000 in Outer channel of the lake. Then after, the fish landings of the lake increased while comparing the fish landings date before opening of mouth.The Pre-interventions period was 1978-79 to 1999-2000 and after the hydrological interventions, the Post-intervention period was 2000-2001 to 2021-2022. For the better comparison, 22 years of data are taken from both Pre- and Post-interventions period.

Pre-interventions Period (1978-79 to 1999-2000)

During the Pre-interventions period, the fish landing was varied between 1056.0 MT to 7446.0 MT/ Year with an average of 4050.0 MT/ Year. Similarly, the prawn landing was varied from a minimum of 136.93 MT/ Year to a maximum of 1927.0 MT/ Year with an average of 926.23 MT/Year. The crab landing data ranged from 3.0 MT to 122.50 MT/Year with an average of 46.81 MT/Ycar during the above period. The total fish landings varied from 1274.0 MT/ Year to 8926.0 MT/Year with an average of 5023.06 MT/ Year. The fish landings data of Chilika increased slowly from 1978-1979 and showed its peak value during 1985-1986(Fig.5). Thereafter, it decreased slowly and showed its minimum value during the year 1995-96. (Fig. 5). Similarly, the prawn landings data varied between 136.93 MT to 1927 MT/ Year recorded the higher value during 1978-79 which decreased till 1980-81. (Fig.5). Then it suddenly reached its peak value during 1981-82. The data decreased slowly till 1993-94, then it abruptly decreased from 1994-1995 onwards and reached its lowest value during 1998-1999. The peak value of crab landings data recorded during 1980-81 being 122.5 MT/ Year which decreased till 1991-92. Then after, it suddenly decreased to record the minimum value of 3.0 MT/ Year during 1994-95. Then again it slowly increased till 1999-2000. (Fig.5). The average base level price during Pre-interventions period of fish was Rs.4.01 crores/year, prawn was Rs.2.86 crores/year, crab was Rs.0.15 crores/year and total landings of fish, prawn and crab was Rs.7.03 crores/year.

Post-interventions Period (2000-01 to 2021-22)

In the post-interventions period, the fishery resources of the lake increased dramatically. After the opening of the lake mouth during September, 2000, the total landings was increased to 4982.75 MT/ Year. During the last 22 years (2000-2001 to 2021-2022), the fish landings varied between 3592.95 MT/ Year to 12187.58 MT/ Year. The minimum value was recorded during 2000-2001 and maximum value in 2021-2022. (Fig. 5). The prawn landings data recorded its minimum value during 2000-2001 being 1296.26 MT/Year to its maximum value during 2021-2022 being 6762.02 MT/ Year. Similarly, crab landings data noticed its lower value of 93.54 MT/ Year during 2000-2001 and the higher value observed during 2021-2022 being 478.70 MT/ Year. The same trend was observed in case of total landings being the minimum value 4982.75 MT/ Year during 2000-2001 and maximum value of 19428.30 MT/ Year during the year 2021-2022. The average landings of fish, prawn, crab and the total landings were 8522.11 MT/ Year, 4214.40 MT/Year, 252.60 MT/Year and 12989.11 MT/Year, respectively. While analyzing the total landings of the fisheries output (fish, prawn and crab), the landings increased from 2001-2002 and reached its peak value during 2003-2004 being 14053.22 MT/Year, then after the value decreased and remained low

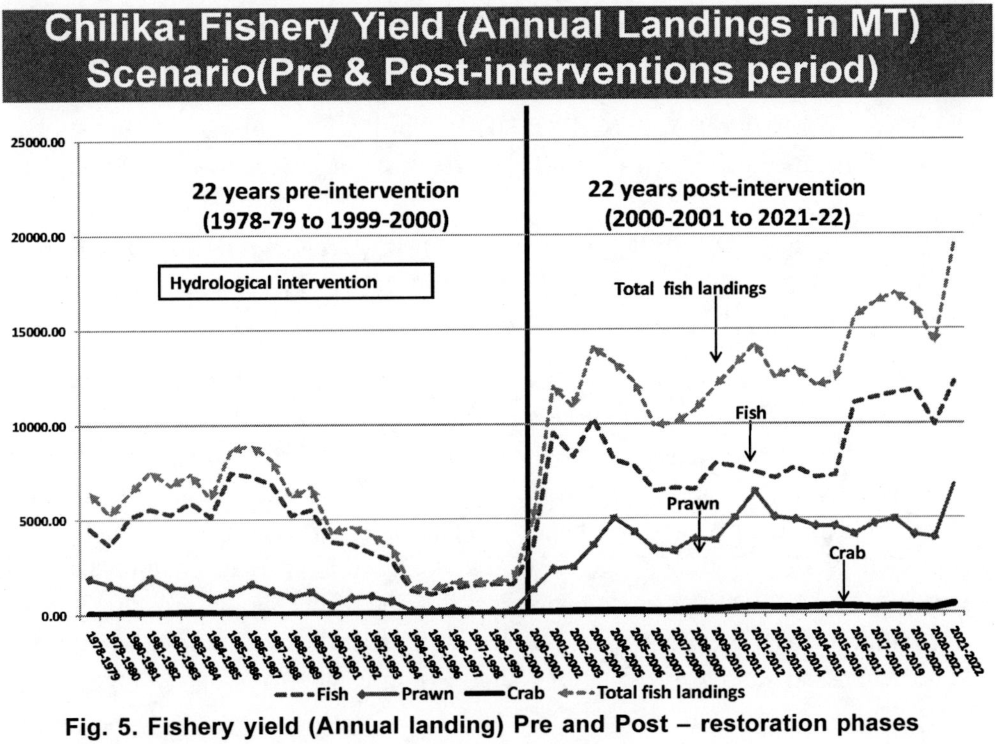

Fig. 5. Fishery yield (Annual landing) Pre and Post – restoration phases

till 2009-2010. The higher value was recorded during 2011-2012 being 14,228.2 MT/Year. From 2012-2013 onwards, the landings decreased and remained almost similar till 2015-2016. From 2016- 2017 onwards, the total landings increased and reached its peak during 2021-2022 being 19,428.30 MT/Year (Fig. 5). It has to be mentioned that the total landings remained more then 12,000 MT/Year from the year 2012-2013 till 2015-2016. But after the demolition of net gherries and zero net Khanda (mesh size less than 10.0 mm) by District Administration, District Fisheries Officer and staffs of Chilika Development Authority from Chilika under OMFR Act 1982 and for Chilika 1988, the total average landing was increased to 12989.11 MT/Year during Post-interventions period. The average base level price for the last 22 years in Post-interventions period (2000-2022) was increased to Rs.138.54 crores/year. This increase in price of Chilika fisheries output was due to increase in fish landings and also increase of price of fish, prawn and crab from year to year.

While comparing the average total landings of fish, prawn and crab during Pre- interventions period (5023.06 MT/Year) and Post-interventions period (12989.11 MT/Year), it has increased by 158.5% as compared to Pre-interventions period. The average fish production potential (capacity) of Chilika lake has been recorded by CIFRI, Barrackpore, 2003 was 27,153.0 MT/Year. The average fish production potential of Chilika lake increased to 32,324.50 MT/Year recorded by CIFRI, 2017. The Maximum Sustainable Yield (MSY) of Chilika lake is 11,760 MT/Year and also reported the average value for last 15 years of their research work was 12,140.0 MT/Year (CIFRI, 2017). Chilika Development Authority has recorded the average value of 12,989.11 MT/Year for last 22 years (2000-2022).

RESPONSIBLE MANAGEMENT AND UTILIZATION OF FISHERIES RESOURSE OF CHILIKA LAKE WITH ACTIVE PARTICIPATION OF LOCAL COMMUNITIES

Among all bio-resources of Chilika lake, fisheries resources are the most important and renewable, but not infinite; hence warrants responsible management and wise use fisheries resources of the lake contribute substantially to the State's economy, livelihood and social well being of the sizeable population of local communities. About 0.2 million fisher population living in 16,710 households are entirely dependent on the fisheries of Chilika lake and the fisheries output accounts for more than 71% of the total economic valuation of the lake (CDA, 2013). The restoration measures with opening of a new artificial lake mouth in September, 2000 witnessed sudden increase of fisheries output due to 'trophic burst'.

The fish landings of Chilika lake fluctuated from time to time. Due to the environmental degradation and anthropogenic pressure, which impact on the Chilika fish landings year to year. There are many factors like siltation, weed infestation, decrease in salinity gradients, shifting in traditional fishing methods to improved modern methods and excessive playing of tourist boats have impact on Chilika fish landings. In Chilika total fish landings data from 1978- 79 to 2021- 22 is enumerated in Fig. no. 5. The Chilika fish landing data from 1978-79 to 2021-22 was analysed in two phases. After the classic hydrological interventions like opening of new mouth, in the year 2000 and excavation of inner channels were undertaken during 2000 to 2002. The first phase was the pre- intervention period (1981-82 to 1999-2000) and second phase was post-intervention period (2000-01 to 2021-22).

The uniqueness of the fisheries resources of Chilika lake with estuarine character is featured by assemblage of marine, brackish and freshwater elements; migratory species constituting more than 86% including juveniles recruitment taking place from both marine and freshwater sources (Mohanty et al 2015). Thus, sustainable management of fisheries resources of such complex nature in a dynamic coastal wetland ecosystem with unique ecological and environmental characteristics, though poses greater challenges, is imperative for the enhancement of livelihood and nutritional supplementation of local communities, largely comprising of impoverished fisher population. The present improved scenario of the lake fishery cannot sustain for longer time if yearly measures to implement an appropriate ecosystem-based fisheries resources management plan based on scientific evidences along with effective enforcement of regulatory Laws/Acts are not taken with people's participation and strong governmental commitment. At present, some comprehensive policy/L to regulate the destructive fishery activities in Chilika lake is not sufficient due to its proper implementation. At present The Hon'ble High Court of Odisha is taking stringent action for proper implementation of OMFRA Act, 1982 and for Chilika, 1988 aquaculture inside the lake Chilika Act, OMFRA(1988). The present non- responsible fishery practices in Chilika are also resulting of inter-community conflicts, which is not desirable for sustainable wetland management with wise use of renewable aquatic resources.

The main objective of fisheries resources management is to make the resources economically productive, socially equitable, scientifically managed and environmentally sustainable. Keeping in view, the influence of social, economic and technological forces on the status and wise use of fisheries resources of Chilika lake, it is imperative that the resource users (fishers) should actively participate in the implementation of resources conservation and management plan in a manner that they can protect the fisheries resources and increase productivity on sustainable basis for their own long-term benefit, particularly to benefit the future generations. Hence, to achieve the greater objectives of responsible fisheries in Chilika lake, the following pragmatic approaches are needed.

Gathering of Scientific Evidences

It is well recognized that sound scientific data are a pre-requisite to assist the management agencies and stakeholders in making appropriate decisions for responsible fisheries. Therefore, it must be ensured that need-based researches/ scientific studies are carried out to bridge the knowledge gaps in the areas of fisheries in Chilika lake. The first holistic fisheries investigation in Chilika was carried out by the erstwhile Chilika unit of the Central Inland Fisheries Research Institute, Barrackpore (Kolkata) during 1957-65, that generated much needed valuable scientific information in published forms. After that, no such holistic study was undertaken during the past four decades. The degradation of the ecosystem and fisheries almost touched the nadir by mid-nineties. However, immediately after the hydrological intervention for restoration of the lake the Central Inland Fisheries Research Institute (CIFRI), Barrackpore was commissioned to assess the fish yield potential and maximum sustainable yield (MSY) level for total fish catch and to evaluate the production function of the ecosystem during 2003/2004 (Panda 2013). Many of the available information from the past investigations have lost relevance in the face of changed ecological and habitat characteristics during the eco-degradation and restoration phases. The MSY as determined by ICAR-CIFRI during 2011 to 2017 was 11,500 t/year. The average catch in last 15 years has been 12,140 t. (CIFRI, 2017).

The Chilika Development Authority (CDA) in cooperation with Japan International Cooperation Agency (JICA) launched a project "Conservation and Wise use of Natural Resources of Chilika lake through Community Participation" in October 2006, being scheduled to be completed in September 2009. With a greater focus on commercially important target species of fish (*M.cephalus, L.calcarifer, E.tetradactylum, D.albida, T.ilisha, and E.suratensis*), prawns (*P.monodon* and *F.indicus*) and mud crabs, the fishery resources survey has been organized with biological, ecological and fishery data collection, data management and stock assessment of the target species (Bhatta et. al, 2009). Besides, baseline survey for socio-economics and sensitization workshops to associate fishers and other stakeholders in the planning process of fisheries resources management were undertaken. All of these have been integrated to give the final shape to the fisheries resources management plan. However, the following studies are suggested to help bridge the knowledge gaps for better conservation and management of renewable, but not infinite fisheries resources of Chilika lake:

- Sufficient time series data-based development of carrying capacity model to harvest each of the target species on a sustainable basis.
- Population structure and breeding biology of Mud crab *Scylla traunquebarica* from Chilika lake.
- Continuance of fish and shellfish biodiversity inventory.
- Long-term monitoring of the data sets used in the stock assessment for target species in the present CDA-JICA project and re-assessment of their stock on annual basis.
- Time series data development on fish, prawn and crab landings, CPUE for target species and overall fishing and environmental variables.
- As Chilika is a dynamic ecosystem, fresh inventorization of leasable fishery sources need to be carried out with categorization on the basis of productivity level for judicious leasing to Primary Fishermen Cooperative Societies (PFCS).
- Survey of fishing gears and crafts currently in use and evaluation of destructive fishing including illegal prawn ghery operation.
- Gear selectivity for responsible fishing.
- Periodical EIA to assess human-induced activities and their impact on sustainable fisher development and ecosystem functioning.
- Documentation of the traditional knowledge of the fishing communities on the fisheries resources and fishing methods for improvisation of traditional technologies which can protect the natural resources and enhance productivity.

Conservation and Management Objectives

Long-term sustainable use of fisheries resources is the basic objective of the management of fisheries resources of the lake. The fishers and other stakeholders should *interalia* adopt appropriate measures, based on the best scientific evidences available which can help restore fish stock so as to achieve maximum sustainable yield. To achieve this goal the following aspects need to be taken care of;

- A pragmatic conservation and management policy to regulate the fishery activities in Chilika with active participation of local communities.
- Regulatory mechanism (including enforcement of laws like suitably amended OMFRA, provision of alternative livelihood options, financial support during fishing ban period, etc.) to avoid excess fishing pressure so that the exploitation of target fish stocks does not exceed the maximum allowable catch limit.

- The traditional rights and interests of fishers and the subsistence fishery in the form of small-scale and artisanal fisheries need to be safe guarded.
- The recovery of depleted fish stock and fish biodiversity to be sustained by ensuring unhindered free recruitment, hassle-free breeding migration and phasing out of destructive fishing gears and fishing methods.
- Eco-sensitive zones of the lake consisting of spawning and nursery grounds, recruitment and breeding migration routes are required to be identified, demarcated and kept free for fish and shellfish stock development.
- Keeping in view the scary sound and obstructions of motorization boats in the main breeding migration and recruitment route (outer channel sector) and oil and air pollution in the lake, further increase in motorized boat population should be restricted. Plying of unregulated motorized tourist boats in the eco-sensitive outer channel need to be critically examined to determine the safe limit.

Community Participation

Fishery practices in Chilika are open use of the common property resources, soothe participation of the local communities at all levels of resources management planning, implementation and enforcement of regulatory laws is imperative. To ensure their active participation the management practices should be made as inclusive as possible. This could be achieved through;

- Assessment of needs and aspiration of the local fishers through extensive stakeholders consultations.
- Stakeholders forum for consultation with communities during planning, decision making and implementation phases.
- Fishers core committees to be constituted for four sectors of the lake to assist in implementation of management plan, law enforcement to prevent illegal fishing and aquaculture practices (Ghery operation) and to resist the entry and activities of vested interest groups. They will also help prevent destructive fishing in eco-sensitive areas of the lake like outer channel, Magarmukh and Palur canal.
- Mechanism development for strong networking with local communities representing all sectors of the lake.
- Dissemination of information on scientific evidences, regulatory laws and rules, new technologies, etc. to the fishers through the fishers core committees.
- Encouraging self-initiated good practices for regulation of destructive fishing practices by the local fishers and intruders.

- Empowerment of fishers through capacity building.
- Strengthening of PFCSs to make the fishers to prevent exploitation by the middlemen and moneylenders and to market their catches through the PFCS.
- The Centrally Sponsored National Schemes of Fishers Welfare (Development of model fishers villages, Group accident insurance for active fishers and saving-cum-relief) could be implemented in Chilika fishing villages with financial support from Government and active involvement of fishers.
- Integration of social, economic, ecological and biological dimensions and incorporation of traditional management practices in the fisheries resources management would be more effective.
- Promotion of alternate source of livelihood like community-based eco-tourism, ornamental fish farming/rearing, aquaculture in village community ponds, mud crab fattening, non-fishery activities (Diary unit, goat rearing, duck rearing, dry fish making using hygienic methods/solar dryer, dry fish marketing, vegetable vending, tent house, vermin-compost production etc.) which may help reduce fishing pressure in the lake and for fishers sustenance during lean and fishing ban period (Government subsidy assistance is available in many cases).
- Formation and capacity development of women Self-Help Groups (SHGs) to adopt income generation activities with value addition, supported by micro-financing programme would help increase family income and women development.
- Organizing training / awareness programmes for the fish workers/traders to cooperate in implementing the fishery conservation measures like refusal to buy small young fish, prawn and crabs (below the prescribed / restricted sizes under OMFR Act) and not to sale small meshed nets like 'O' net, mantle net, etc in Chilika area.

Institutional Development

- Developing policy for biodiversity conservation of Chilika lake.
- Enforcement and implementation of laws related to fisheries, wildlife and water resources development.
- Monitoring, implementation and impact assessment.
- Awareness generation about biodiversity conservation and its sustainable use by all stakeholders.
- Developing scientific basis/evidences for various problem solutions

Regulation of Fishing Practices

Responsible fishing practices could be promoted in Chilika by regulating the registered boat population in consonance with the maximum number as determined to achieve maximum sustainable catch and phasing out of fishing gears which are not commensurate with the responsible gears. Thus responsible fishing practices can be promoted by adhering to the following measures;

- All fishing boats must be registered under OMFR Act, 1982 and Rules, 1983/84 for the purpose of fishing in the lake and obtain certificate of registration to strictly follow up the conditions as laid down under clause 5(1) of OMFRA Rules. All other boats plying in the lake for general purposes should be registered under Orissa Boat Rules and should operate/ply in the lake with valid licenses.
- The population of fishing boats should be determined on the valid scientific basis to harvest fish at MSY level, which is termed as maximum sustainable effort (Fmsy). Recently, ICAR-CIFRI has assessed this limit to be 9000 fishing boats which should not be exceeded. The local fishers need to be sensitized in this regard.
- The existing gears need to be evaluated and those which are not commensurate with the responsible fisheries should be phased out. Based on the results of gear selectivity study, the best suited and acceptable (expectable to fishers) new fishing gears should be introduced with regular performance evaluations. Due attention need to be given to the ability of the fishers to effectively use the new fishing gears.

Post-harvest Practices Including Value Addition

Government of India (GoI) have been emphasizing on proper sanitation in fishing vessels and landing centres in accordance with the EU specifications as a part of responsible fisheries to boost production of seafood. This is quite relevant to the landing centres of Chilika lake, since 40-50 % of total shrimp catch and about 45% of live mud crabs from Chilika are exported to international markets. Besides, responsible utilization of fish emphasizes to ensure the right of consumers to safe, wholesome and unadulterated fish and fish products. Minimization of post-harvest loss and value addition to low value fishes would enhance return from catches. Post-harvest practices at the landing centres would provide employment to local fish workers, retail ice sellers, porters, artisans, fish transporters etc.; hence the post-harvest practices need to focus on the following aspects;

- All the fish landing centres in Chilika, particularly the four major centres i.e., Balugaon, Sorana, Kalupadaghat and Bhusandpur should be up-graded with infrastructure facilities in conformity with EU prescribed standards.

- Proper hygienic and sanitation need to be extended to all the fish landing centres of the lake with active cooperation of fishers and local fish merchants associations to maintain fresh fish quality for food safety.
- Infrastructure like road connectivity, freshwater supply, waster management, ice supply and navigable approach channels need to be provided to all the established fish landing centres in Chilika.

Education and Capacity Development

Education and capacity development are key to achieve the goal of sustainable fisheries management. Empowerment of the local fishers through capacity development and awareness building would be crucial for promotion of responsible fisheries in the lake. This can be achieved through;

- Development of curricular on important aspects of responsible fisheries i.e., good practices, conservation and fishing regulations, management frame work and procedures, rules and regulations of existing Acts, perceivable elements of scientific evidences, destructive and responsible fishing gears, environmental education etc.
- Training modules need to be developed for need-based training for the fishers and other stakeholders.
- A carefully planned outreach programme focused on responsible fisheries involving the local NGOs & CBOs for different target groups i.e., fishers, catchment communities, school children, youth and women.

CONCLUSION

A holistic management and conservation plan needs to be implemented for Chilika lake, taking care of the livelihood of fishing communities, sustainable fisheries and biodiversity of the lake. Regular maintenance measures of hydrological intervention should be continuing along with dredging of major river mouths in the northern sector of the lake. Maintaining topographic and hydrographic characters of outer channel sector and sea mouth.

ACKNOWLEDGMENTS

The authors are thankful to Ministry of Environment Forest and Climate Change, (NPCA) Govt. of India, New Delhi for financial support. Our thanks are due to Mr. Bibhuti Bhusan Dora, JRF, GIS cell of CDA for preparing of the maps reproduced in this book chapter. We are also thankful to Shri Padma Charana Pradhan, Shri Manas Ranjan Moharana, Shri Sadananda Sahu & Shri Bhikari

Charana Jena, fish landing data collection supervisors of Kalupada, Balugaon, Palur & Satapada centres respectively and their supporting field staff for the collection of fish landing data under systematic sampling method from 35 data collection points in Chilika since 2000-2001. We would like to thank all the staff members of Chilika Development Authority who have directly or indirectly rendered their help unconditionally to accomplish this work. The authors are thanking to the fishers of Chilika lake for sharing valuable knowledge.

REFERENCES

Annandale N and Kemp S (1915) Fauna of the Chilika lake . *Mem Indian Mus* 5(1): 1-118.

Annandale N (1915) Fauna of the Chilka Lake: Mammals, Reptiles and Batrachiana. *Mem. Indian Mus.,* 5(2): 163-174.

Bhatta KS, Samal RN, Karna SK, Sahoo D, Panda S, Pattnaik AK, Hiramatsu K and Ito K (2009) The biological & ecological characteristics and the current status of fisheries & resources of commercially important species in Chilika Lake. 174.

Blanford WT (1872) Sketch of the Geology of Orissa, *G. S. I. Records*, 5(2), 41-72.

CDA (2013) Annual report of Chilika Development Authority,Bhubaneswar,1-53.

CIFRI (2017) Post-restoration assessment of the ecology and fisheries diversity of Chilika Lake – ICAR-CIFRI/CDA-ICZM Consultancy Project (October 2011 to April 2017) (Final Report) ICAR-CIFRI, Barrackpore, Kolkata: 498p.

Chaudhuri BL (1916) Fauna of the Chilika Lake: Fish, Part I. *Mem. Indian Mus.,* 5(4): 403-439.

Chaudhuri BL (1916) Fauna of the Chilika Lake: Fish, Part II. *Mem.Indian Mus.,* 5(5): 441-458.

Chaudhuri BL (1917) Fauna of the Chilika Lake: Fish, Part III. *Mem.Indian Mus.,* 5(6): 491-508.

Chaudhuri BL (1923) Fauna of the Chilika Lake: Fish, Part IV. *Mem.Indian Mus.,* 5(11): 711-736.

Das,C.R.(1996) Conservation and Management of Chilika Lake. In M .Ando (ed.),Conservation and Management of Lakes/Reservoirs in India, ILEC,Susastu, Japan,184p.

Devasundaram MP (1951) Fishing methods for Chilika mullets. *Indian Famg.,* 12 (1-2): 22-25.

Devasundaram MP (1954) A report on the fisheries of the Chilika Lake from 1948 – 1952. Orissa Govt. Press, Cuttack: 34p.

Dujovny E (2009) The deepest cut: political ecology in the dredging of a new sea mouth in Chilika Lake, Orissa, India. Conservat Soc 7(3): 192–204.

Government of Orissa (1988) Conservation of Chilika Fishery under Orissa Marine Fisheries Regulation Act and Rules, Order no 19933/FFAH dated 31.08.1988 FFAH Dept., 54-55.

Ghosh AK and Pattanaik AK (2005) Chilika Lake: Experience and lessons learned brief. In UNEP International Waters Learning Exchange and Resource Network. http://iwlearn.net/iw-projects/1665/experience-notes-and lessons-learned/chilika/ lake 115–132.

Hiddink JG, Mackenzie BR, Rijnsdrop A, Dulvy NK, Nielsen EE, Bekkevold D, Heino M, Lorance P and Ojaveer, H (2008) Importance of fish biodiversity for the management of fisheries and ecosystems. *Fisheries Research*, 90, 6-8.

Hora SL (1923) Fauna of the Chilika Lake: Fish, Part V. *Mem. Indian Mus.,* 5 (11): 737-769.

Jhingran VG (1963) Chilika can produce more fish for the nation. *Indian Livestock,* 1(1): 47-49.

Jhingran VG and Natarajan AV (1966) Final report on the Chilika Lake (1957-65) Bulletin No. 8, Central Inland Fisheries Research Institute, Barrackpore (India): 40p.

Jhingran VG and Natarajan AV (1969) A study of the fisheries and fish population of the Chilika Lake during the period 1957-65.*J. Inl. Fish. Soc. India*, **1**: 49-126.

Job TJ and VR Pantulu (1953) Fish trapping in India. J. Asiatic Soc. Sci., 19: 175-196.

Jones S and KH Sujansinghani (1952) The Mani jaal of the Chilika Lake – a special net for Beloniform fishes. *J. Bom. Nat. Hist.Soc.,* 51 (1): 288-289.

Jones S and KH Sujansinghani (1954) Fish and fisheries of the Chilika Lake with statistics of fish catches for the years 1948-1950. *Indian J. Fish.,* 1 (1&2): 256-344.

Kemp S (1915) Crustacea: Decapoda, Fauna of the Chilika Lake. *Memoir of Indian Museum*, 5, 199-325.

Mangla B (1989) Chilika Lake: Desilting Asia's largest brackish water lake. *AMBIO - A Journal of the Human Environment*, 18(5): 298p.

Mitra GN (1946) Development of the Chilika Lake, Orissa Govt. Press, Cuttack: 26p.

Mitra GN and Mohapatra P (1957) Bulletin on the development of Chilika Lake – survey report on the fishing industry. Orissa Govt. Press, Cuttack: 51p.

Mohanty PK and Otta SN (2008) Population of Irrawaddy Dolphin in Chilika Lake. Dolphins of Chilika Odisha Review, India, pp. 24-25.

Mohanty SK, Mohanty SK and Mohanty RK (2004) Evaluation of Gherry prawn culture in Chilika lake with special reference to social issues. *J. Indian. Soc. Coastal Agric. Res.,* 22 (1&2): 293-297.

Mohanty SK, Mishra SS, Khan M, Mohanty RK, Mohapatra A and Pattnaik AK (2015) Icthyofaunal diversity of Chilika Lake, Odisha, India: an inventory assessment of biodiversity status and comprehensive systematic checklist(1964-2014) *Checklist,* 11(6),1-19.

Mohanty RK, Mohapatra A and Mohanty SK (2009) Assessment of the impact of a new artificial lake mouth on the hydrology and fisheries of the Chilika Lake, India. *Lakes, Reservoir: Research and Management* 14, 231–245.

Mohanty SK (2002) Fisheries biodiversity of Chilika lake. *Chilika news letters,*3,11-12. Odisha Economic Survey (2016-17)

Mohapatra, P. (1955) The Thatta-Khonda-A screen trap of the Chilika Lake. *J. Bombay Nat. Hist. Soc.,* **53** (2): 277-279.

Mohapatra A, Mohanty, RK Mohanty, SK, Bhatta KS and Das NR (2007) Fisheries enhancement and biodiversity assessment of fish, prawn, and mud crab in Chilika Lake through hydrological intervention. *Wetlands Ecol. Manage.,* 15: 229-251.

O Malley LSS (1929) Bihar and Orissa District Gazetters. Puri, Rev. P.I. Mansfield (edi.) Supt. Govt. printing, Bihar and Orissa, Gulzarabagh, Patna.

Panda D (2013) Final report on fisheries and fish yield potential of Chilika lake during the post-engineering intervention. ICAR- Central Inland Fisheries Research Institute, Barrackpore,Kolkatta.104p.

Parida S, Mohapatra D and Bhatta KS (2013) Recent trends in fisheries of Chilika lake: A Ramsar site at East coast of India. *Fishing Chimes* 32 (10), 72–75.

Parida S and Bhatta KS (2023) "Environmental Management and Fisheries Restoration in the tropical Lake Chilika" : 235-251 (eds) B. P. Mohanty, S.D. Sarkar, S. Mohanty, S. C. Parija, S. K. Nag & B.K. Bhattachariya Advances in Fish Research: Vol.-VIII, Narendra Publishing House, Delhi, India

Parida S, Parida P, Bhatta KS and Karna SK (2019) Shifting Paradigm shift in fishing technology of Chilika: The largest brackish water wetland of Asia. *Proc Zool Soc,* 72(3), 263–272 https://doi.org/10.1007/s12595-018-0267-x.

Phleger FB (1981) A review of some general features of coastal lakes 7-14. In:Coastal lake Research present and Future. UNESCO Technical papers in Marine Science, p, 33.

Rajan S, Patnaik S and Basu NC (1968) New records of fishes from Chilika Lake. *J. zool. Soc. India,* 20 (1&2): 80-93.

Remesan MP, Pravin P, Pradhan BK and Meenakumari B (2011) Fishing methods of Chilika Lake, CIFT Information Series, Central Institute of Fisheries Technology, Cochin: 24p.

Sekhar NU (2004) Fisheries in Chilika Lake: how community access and control impacts their management. *J. Environ. Manage*. 73, 257–66.

Suresh VR, Mohanty SK, Manna RK, Bhatta KS, Mukherjee M, Karna SK, Sharma AP, Das BK, Pattnaik AK, Nanda S and Lenka, S (2018) Fish and shellfish diversity and its sustainable management in Chilika lake, ICAR-Central Inland Fisheries Research Institute, Barrackpore, Kolkata and Chilika Development Authority, Bhubaneswar. 376pp.

Taylor JH (1898) Final report on the Khurda settlement of 1897-98, p, 8-9.

Venkataratnam K (1970) Formation of the barrier spit and other sand ridges near Chilika lake on the east coast of India. *Marine Geology*, 9, 101-116.

Wetzel RG (1975) *Limnology* .W.B. Saunders Co. Philadelphia, USA.,743 pp.

World Bank (2005) Scenario assessment of provision of environmental flows to Lake Chilika from Naraj Barrage, Orissa, India. Reports from the Environmental flows window of the bank Netherlands Water partnership programme (World Bank) to the Government of Orissa, India, 1–40.

Sustainable Agriculture and Biodiversity: Innovations and Insights, Pages: 312–330
Edited by: Bibhuti Bhusan Sahoo, Harvinder Kumar Singh, Niranjan Chinara and Amitabh Panda

CHAPTER - 13

HYDROPONICS: A CONTEMPORARY APPROACH TO SOILLESS PLANT CULTIVATION

Amitabh Panda*, Swaraj Meher and Subhashri Sahoo

Krishi Vigyan Kendra, Kalahandi, Bhawanipatna, Kalahandi-766001, Odisha
**Corresponding email: amitabhpandap70@gmail.com*

INTRODUCTION

Hydroponics is a versatile and efficient method for growing plants, offering benefits in terms of resource use, space utilization, and the ability to customize growing conditions for optimal plant health and productivity. Soil, while a common growing medium for plants, has limitations such as disease, nematodes, poor drainage, and unsuitable soil conditions. Conventional open-field agriculture requires extensive space, labour, and water. In metropolitan areas, soil may be unavailable, and fertile lands may be scarce due to unfavourable conditions. Additionally, hiring labour for traditional agriculture has become challenging. In response, soil-less cultivation, like hydroponics, can be a successful alternative, overcoming these challenges.

Soil-less cultivation involves Hydroponics and Aeroponics. "Hydroponics" comes from Greek language, "Hydro" means "water" and "Ponos" means "labour" (Shrestha A. *et al.*, 2017). It's a method of growing plants without soil, using nutrient solutions. Plants can be grown with roots directly in the nutrient solution or in inert mediums like perlite. Hydroponics helps address climate challenges, improves production system management, and combats malnutrition. "Aeroponics" is similar to hydroponics, but plants are grown with fine drops of nutrient solution.

Thus, hydroponics involves growing crops in a nutrient-containing medium, whether solid or liquid, excluding soil. The medium can range from mineral nutrient solutions to inert substances like gravel, sand, wool, perlite, or rice husks. Aeroponics, on the other hand, employs a fine mist of liquid nutrient solution instead of a traditional medium, closely resembling hydroponics.

Hydroponics is a popular method for growing vegetables like tomatoes, lettuce, cucumbers, and peppers, as well as ornamental crops such as herbs, roses, freesia, and foliage plants. The ban on methyl bromide in traditional soil farming has boosted the demand for hydroponically grown produce in recent years. (Shrestha A. *et al.*, 2017).

1 (A) Historical Background: A brief history of hydroponics and its development over time.

The term 'Hydroponic' was coined by William Frederick Gericke in 1937. (Singer J., 2021)

- **Ancient Roots:** The Hanging Gardens of Babylon, considered one of the Seven Wonders, possibly used a form of hydroponics on terraced platforms irrigated with water.
- **17th Century:** English naturalist John Woodward's late 17th-century experiments hinted at plant growth relying on waterborne substances, laying the groundwork for later hydroponic developments.
- **19th Century:** In the late 19th century, German botanist Julius von Sachs and American scientist William Frederick Gericke independently researched soilless culture. (Senn's 1860 experiments involved growing plants in water solutions.)
- **20th Century:** In the 1920s, Gericke popularized "hydroponics" and conducted significant research at the University of California, Berkeley, showcasing successful soil-less plant growth. In 1946, English scientist W. J. Shalto Duglas introduced hydroponics in India, establishing a lab in Kalimpong, West Bengal, contributing to its early development. In the 1960s and 70s, commercial hydroponic farms emerged globally, from Abu Dhabi to the Russian Federation. Automation and computerization of hydroponic systems became prevalent in the 1980s. (Singer J., 2021)
- **21st Century:** Hydroponics has gained global popularity, with a surge in home hydroponics kits during the 1990s. Ongoing technological advancements and a focus on sustainable agriculture shape the future of hydroponics.

HYDROPONIC SYSTEM

Hydroponics can be straightforward with just plants, a container, water, support for plants, fertilizers, and a light source. A growth medium like perlite or rockwool is necessary for the plant's roots, enhancing water absorption. Outdoor hydroponics in the sun on a balcony is common, while indoor farmers often use LED or fluorescent lamps. Plants get nutrients from a solution, delivered through methods like drip systems or aeroponics, depending on factors like weather and crop. Maintaining the nutrient solution's pH balance is crucial for plant health. Hydroponic systems offer benefits like faster growth, higher yields, and the ability to grow in limited spaces or poor soil. Proper setup and maintenance are key for success.

ADVANTAGES OF HYDROPONICS

Hydroponics offers several advantages for plant cultivation, making it an increasingly popular method of farming. Here are some key advantages of hydroponics:

- **Increased Growth Rates:** Plants in hydroponic systems often grow faster compared to traditional soil-based methods. The controlled environment allows for optimal nutrient absorption and faster metabolism.
- **Water Efficiency:** Hydroponic systems use less water compared to traditional soil farming because water is recirculated within the system, and there is reduced water wastage through evaporation or runoff.
- **Space Efficiency:** Hydroponics is well-suited for urban farming and areas with limited space. Plants can be grown vertically, maximizing the use of available space.
- **Optimized Nutrient Control:** Hydroponic systems provide precise control over nutrient levels, ensuring that plants receive the right balance of essential elements. This helps in avoiding nutrient deficiencies or excesses.
- **Reduced Dependency on Soil Quality:** Hydroponics eliminates the need for fertile soil, making it possible to grow crops in areas with poor or contaminated soil. This is particularly beneficial in urban or industrialized areas.
- **Year-Round Crop Production:** Hydroponic systems allow for year-round cultivation, independent of seasonal changes. This feature is especially valuable in regions with extreme climates.
- **Reduced Pest and Disease Risks:** Hydroponic environments can be more easily controlled and monitored, reducing the risk of pests and diseases. Without soil, there are fewer opportunities for soil-borne pathogens to affect plants.
- **Energy Efficiency:** Although hydroponic systems may require initial energy inputs, they can be designed for energy efficiency. Additionally, the controlled environment allows for optimized use of resources.
- **Customization and Adaptability:** Hydroponic systems are highly adaptable and can be customized to suit the specific needs of different plant varieties.

This flexibility is valuable for experimenting with and optimizing growing conditions.

- **Higher Crop Yields:** With optimal growing conditions and nutrient control, hydroponic systems often result in higher crop yields per square foot compared to traditional farming methods.
- **Environmental Sustainability:** Hydroponics can be environmentally sustainable, especially when coupled with renewable energy sources and water recycling. It reduces the need for large-scale land use and minimizes environmental impact.
- **Research and Innovation:** Hydroponics provides a controlled environment that is conducive to research and experimentation, driving innovation in agriculture and plant science.

TECHNIQUES IN HYDROPONIC SYSTEM

Managing nutrient solutions is crucial in soilless production for ensuring healthy plant growth. Various techniques are employed in hydroponics, selected based on factors like plant type, climate, environment, and cost. The primary difference lies in the structural setup of each method. In hydroponics, plants receive nutrition through circulating a nutrient solution through the system, crucial for their growth. One such technique is Continuous Flow Solution Culture. In this system, the nutrient solution circulates through the plant roots, with excess solution collected and recycled back to the storage tank. The solution is changed when the electrical conductivity (EC) changes. Continuous flow systems are regarded as excellent nutrient delivery systems, where maintaining a consistent EC is vital for plant yield. Two types of systems, namely the nutrient film technique and the deep flow technique, further illustrate this principle.

Nutrient Film Technique (NFT)

The Nutrient Film Technique (NFT) is a widely adopted hydroponic method. In this system, plants grow in channels where a nutrient solution is consistently pumped. Plant roots receive nourishment from a thin film of this nutrient solution, eliminating the need for a timer or a growing medium due to the continuous process. Unlike Deep Water Culture, the roots are kept moist but not fully submerged in water. It's crucial to maintain the electricity and pump system to prevent system failure or stoppage. The roots absorb nutrients directly from the nutrient film. Typically, NFT systems circulate the solution at a rate of 1 liter per minute. (Karne H. *et al.*, 2023) This technique is particularly suitable for short-term crops like lettuce and herbs. If the channel size is increased, NFT systems can also accommodate long-term crops. Regular maintenance is essential for the reliable performance of the system.

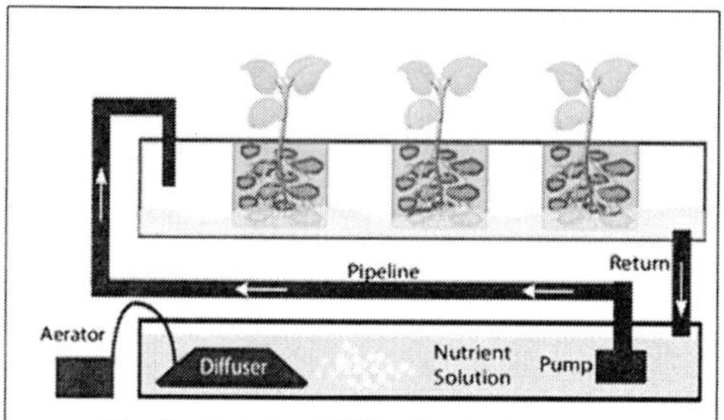

Fig-1: Nutrient Film Technique

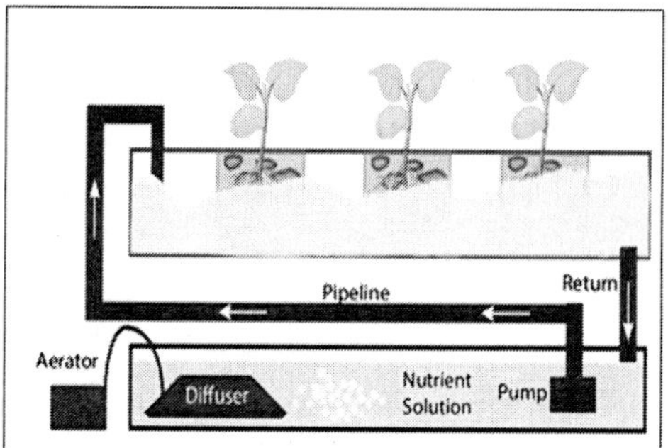

Fig-2: Deep Flow Technique

Deep Flow Technique (DFT)

The Deep Flow Technique (DFT) is a hydroponic system suitable for growing herbs like oregano, basil, and thyme. It works in open or protected spaces. In this system, plants are placed in plastic net pots, and PVC pipes are arranged in a zig-zag pattern to deliver the nutrient solution to the plant roots. This setup is designed for space efficiency, especially for low-growing crops. The water, after collecting in a storage tank, is oxygenated before being recirculated through the system. Similar techniques include Root Dipping, Capillary Action, Trench or Trough, Pot, Deep Water Culture (DWC), among others. DFT is ideal for cultivating herbs in a controlled environment, providing an efficient and space-saving solution for hydroponic herb cultivation.

Deep Water Culture (DWC)

In this system plant roots are submerged in a nutrient solution while resting on a floating platform. This method ensures a continuous supply of water, oxygen, and nutrients to the plant roots. An air pump oxygenates the water, facilitating root breathing. Deep Water Culture is commonly used for growing fruiting plants such as tomatoes and cucumbers. In essence, it's a simple and effective hydroponic technique for nourishing plants directly in nutrient-rich water.

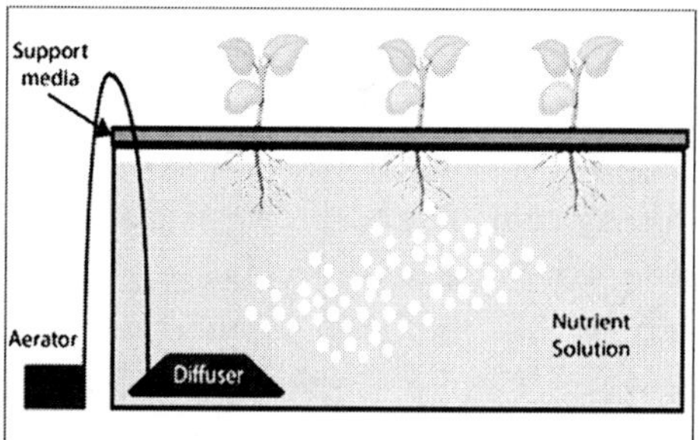

Fig-3: Deep Water Culture

Ebb and Flow (Flood and Drain)

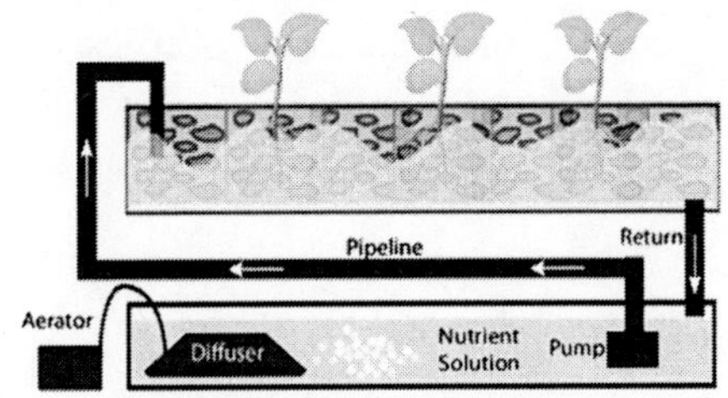

Fig-4: Ebb & Flow Technique

Ebb and Flow, also known as flood and drain, is a hydroponic technique that involves periodically flooding plants with nutrient solution for a few minutes and then draining it away. The nutrient solution is stored in a reservoir. While it might not be the top choice for large-scale soilless production, it works well for smaller hydroponic setups. The system uses gravity to drain water and a pump to recirculate it. Crops like strawberries, beans, radishes, cucumbers, and tomatoes can thrive in this system. The frequency of the flooding process varies based on the plant, growing medium, and environmental factors. Although not ideal for commercial use, Ebb and Flow is a practical and efficient method for smaller hydroponic setups.

Drip System

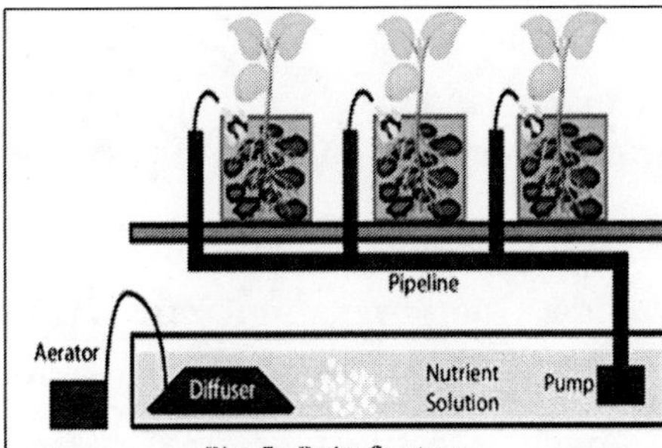

Fig-5: Drip System

Drip systems are popular in soilless production for their effective moisture control. Ideal for long-term crops like tomatoes, peppers, and onions, this system pumps nutrient solution from tanks to plant roots via a drip mechanism. The growing medium, flushed during the drip cycle, retains essential nutrients, water, and oxygen for plants. The drip rate is adjustable, allowing diverse plant growth. Continuous drip systems fall into two categories: recovery and non-recovery. Recovery systems recycle the nutrient solution, offering a cost-effective approach. In contrast, non-recovery systems maintain a constant solution strength without reuse. Drip systems are known for their simplicity and efficiency in delivering precise nutrition to a variety of plants.

Wick System

It operates without electricity or pumps. Nutrient solution from reservoirs is delivered to plant roots through wicks using capillary action. Preferred growing mediums for this method include perlite, vermiculite, Pro-Mix, and coconut fibre.

This is suitable for plants that don't need frequent watering, making it useful for small plants, herbs, and spices. However, a notable drawback is the limited oxygenation of plant roots (Karne H. *et al.*, 2023). Due to this limitation, it is best suited for smaller plants, herbs, and spices rather than larger crops.

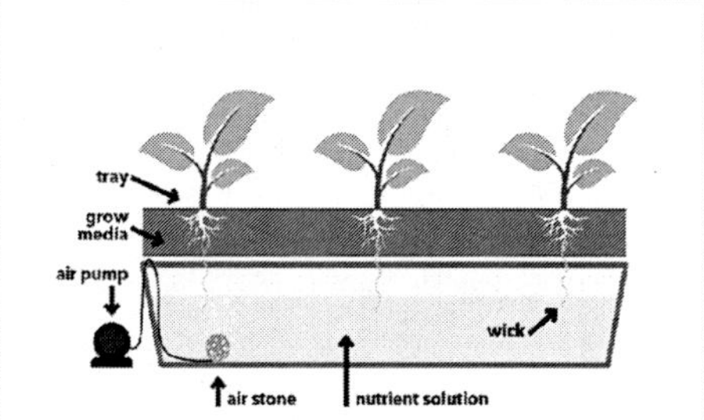

Fig-6: Wick System

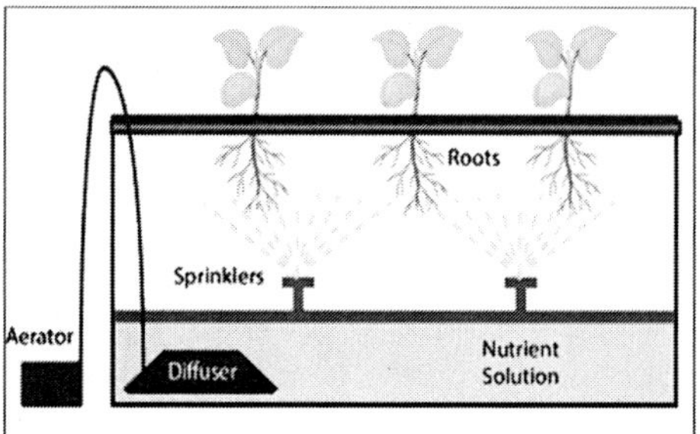

Fig-7: Aeroponics

Aeroponics

Aeroponics is an ideal method for growing tubers and roots. In this setup, plants suspend their roots in the air and receive nutrients through periodic spraying by sprinklers. Unlike other methods, aeroponics doesn't need a separate air system since oxygen is naturally delivered with the sprayed nutrient solution.

DIFFERENT GROWING MEDIUM USED IN HYDROPONICS

An ideal growing medium should:

Provide Good Aeration and Drainage:

- Balance water retention with proper drainage.
- Avoid excessively fine materials to prevent waterlogging and ensure sufficient aeration.

Be Durable:

- Maintain structural integrity over time.
- Avoid soft aggregates that break down easily.

Have Adequate Porosity:

- Retain moisture long enough for plants to absorb nutrients between watering cycles.

Be Sterile:

- Minimize disease and pest spread.
- Ensure cleanliness and avoid introducing additional nutrients to the roots.
- Some media can be reused after pasteurization or a bleach soak followed by rinsing.

Possess Neutral pH and Good Cation-Exchange Capacity:

- Maintain a neutral pH.
- Exhibit good cation-exchange capacity for nutrient retention.

Be Functional:

- Lightweight, making it easy to handle.
- Reusable and durable for sustained use.

In essence, an ideal growing medium balances water-related factors, durability, porosity, sterility, chemical properties, and functionality to support optimal plant growth in a hydroponic or soilless environment.

5 (A) Summary of common hydroponics growing mediums:

Mineral wool: Mineral wool, like Rockwool, is a sterile, porous, and non-degradable material made mainly from heated and melted granite and limestone. It's formed into blocks, sheets, cubes, slabs, or flocking. This medium easily absorbs water, has good drainage, and is commonly used for starting seeds, rooting cuttings, and growing large crops like tomatoes.

Coconut Coir: Coconut coir, known by names like Ultrapeat®, Cocopeat®, and Coco-tek®, is an organic medium made from shredded coconut husks. It comes in various forms due to different sources and production methods. The compressed briquette form is popular, needing soaking before use. When soaked, the coir expands up to six times its original size, making it ready for gardening use.

Expanded clay pellets: Expanded clay pellets, also called LECA, grow rocks, or Hydroton®, are created by heating dry, dense clay to form round, porous balls. They offer sturdy plant support while remaining lightweight. Their spherical shape and porosity maintain a balanced environment for roots, preventing excessive drying or drowning by promoting a good oxygen-to-water ratio.

Perlite: Perlite is a natural volcanic mineral that expands when heated, becoming lightweight, porous, and absorbent. It comes in different grades, with the most common diameters being 0 to 2 mm and 1.5 to 3 mm. It can be used alone or mixed with other growing media.

Vermiculite: It's a micaceous mineral heated to about 2,000 F, expanding into pebbles. Widely used for rooting, it's often combined with other media like coconut coir or peat moss for starting seedlings. Comes in various grades, with common diameters of 0 to 2 mm, 2 to 4 mm, and 4 to 8 mm.

Peat: Peat is made from partially decomposed marsh plants like sedges, grasses, and mosses. In horticulture, there are three types: Sphagnum peat moss, hypnum peat moss, and reed and sedge peat moss. Sphagnum peat moss is the preferred type due to its higher moisture retention and slower breakdown compared to other peat types.

Sand: Sand is one of the oldest and widely used hydroponic mediums. It's often mixed with other substrates like vermiculite, perlite, and coconut coir. Coarse sand is preferred for better root aeration, increasing the size of air pockets between the sand grains when used as a growing medium.

Gravel: Gravel, derived from rocks like sandstone, limestone, or basalt, is successfully used in ebb and flow systems. It has large spaces between particles, providing ample air to the roots. However, it doesn't retain water well, making it prone to quick root drying.

Expanded Shale: Expanded shale is made by heating quarried shale above 2,000°F, making it chemically and biologically inert. The process removes water, causing the shale to expand. This lightweight material is excellent for aquaponics grow beds, providing a large surface area for bacteria to convert ammonia into nitrates.

NUTRIENT SOLUTIONS USED IN HYDROPONICS

The nutrient solutions in hydroponics must provide a balanced mix of essential elements necessary for plant growth, which can be categorized into macronutrients and micronutrients.

Macronutrients: Primary nutrients, including nitrogen (N), phosphorus (P), and potassium (K), crucial for overall plant development.

Micronutrients: Trace elements like iron (Fe), zinc (Zn), and manganese (Mn), required in smaller amounts but vital for specific plant functions.

pH and EC Levels of Nutrient solutions: Monitoring and adjusting pH and EC levels are crucial for ensuring optimal nutrient absorption and promoting healthy plant growth in hydroponic systems.

Importance of pH and EC:

- **pH Levels:** Maintaining the correct pH is essential for nutrient availability. Most hydroponic plants thrive in a slightly acidic to neutral range (pH 5.5-6.5). Incorrect pH levels can impact nutrient uptake.
- **Electrical Conductivity (EC) Levels:** EC measures the nutrient concentration in the solution. Proper EC levels ensure plants receive an optimal nutrient balance, preventing overfeeding or nutrient deficiencies.

Source of Nutrient Element with their characteristics:

Table 1: Sources of elements with respective characteristics (Singh S. *et al.*, 2012)

Source	Element	Characteristics
Potassium Nitrate (KNO_3)	N, K	Very soluble salt
Potassium Phosphate monobasic (KH_2PO_4)	P, K	Corrects phosphorus deficiency
Magnesium sulphate ($MgSO_4$)	S, Mg	Cheap, highly soluble, pure salt
Iron chelate	Fe Cit	Best sources of iron
Boric acid (H_3BO_3)	B	Best source of boron
Calcium nitrate $Ca(NO_3)_2$	N, Ca	Very soluble salt

CROPS GROWN IN HYDROPONICS

A wide variety of crops can be successfully grown using hydroponic systems. The suitability of hydroponics for different crops depends on factors such as the specific system used, the crop's growth characteristics, and the grower's expertise. Hydroponics provides an environment where plants receive a controlled and optimized balance of nutrients, water, and oxygen, leading to faster growth and potentially higher yields compared to traditional soil cultivation. Growers often choose hydroponics for its efficiency, resource conservation, and the ability to grow crops in locations with limited access to arable land. Here are some common crops grown in hydroponics:

Table 2: List of crops suitable for commercial production under Hydroponics. (Singh S. *et al.*, 2012)

Type of crops	Common Name	Botanical Name
Fruits	Strawberry	*Fragaria ananassa*
Vegetables	Tomato	*Lycopersicon esculentum*
	Chili	*Capsicum frutescens*
	Brinjal	*Solanum melongena*
	Green bean	*Phaseolus vulgaris*
	Beet	*Beta vulgaris*
	Winged bean	*Psophocarpus tetragonolobus*
	Bell pepper	*Capsicum annum*
	Cabbage	*Brassica oleracea var. capitata*
	Cauliflower	*Brassica oleracea var. botrytis*
	Cucumbers	*Cucumis sativus*
	Melons	*Cucumis melo*

[Table Contd.

Contd. Table]

Type of crops	Common Name	Botanical Name
	Radish	*Raphanus sativus*
	Onion	*Allium cepa*
Leafy vegetables	Lettuce	*Lactuca sativa*
	Kang Kong	*Ipomoea aquatica*
Cereals	Rice	*Oryza sativa*
	Maize	*Zea mays*
Condiments	Parsley	*Petroselinum crispum*
	Mint	*Mentha spicata*
	Sweet basil	*Ocimum basilicum*
	Oregano	*Origanum vulgare*
Flower / Ornamental crops	Marigold	*Tagetes patula*
	Roses	*Rosa berberifolia*
	Carnations	*Dianthus caryophyllus*
	Chrysanthemum	*Chrysanthemum indicum*
Medicinal crops	Indian Aloe	*Aloe vera*
	Coleus	*Solenostemon scutellarioides*
Fodder crops	Sorghum	*Sorghum bicolor*
	Alphalfa	*Medicago sativa*
	Barley	*Hordeum vulgare*
	Bermuda grass	*Cynodon dactylon*
	Carpet grass	*Axonopus compressus*

PLANT SELECTION AND ADAPTATION

Successful hydroponic plant selection involves knowledge, environmental attention, and continuous monitoring for optimal growth conditions. Choosing plants for hydroponics involves selecting varieties that thrive in soilless environments and adjusting cultivation practices for optimal growth. Considerations include:

Crop Suitability:

- Leafy Greens: Lettuce, spinach, kale.
- Herbs: Basil, mint, cilantro.
- Fruit-bearing Plants: Tomatoes, peppers, cucumbers.
- Berries: Strawberries, blueberries.
- Microgreens and Sprouts: Well-suited for hydroponics.

Adaptation of Root Systems:

- Choose plants with fibrous, shallow roots.
- Use support structures for vining crops like tomatoes.

pH and Nutrient Preferences:

- Choose plants within the pH range of 5.5 to 6.5.
- Select crops with similar nutrient requirements for efficient management.

Temperature and Humidity:

- Consider plant preferences for temperature and humidity.
- Choose plants adapted to hydroponic conditions.

Disease Resistance:

- Prioritize varieties resistant to common diseases and pests.

Growth Habit and Space:

- Adapt hydroponic systems to accommodate plant growth habits.
- Plan for spacing and support structures.

Light Requirements:

- Consider lighting needs and ensure adequate exposure.

Crop Rotation:

- Implement rotation to prevent disease buildup and optimize nutrients.

Experimentation and Monitoring:

- Experiment with different plant varieties and hydroponic techniques.
- Regularly monitor plant health and growth.

Adaptation to System Types:

- Choose plants based on compatibility with hydroponic systems (NFT, DWC, aeroponics).

ENVIRONMENTAL CONTROL IN HYDROPONIC SYSTEM

Precise control of the prime environmental components determining plant growth and productivity viz. temperature, humidity, and lighting is essential in hydroponics to create an optimal environment for plant growth, development, and productivity.

Temperature Regulation: Ensures an ideal range for enzymatic processes, nutrient absorption, and overall metabolic activity. Typically, 70-75°F (21-24°C) for most plants. Variations exist based on specific crops. High Temperature can lead to heat stress, accelerated transpiration, and nutrient deficiencies. (Resh H.M., 2015)

Humidity Control: Prevents issues like mold, mildew, and diseases. Optimizes transpiration rates, impacting water and nutrient uptake. Generally, 60-70% for the vegetative phase, lower (around 40-50%) during flowering to prevent mold. Low Humidity may result in excessive transpiration, wilting, and nutrient concentration in plant tissues. (Resh H.M., 2015)

Lighting: In hydroponics, especially in indoor environments, artificial lighting plays a critical role in providing the necessary energy for photosynthesis and influencing plant development.

Role of Artificial Lighting:

- **Photosynthesis:** Supplies light energy crucial for the conversion of carbon dioxide and water into glucose.
- **Growth Stimulation:** Influences plant morphology, flowering, and fruiting.
- **Photoperiod Control:** Regulates day and night cycles for specific crops, impacting flowering and fruiting.

Types of Artificial Lighting:

- **Fluorescent Lights:** Suitable for seedlings and early growth stages.
- **High-Intensity Discharge (HID) Lamps:** Provide intense light suitable for various growth phases.
- **Light Emitting Diodes (LEDs):** Energy-efficient and customizable for different plant needs.

Optimal Light Spectrum and Duration:

- **Spectrum:** Adjust light spectrum based on growth phases; blue light for vegetative, red for flowering.
- **Duration:** Mimic natural day and night cycles; typically 14-18 hours of light for vegetative and 12 hours for flowering. (Resh H.M., 2015)

Effects of Inadequate Lighting:

- **Insufficient Light:** Stunted growth, poor flowering, and reduced yields.
- **Excessive Light:** Can lead to photoinhibition, affecting photosynthetic efficiency.

COMMON CHALLENGES AND SOLUTIONS IN HYDROPONICS

Although the hydroponic system is highly efficient, but there are some common challenges and their solutions, which need to be addressed, viz. control of plant diseases and nutrient imbalance. Disease control in hydroponics involves preventive measures and prompt responses to symptoms, while addressing nutrient imbalances requires ongoing monitoring, diagnostic tools, and appropriate adjustments to the nutrient solution.

Disease Control:

Potential Issues:

- Root Diseases: Pythium and Fusarium can thrive in the nutrient-rich, water-saturated conditions of hydroponic systems.
- Foliar Diseases: Powdery mildew and other foliar diseases can affect plant health.

Solutions:

- Hygiene Practices: Maintain cleanliness in the system to reduce the risk of pathogens.
- Beneficial Microorganisms: Introduce beneficial bacteria and fungi to outcompete harmful pathogens.
- Proper Ventilation: Ensure good air circulation to minimize conditions conducive to disease.
- Isolation: Isolate infected plants promptly to prevent the spread of diseases.
- Integrated Pest Management (IPM): Implement a comprehensive approach that includes biological controls, physical barriers, and, if necessary, targeted pesticide use.

Nutrient Imbalances:

Diagnosis:

- Visual Inspection: Monitor plant appearance for signs of deficiencies or excesses (yellowing, stunted growth, leaf discoloration).
- Nutrient Solution Analysis: Regularly test the nutrient solution to identify specific imbalances.

Correction:

- Adjusting pH: Maintain the appropriate pH (usually 5.5-6.5) to optimize nutrient availability. (Sardare M. D. *et al.*., 2017)
- Balanced Formulas: Use well-balanced nutrient solutions with the correct ratio of macronutrients and micronutrients.
- EC/TDS Monitoring: Regularly check and adjust electrical conductivity (EC) or total dissolved solids (TDS) to ensure the right nutrient concentration.
- Flush Systems: If imbalances persist, flushing the system with pH-balanced water can help reset nutrient levels.

Prevention:

- Regular Monitoring: Keep a consistent schedule for checking and adjusting nutrient levels.

- Proper Dilution: Follow recommended dilution rates to prevent over-fertilization.
- Water Quality: Ensure water quality is suitable, as poor water quality can contribute to nutrient imbalances.

FUTURE SCOPE OF HYDROPONICS ACROSS THE GLOBE

Hydroponics is experiencing remarkable growth and is considered the fastest-growing sector in agriculture. Forecasts suggest that it could eventually dominate global food production. This trend is driven by factors such as poor farmland management, overuse, and the need for innovative farming methods.

Countries like Japan, facing limited arable land and high land prices, are proactively adopting hydroponic technologies, primarily using methods like NFT or sand/gravel. In Japan, hydroponics has led to the development of more productive plants for rice production, allowing for multiple harvests in a year compared to traditional methods.

Urban areas, exemplified by cities like Indianapolis, are investing in local food production systems to boost economic development and reduce carbon footprints. Hydroponics in urban settings uses significantly less water compared to traditional farming methods.

In arid regions like Israel, hydroponics is solving water scarcity issues by growing fruits like berries and bananas in shipping containers. Despite the challenging climate, hydroponics in Israel yields produce at a rate 1,000 times greater than traditional methods. (Jan S. *et al.*, 2020)

Even large businesses, such as Target, have recognized the value of hydroponic systems. Target has implemented hydroponic gardens in selected locations, providing customers with fresh vegetables and herbs while minimizing water usage.

In conclusion, the future of hydroponics holds promise, with its potential recognized globally. It offers solutions to challenges like land scarcity, water shortages, and the need for sustainable agriculture. As innovation continues, hydroponics is set to play a significant role in shaping the future of food production.

INDIAN SCENARIO

Hydroponic farming is gaining momentum in India due to technological advancements that reduce costs and facilitate scalable operations. The integration of Artificial Intelligence and Data Analytics enhances food/crop planning, contributing to the food security ecosystem. The push for smarter and greener markets in densely populated nations creates opportunities for hydroponic farming, including rooftop and vertical farming. Despite the annual growth of hydroponics in Indian agriculture, there remains a considerable portion of farmers unaware of hydroponic systems.

Global demand for food continues to rise, and India is no exception. Hydroponics presents an alternative to traditional agricultural methods. However, low awareness, particularly due to the low literacy rate among farmers, poses a significant challenge. Educating farmers is crucial to the widespread adoption of hydroponic systems. Greenhouses offer a solution for farmers with limited land, enhancing crop productivity. (Solanki S. *et al.*, 2017)

In the face of natural disasters causing crop loss, hydroponic farming provides an alternative for farmers to grow organic, pesticide-free vegetables, contributing to economic growth. (Kaur S. *et al.*, 2023)

Despite the positive trajectory, public acceptance of hydroponic farming in India is hindered by factors such as poverty, lack of education, and resistance to change among rural farmers. However, the potential benefits, including significant water savings and the production of chemical-free crops, make hydroponics a promising solution for India's agricultural challenges. (Solanki S. *et al.*, 2017)

In conclusion, while challenges persist, the increasing demand for organic products and growing interest from farmers and researchers suggest that hydroponic crops may become more widespread in India over time. Addressing the need for financial and technical assistance for farmers is crucial for the successful adoption of hydroponic technology.

COMPARISON BETWEEN HYDROPONICS AND TRADITIONAL FARMING

Hydroponic farming has several great advantages over traditional methods. One key benefit is the ability to grow crops in areas with poor or contaminated soil. Plants grown hydroponically are more resilient to high-salt water. Besides the hassle of dealing with insects, animals, and diseases present in soil. If automated, significantly cuts down on labour costs.

In terms of efficiency, hydroponic plants use only a fraction of the water compared to soil-grown plants. Traditional farming often results in water quickly passing through the soil, but hydroponics conserves water by delivering nutrients directly to the roots. The nutrient solution needed for hydroponics contains just 25% of the essential elements found in solid fertilizers. Without the need to compete for soil nutrients, hydroponic plants can be grown closer together, maximizing space. (Prakash S. *et al.*, 2020)

Hydroponic systems also promote faster and larger plant growth, leading to higher yields per unit area compared to traditional farming. The only limitation is the amount of available light, allowing for more flexibility in spacing. In a nutshell, hydroponics offers a smart, efficient, and space-saving way to grow crops.

14. CONCLUSION

Hydroponics offers a sustainable solution to the increasing challenges of water scarcity and food production. As traditional soil-based agriculture faces difficulties, the hydroponics industry is expected to grow significantly. In rapidly urbanizing countries like India, hydroponics is poised to surpass conventional agriculture, enhancing both quantity and quality of produce for future food security. Despite initial capital investment, its minimal operating costs, reduced staff, and maintenance expenses make it a cost-effective alternative to conventional farming. The technology effectively tackles food safety issues related to soil contamination and pesticide residues. The global hydroponic production is expected to experience exponential growth in the coming years. While hydroponic systems are efficient, they demand specialized knowledge, and the initial setup can be expensive and complex. To make the system more economical, there is a need for innovation and equipment development. Although hydroponics requires fewer workers, it can significantly boost productivity. Nations can enhance food production by implementing hydroponics, whether through urban farms, government-supervised food parks, or village setups. When used wisely, hydroponics has the potential to contribute to self-sustainability. Accelerating the progress of hydroponics requires increased interest, governmental support, and focused research.

REFERENCES

Jan S, Rashid Z, Ahngar TA, Iqbal S, Naikoo MA, Majeed S, Bhat TA, Gul R and Nazir I (2020) *Hydroponics – A Review, International Journal of Current Microbiology and Applied Sciences*, ISSN: 2319-7706 Volume 9 published on 10th August 2020. https://doi.org/10.20546/ijcmas.2020.908.206

Karne H, Iyer V, Joshi S, Diwan S, Gole M, Sunthankar S and Phansalkar S (2023) Hydroponics: A Review of Modern Growing Techniques. European Chemical Bulletin. Special Issue 4, doi: 10.48047/ecb/2023.12.si4.1016

Kaur S and Dewan B (2023) Hydroponics: A Review on Revolutionary Technology for Sustainable Agriculture, *Asian Journal of Agricultural and Horticultural Research*, Volume 10, Issue 4 published on 9th Aug 2023. DOI: 10.9734/AJAHR/2023/v10i4270

Prakash S, Singh R, Kumari AR and Srivastava AK (2020) Role of Hydroponics towards Quality Vegetable Production: An Overview, *International Journal of Current Microbiology and Applied Sciences*, ISSN: 2319-7706 Special Issue-10 pp. 252-259. https://www.ijcmas.com/special/10/Satya%20Prakash 2,%20et%20al.pdf

Resh HM (2015) Environmental Control Components for Hydroponic Systems, Hydroponics for the Home grower. eBook ISBN:9780429156458, published on 20th July 2015, https://doi.org/10.1201/b18069

Sardare MD and Admane SV (2013) A Review on Plant Without Soil – HYDROPONICS, *International Journal of Research in Engineering and Technology*, ISSN: 2319 - 1163 Volume: 2 Issue: 3 published on March 2017. DOI: 10.15623/ijret.2013.0203013

Shrestha A and Dunn B (2017) Hydroponics, Extension, Oklahoma State University, Oklahoma Cooperative Extension Service, HLA-6442, published on April 2017. https://extension.okstate.edu/fact-sheets/print-publications/hla/hydroponics-hla-6442.pdf

Singer J (2021) A Brief Overview of the History of Hydroponics. Garden Culture Magazine. Issue 37, published on 8 Feb 2021. https://gardenculturemagazine.com/a-brief-overview-of-the-history-of-hydroponics/

Singh S and Singh BS (2012). Hydroponics – A technique for cultivation of vegetables and medicinal plants. In Proceedings of 4th Global conference on ¯Horticulture for Food, Nutrition and Livelihood Options Bhubaneshwar, Odisha, India. p.220.

Solanki S, Gaurav N, Bhawani G and Kumar A (2017) Challenges and Possibilities in Hydroponics: An Indian Perspective, *International Journal of Advanced Research*, published on Nov 2017, DOI URL: http://dx.doi.org/10.21474/IJAR01/5752

Sustainable Agriculture and Biodiversity: Innovations and Insights, Pages: 330–344
Edited by: Bibhuti Bhusan Sahoo, Harvinder Kumar Singh, Niranjan Chinara and Amitabh Panda

CHAPTER - 14

ROLES OF NEMATODES IN MUSHROOM CULTIVATION: BENEFICIAL, SAPROPHYTIC AND PARASITIC

Harjot Singh Sidhu

Regional Mushroom Research Centre
Maharana Pratap Horticultural University, Karnal-132001, Haryana
**Corresponding email: hss.apn@mhu.ac.in*

INTRODUCTION

Nematodes thrive in raw compost and can exist in excessive numbers during the mushroom growing process. While some mushroom growers believe that nematodes are merely an indicator that compost and casing preparation has not gone orderly, but somehow it is good to assume that nematodes can represent the risk of yield losses, and it is necessary to take precautions against their proliferation.

NEMATODES

Nematodes are tiny, very primitive roundworms. They appeared early on the evolutionary stage, being the first animals to evolve a body cavity. They are extremely abundant in both types and numbers. There are about 25,000 species currently known, but scientific opinion holds that the number of species actually could be 100 times greater (Blaxter, 2011). Typically, nematodes range in size from 0.2 mm to over 6 m in length, though some may be much longer (Blaxter and Denver, 2012). Nematodes are found in marine, freshwater, and soil habitats. It has been estimated that there are 8 billion nematodes in an average acre of field soil. One square meter of garden soil probably contains approximately 3.5 to 5 million nematodes (Yeates, 2003). Many are parasites; in fact, almost all types of creatures studied by scientists have at least one species of nematode that parasitizes them. Roughly 100 species parasitize humans (Ashford & Crewe, 2003). *Caenorhabditis elegans,* one of the saprophytic nematodes, has become an important tool for genetic and developmental researchers. This organism is made up of only 1,000 cells. It matures in 3 days and has a transparent body that allows scientists to watch the dividing cells.

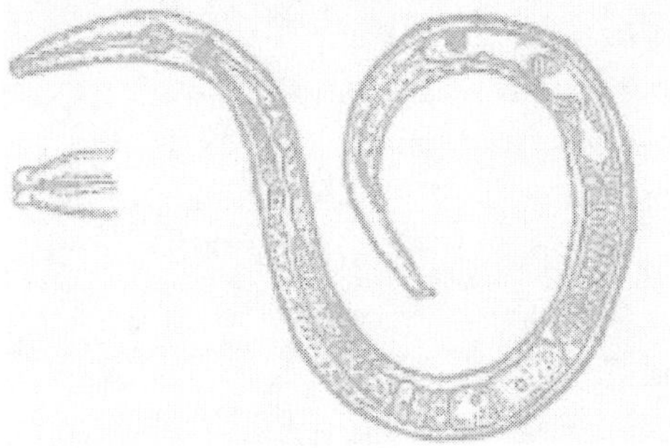

Figure 1. A pictorial representation of a general nematode.

NEMATODES IN MUSHROOM GROWING

It is fortunate that nematodes do not harm in raw compost, because they are ubiquitous in the materials used to prepare the compost mix, and their complete removal, if possible, would be extraordinarily expensive. The richness of the compost environment in terms of food, water, and oxygen provides nematodes with an excellent habitat, at least until composting temperatures reach lethal ranges. The cooler outer portions of the compost pile, if not mixed and turned into the interior, will continue to support nematode populations into Phase II.

There are four general types of nematodes: parasitic, saprophytic, predatory, and animal parasitic. Mushrooms nematodes can be grouped into at least two broad categories: beneficial and harmful. Mushroom growers have typically thought of nematodes as being harmful to the crop. Parasitic (mycophagous) and saprophytic nematodes are associated with commercial mushroom yield loss. The occurrence of mycophagous nematodes in mushroom culture is rare under modern mushroom farming techniques. Saprophytic nematodes are common. Generally, their overall economic impact on production is minimal; however, under heavy populations significant individual crop reduction is experienced. Beneficial nematodes, on the other hand, control insect populations. They may be either natural to the population or introduced to the mushroom crop to control mushroom pests.

For mushroom growers, the primary difference between these two groups lies in their feeding habits. The parasitic nematode feeds directly on mushroom mycelium, whereas the saprophytic nematode feeds on bacteria, protozoa, fungal spores, and other bits of organic matter, but does not attack the mycelium.

NEMATODE TYPES, SPECIES, BIOLOGY AND CONTROL

1. SAPROPHYTIC NEMATODES

These nematodes, often referred to as "free-living" now are more commonly associated with mushroom farming than the parasitic species. They characterize poorly prepared compost and/or casing and cause severe deterioration of mycelium in their own right. There is inconsistency as to the correlation of nematode populations and mushroom yield reductions. Nevertheless, all commercial mushroom operations attempt to minimize these free living nematode populations.

1.1 Symptoms and Signs

At low levels, these nematodes have little effect on mycelium. As the nematode numbers increase, mycelium begins to grow slowly and weakly. At high infestation levels, the mycelial strands completely degenerate. Inside the cropping rooms, black necrotic areas may be visible on the spawned compost surface prior to laying of the casing. These spots may show some colonized compost but the mycelium will be fragmented and the compost appears wet. These areas will not be re colonized by the mushroom mycelium. The surrounding colonized compost will degenerate, and the nematodes will migrate into the casing layer. With careful observation and a bright light, the nematodes can be recognized as they flicker (reflecting the light) on the compost straws.

Often the casing is well colonized by the mycelium but after ruffling or scratching, the mycelium does not re knit well in spots or the whole shelf. Sometimes the casing will initially be slow to be colonized by the mycelium. The mycelium is fragmented and the casing does not hold together well. As a consequence, yield is reduced (Kaufman et al., 1984).

The whiteness of the mushroom may be negatively affected. Bacteria on which the nematodes feed reproduce well in the moist environment. The nematodes and their associated bacteria lower the quality of the fresh mushrooms (Grewal, 1991).

Sometimes brown fungal colonies develop on the compost or casing. This fungus, *Arthrobotrys* spp., is a parasite of nematodes. Its presence is an indicator of high nematode populations (Fletcher and Gaze, 2008)

1.2 Causal Agents

The majority of saprophytic nematodes recovered from mushroom casing samples belong to the order Rhabditida. They are all bacteria feeders, characterized by having three or six lips fused or replaced by other structures. Their stoma lacks a stylet and their cuticle is annulated or smooth. Amphids are in conspicuous and the oesophagus has a terminal bulb. The tail of the male usually possesses a bursa supported by rays. There are no caudal glands.

They possess a muscular pharyngeal bulb, which creates the suction to draw in food particles and liquids. The saprophytic nematodes multiply even faster, one hundred-fold in 3 days than those that are parasitic. Many of these nematodes are parthenogenetic (self-fertile).

The common saprophytic species are listed below. In most cases of infestation in India, genera of common occurrence are *Rhabditis*, *Caenorhabditis*, *Panagrolaimus* and *Acrobeloides*.

Common Saprophytic Nematode Species

S. No.	Nematode species	Order
1.	*Acrobeloides apliticus*	Rhabditida
2.	*Acrobeloides buetschii*	-do-
3.	*Caenorhabditis elegans*	-do-
4.	*Cruzenema lambdiensis*	-do-
5.	*Panagrolaimus rigidus*	-do-
6.	*Pelodera (Pelodera) strongyloides*	-do-
7.	*Rhabditis (Cephaloboides) oxycera*	-do-
8.	*Rhabditis (Choriorhabditis) longicaudatus*	-do-
9.	*Rhabditis (Rhabditis) terricola*	-do-
10.	*Rhabditis (Pellioditis) pellio*	-do-
11.	*Rhabditis (Cephaloboides) oxycera*	-do-

1.3 Disease Cycle and Epidemiology

Saprophytic nematodes are common inhabitants of compost and casing mixtures. Saprophytic nematodes' feeding habits differ markedly from their parasitic counterparts. They suck in and chew the particles of food they consume. Under excessively wet compost conditions, bacteria have an advantage over mycelium, and as the nematode food source, their increase in numbers encourages the expansion of the nematode population. The high numbers of bacteria also inhibit normal growth of mycelium. The compost deteriorates and becomes wet and increasingly anaerobic. Under slow drying conditions, especially during the pre pasteurization phase of compost, some nematodes can form resistant stages which can enable them to survive pasteurization temperatures. Insects, equipment, workers, and irrigation of the casing can disperse the nematodes. In older farms with wooden ceilings, nematodes can reproduce in the wet insulation and drop onto the compost through the condensation on the ceiling.

Research suggests that the extent of the saprophytic nematode damage to mycelium is closely tied to the number of bacteria, the nematodes' primary food source, present in the spawned compost or casing. Their detrimental effects on the mycelium appear to be linked to the release of a toxin or byproduct into the compost. Extracts taken from diseased compost and casing show there is greater crop damage when both bacteria and nematodes are present in high numbers than when only bacteria are present. The enhanced crop injury may be the result of increased production of toxins when both are present, or may reflect some way in which the nematodes make possible a more rapid or thorough bacterial colonization of the compost. There is an interesting ecological relationship among the nematodes, bacteria, and mycelium.

1.4 Management Strategies

Good sanitation and hygiene practices reduce the spread of nematodes on the farm. Thorough post crop pasteurization and cleaning of the production room, netting, and equipment will reduce carry over from one crop to another. Farms with wooden shelving or trays need to do an especially thorough post crop heat treatment since the nematodes may be located in crevices of the wooden boards and survive the post crop heat treatment.

Under less wet conditions, the mycelium can spread, use the water for its own growth, and dry out the compost to a point that inhibits the bacterial proliferation. The nematodes remain in low number because of the dry conditions and the limited food source. The environment remains favorable for mushroom production.

Maintenance of proper temperatures during pasteurization of compost is critical to reduce the nematode threat. If the surface of the compost becomes dry during the pre pasteurization phase, resistant stages of the nematodes may survive the pasteurization temperatures. Changes in the length of the pre pasteurization period, air flows, and humidity control may be necessary.

The casing material can be a source of nematodes. Once nematodes are noticed on the compost or casing the best control is to reduce their spread in the room and on the farm and to determine the source or cause of infestation. Tools and equipment should be sanitized between shelves during the spawning operation. Ruffling or scratching should be avoided where nematodes are visible on the casing. There are no registered chemical controls.

2. PARASITIC NEMATODES

Parasitic or mycophagous nematodes were serious pests of commercial mushroom production. These nematodes, also referred to as fungal-feeding or mycophytic nematodes, are increasingly rare in commercial mushroom farming today. Presently, mushroom growers choices of casing materials or pasteurization of casing usually avoid outbreaks. In the past, however, they were responsible for disastrous crop losses. Extensive sampling of commercial mushroom facilities in India over the years has not revealed the presence of these nematodes. However, they remain problematic in low cost thatched houses of the country.

2.1 Causal Agents

The mycophagous nematodes like *Aphelenchus avenae*, *Aphelenchoides spp.* (particularly *Aphelenchoides composticola*), *Ditylenchus myceliophagus* are common in mushroom beds (Bajaj and Kanwar, 2011). In India, 8 spp., of Aphelenchoides (*A. agarici*, *A. asterocaudatus*, *A.composticola*, *A. minor*, *A. myceliophagus*, *A. neocomposticola*, *A. sacchari*, *A. swarupi*) and *Ditylenchus myceliophagus* have been frequently reported from mushroom beds.

The common parasitic species are listed below. In most cases of infestation, two genera of these nematodes are present.

S. No.	Nematode species	Order
1.	*Ditylenchus myceliophagous*	Tylenchida
2.	*Aphelencoides agarici**	Aphelenchida
3.	*A. asterocaudatus*	-do-
4.	*A. composticola**	-do-
5.	*A. dactylocerus*	-do-
6.	*A. radicicolus*	-do-
7.	*A. neocomposticola*	-do-
8.	*A. sacchari**	-do-
9.	*A. saprophilus*	-do-
10.	*A. swarupi**	-do-
11.	*Aphelenchus avenae*	-do-
12.	*Paraphelenchus myceliophthorus*	-do-
13.	*Seinura winchesi*	-do-

* Species reported from India

2.2 Symptoms and Signs

The symptoms and signs of mycophagous nematodes are similar to that of the saprophytic ones. They feed on the mushroom mycelia by using their stylet (a needlelike mouthpart) to pierce the mycelial wall with their hollow stylet and suck the contents of the pierced cells. They may attack the mycelium at any time from spawning onwards. The fine strands of the mycelium are destroyed (Figure 2) and the casing/compost becomes soggy and smells with little or no production in the area. Effects on mushrooms can range from little damage to total elimination of the crop. The appearance of the compost can gives clues to the damage to come; when dark, watery, barren patches develop, production will be severely affected.

Figure 2. Patch of wet unproductive casing on a nematode infested bed.

2.3 Disease Cycle and Epidemiology

As nematodes move through the mycelium-filled compost, they first destroy the fine hyphal structures and leave the mycelium looking stringy. Thereafter, larger mycelium is destroyed, leaving small barren bed areas that grow progressively larger as the nematodes moves outward into healthy compost. If the conditions are optimal for the nematodes, moderate temperature (20-25°C) and wetness entire beds can be denuded of their mycelium. Depending on the number of nematodes on the bed, the mushroom crop will be reduced or eliminated. Under good conditions, nematodes can multiply 30- to 100-fold in 2 weeks. When their rapidly growing population exhausts the compost of its nutrients, the nematodes respond to the changing environment by swarming to the surface. Exposed there, they can be picked up easily by vectors such as humans and flies. If dried slowly, the nematodes become dormant and can be distributed by even slight air movements. Compost infested with nematodes has a characteristic appearance:

soggy, sour smelling, and depressed. The nematode trapping gray mold, *Arthrobotrys superba,* may appear in areas where the mycelium has been destroyed. This soggy mess is apparently good habitat for the saprophytic nematodes.

These mycophagous nematodes may be found in soil and cocopeat. One female can produce up to 500 eggs. They can increase rapidly and reproduce as much as 2500 times in one week.

As the food source disappears the nematodes group together in clumps or swarms. As with the saprophytic ones, a light on the compost or casing surface will reveal their presence. This behaviour permits "easy" access to flies that can transport them to other locations. These nematodes can survive for six weeks without feeding and in a desiccated state for upwards of three years. Poorly post crop heat treated spent compost could be a good source of these nematodes.

2.4 Sources of Inoculum

Nematodes are carried into the growing process in a number of ways; the most obvious of which is in the compost. As noted above, nematodes are associated with raw materials entering the compost yard and respond by proliferating in the initially favorable compost environment. On the cooler parts of the compost pile, inside clumps of compost materials, or in excessively wet compost, they may survive Phase I and enter Phase II. The vast majority of their population may be destroyed in a Phase II room during pasteurization at about 60°C, but survivors can persist in wet areas, dry areas, and clumps, and can continue on to spawning. There they encounter a more favorable environment: temperatures around 24°C, moist conditions, and near-neutral pH (7.5). Mixing at spawning distributes the nematodes. The spawning machine can contaminate many subsequent beds or trays after spawning a single infested batch of Phase II compost. When cocopeat is introduced into the growing process, is dry, has a low pH, and usually does not contain nematodes. Once in place as casing, however, as conditions become more moist, cocopeat provides a favorable environment. Compared to compost, casing provides a habitat with less interference by mycelium.

Another source of nematodes in a crop is a preceding infested crop. In growing rooms, woodwork and ceiling insulation contaminated with nematodes can inoculate successive crops. If high temperatures during pasteurization do not penetrate into the wood, especially into the cracks and crevices, nematodes will not be destroyed. Likewise, moisture dripping from contaminated ceiling insulation spreads nematodes to new beds. Nematodes can invade growing rooms from other areas of production in a variety of ways. Dust can carry dormant nematodes between rooms, and flies, mites, hands, boots, and tools can carry nematodes acquired from contact with swarms or contaminated materials. Equipment such as spawning machines, if not designed for easy cleaning and if not routinely sanitized, provide nematodes with an effective means of distribution.

Sampling, Separation, and Identification

Compost piles, trays, or beds that are suspected of infestation can be tested for presence and relative quantity of nematodes. In any testing procedure, the integrity of the sample is critical; in this case, the location at which the sample is collected can strongly influence the results. Collecting samples from the hottest portions of the materials usually will give negative results because nematodes rarely survive there. Sampling at cool locations or areas where heating has been non uniform in the past is more likely to produce detection of the pests. Incubation of samples sometimes is necessary to provide adult nematodes for identification. Nematodes can be separated from compost or casing and identified visually. A Baermann funnel (Figure 3) is a convenient tool for collecting nematodes for further investigation. The apparatus consists of a support system that holds a funnel with a bottom tap closure. The funnel is filled with fresh water. A cloth or strong, fine

Figure 3. A Baermann funnel is a simple tool for extracting nematodes from material.

mesh bag is suspended over the water at the top of the funnel. Strong commercial tissues will work. A sample of compost or casing is placed into the bag. The nematodes move out of the sample, and since they are slightly denser than water, sink down the funnel until they are stopped by the tap closure. The tap closure is opened after the migration has proceeded for several hours, and the nematodes are collected in a shallow glass dish.

Nematode identification by non-experts is limited to differentiation of general type or genus level. The most useful distinguishing characteristic is the appearance of the anterior (front) end where the mouthparts are located. A blunt end from which the needle-like stylet can extend shows the organism to be parasitic. Saprophytic nematodes lack the stylet and appear to have liplike bulbs stuck on their anterior ends. By noting the structures of the anterior end, the shape and size of the internal structures, the body length and other features, and contrasting them to published illustrations of nematode types (rhabditoid, aphelenchoid, etc.), the nematodes' identity can be determined well enough to make control judgments or to process comparisons. But, for the grower, simply the existence of nematodes, regardless of the species, is a problem.

2.4 Management Strategies

Because nematodes are ubiquitous, total prevention of nematode invasion and total eradication during infestations is unlikely. These mycophagous nematodes may be found in soil. Avoiding incorporation of soil into compost, avoiding runoff from soil onto the compost or compost ingredients, or preparation on soil will minimize the risk from infection.

The Phase I compost pile provides the grower with his first opportunity for nematode control. Good Phase I temperatures, uniform mixing of the raw materials, and ensuring that cool shoulders are moved to the interior positions are measures that will reduce the number of nematodes moving on to Phase II.

Phase II compost should be pasteurized at 60°C for 2 hours to subject the nematodes to killing temperatures. This temperature range is adequate to kill wet nematodes, but if the compost and nematodes dry out, temperatures as high as 71°C would be required for lethal effect. Maintaining adequate uniform phase II pasteurization temperature will assist in elimination of this nematode. If the compost dries during warm up to pasteurization, then some nematodes may go into a resistant stage and infest the crop at spawning.

Pasteurization temperatures, as noted above, and careful watering (moist but not wet) can reduce or eliminate nematode populations; but failure to manage these environmental conditions can allow nematodes to move further into the growing process. At spawning, any compost suspected of harboring significant nematode populations should be processed last and followed by careful cleanup of the machinery. During cropping, the grower can do little to control an infestation other than to prevent further spread. A diligent attention to sanitation throughout the growing process, will contribute greatly to nematode control.

Casing should be pasteurized at 60°C unless the grower is confident that each lot is free of nematodes. Recontamination of the casing should be avoided.

Once an infection site is identified, it is quite important to maintain vigilant control of flies. Covering the infected area with plastic will minimize movement by personnel, insects, or irrigation. Adequate management of temperatures during post crop heat treatment, disposal of the spent material away from the farm, a thorough cleaning of the room prior to reuse and a heat treatment of the cleaned empty room before fresh substrate is added are critical.

In addition to typical sanitation practices, the following should be considered:

- Develop and enforce rules restricting personnel movement between compost areas and growing rooms.
- If trays, shelf boards, or other such itcms cannot bc pasteurized, they can be washed with a steam pressure washer, though this treatment may be insufficient to kill all nematodes. Sanitizers should be used in rooms and on equipment. This is especially important on floors, since they act as a heat sink during pasteurization and rarely can have their temperatures raised sufficiently to kill nematodes.
- Redouble sanitation efforts at spawning. Review spawning machine design and positioning to determine if better cleaning is possible with adjustments or retrofitting.
- Take precautions against nematode spread by bits of spilled compost from infested trays. Clean up any dropped materials before they are carried into non-infested areas.

No chemicals are registered for control although some literature suggests Thionazin (80 ppm) (Singh and Sharma, 2016) and Dichlorvos (0.04%) (Keshari and Kranti, 2020) may reduce population growth of *Aphelenchoides*. Further, since nematicides are not available to mushroom growers, measures to prevent or control infestations are limited to ensuring that normal temperature and sanitation safeguards are followed, and that enhanced measures are instituted when necessary. A well-prepared IPM program should outline clearly these measures and help keep all members of the growing team on track.

3. BENEFICIAL NEMATODES

Beneficial nematodes are new to the thought processes of mushroom farmers. Beneficial nematodes can kill the host directly, reduce its ability to produce offspring, or be a carrier of disease to the insect. These nematodes fit well into biological control strategies for insects.

3.1 Endoparasitic Nematodes

The phorid fly (*Megaselia halterata*) has within this insect's population a naturally occurring nematode, *Howardula husseyi*. The nematode enters the fly larva and reproduces within the fly. As the nematodes develop inside the fly, they consume the fly's reproductive system. Thus, an infected female produces few, if any, offspring. When she attempts to lay her eggs, the nematodes are released back into the compost or casing, ready to infect other phorid larvae (Figure 4). Commercialization of this nematode for biological control has not been successful.

3.2 Entomopathogenic Nematodes

Another more successful commercial nematode product for mushroom fly control has been the development of nematodes that enter the fly larva and release infective bacteria that kill the fly larva.

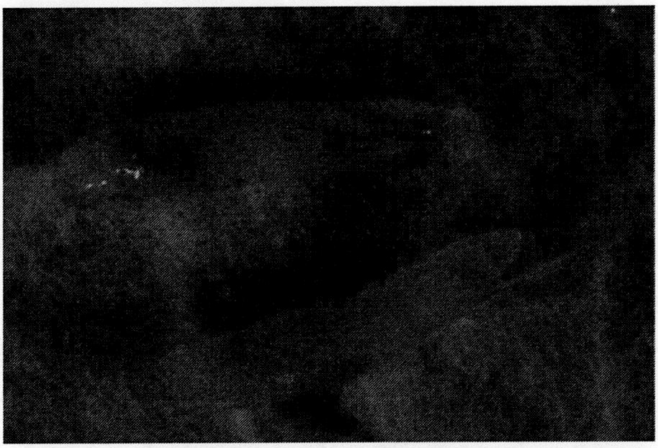

Figure 4. Phorid larvae

These nematodes either bore through the body of the insect or go through natural openings in the insect larva. Once the infective nematode enters the fly larva, it releases bacteria that it carries in its digestive system. As the fly larva dies the nematode uses the cadaver to reproduce more young.

One of the more common nematode species for mushroom fly control is *Steinernema feltiae*. It is one of most effective species against the sciarid fly,

Lycoriella ingenua (syn. *Lycoriella mali*) with 70% average control at the farm level. However, no entomopathogenic nematode species has been reported to control the phorid fly. The nematodes are typically applied to the casing material in the irrigation water with the timing of application important to maximize control. These nematodes prefer to attack large fly larvae or young pupae, about 12-16 days from egg laying. This necessitates appropriate monitoring of invading adult populations. The nematodes may reduce the mycelial growth on the casing surface with a slightly weaker mycelial growth in the casing at pinning time when applied at casing time and first harvesting may be delayed up to one day (Rinker *et al.*, 1995, 1997).

Survival Characteristics

Nematodes owe their abundance and widespread distribution in part to their remarkable survival abilities. If dried slowly, they enter a heat-resistant dormant state that can persist for years until they contact enough moisture to break dormancy. In the dried state, they are distributed easily by air currents. They also can survive without food for months. They are not susceptible to cold or freezing, and they regain their vigor once temperatures are more moderate. When available food in compost begins to deplete due to their high numbers, they show a collective swarming behavior that brings them to the compost surface for a greater chance of dispersal. The saprophytic nematodes take this group behavior a step further and form into columns of living nematodes, hundreds strong, that wave about on the surface of the mushroom bed, ready to adhere effect is to hands, tools, flies, or other objects. This phenomenon is called winking. The waving strands of winking nematodes can be observed by holding a flashlight at a 45-degree angle to the bed (Figure 5). A grower's IPM plan should take into account these survival traits to minimize the opportunities for nematode dispersal.

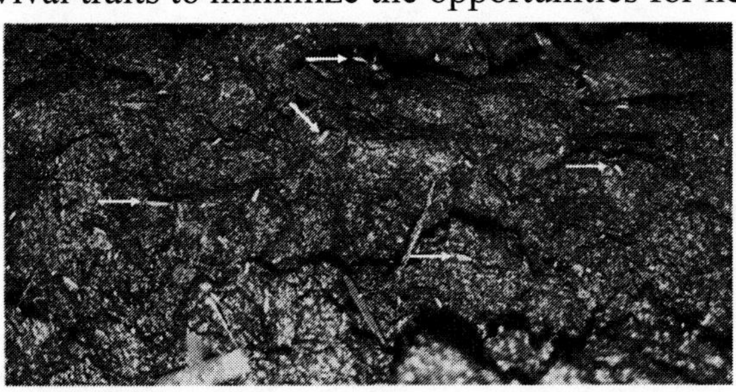

Figure 5. Nictating ('winking') columns of saprophagous nematodes at the surface of nematode-infested casing. Some of the nictating columns are indicated (yellow arrows). The glistening casing surface is evidence of free water and very wet conditions. Note how far above the casing surface some of the columns extend (white arrow) to adhere to flies and other vectors.

THE FUTURE

The future may hold new options for nematode control. At the present, we have few ways of dealing with nematodes once they are infesting a bed. But new applications of biological control may change that. Nematode-trapping fungi, *Arthrobotrys irregularis*, hold promise for pest control. These fungi can pierce the outer surface (cuticle) of nematodes, invade their bodies, and lay spores on the inside. In the process, the nematode is killed, but before death comes, its movements can disperse the fungi by carrying spores through the compost or casing. Research has shown that NSKWE (4 % @ 7.5 litre/q compost) has been found effective and recommended to control the myceliophagous nematodes in mushroom without affecting yield, but when NSKWE was tested against predatory nematodes, it killed the predator after 24 hours, showing its incompatibility with predator, so not developed a wide following (Keshari and Kanwar, 2021). A great aid to the grower would be the development of strains resistant to the toxins that are suspected of being produced in nematode infestations. Off-white and white strains were shown years ago to react differently to extracts prepared from nematode-infested compost and casing, giving rise to hopes for commercially effective levels of resistance in the hybrid mushroom strains used today. Unfortunately, such strains are not available yet, but their pursuit is a worthwhile venture for researchers.

CONCLUSION

This chapter apprises farmers to avoid the belief that nematodes are "not really harmful." A complacent opinion that will not serve the grower well is that nematodes, especially saprophytic nematodes, are not really pests, or that they are only indicators of other problems. A valid IPM program must consider the potential for nematode infestation and detail all farm specific anti-nematode measures to avoid expensive crop loss, especially since pesticides are not an option.

REFERENCES

Ashford R and Crewe W (2003) Parasites of Homo sapiens: An Annotated Checklist of the Protozoa, Helminths and Arthropods for which we are Home (1st ed.). CRC Press. Pp 152.

Bajaj HK and Kanwar RS (2011) Nematode infestation in mushrooms. In: M.R. Khan and M.S. Jairajpuri (Ed.) Nematode Infestations Part II: Horticultural Crops, 357-382pp.

Blaxter M and Denver DR (2012) The worm in the world and the world in the worm. BMC Biology 10: 57.

Blaxter M (2011) Nematodes: The Worm and Its Relatives. *PLOS Biol. 9*, e1001050.

Fletcher JT and Gaze RH (2008) Mushrooms: Pest and Disease Control: A Colour Handbook. Manson Publishing Ltd.: London. 192 pp.

Grewal P (1991) Relative contribution of nematodes (*Caenorhabditis elegans*) and bacteria towards the distribution of flushing patterns and losses in yield and quality of mushrooms (*Agaricus bisporus*). *Annals of Applied Biology* 119:483-499.

Kaufman T, Lukezic F and Bloom J (1984) The effect of free living nematodes and compost moisture on growth and yield in *Agaricus brunnescens*. *Canadian Journal of Microbiology* 30:503-506.

Keshari N and Kanwar RS (2021) Bio-management of Nematodes in *Agaricus bisporus* by *Fictor composticola* and its compatibility with Neem Seed Kernel Water Extract. *Journal of Agri Search* 8(3): 255-259.

Keshari N and Kranti KVVSK (2020) Integrated Management of Phytopathogenic Nematodes Infesting Mushroom. In: Ansari, R., Rizvi, R. and Mahmood, I. (eds) Management of Phytonematodes: Recent Advances and Future Challenges. Springer, Singapore. 141-170pp.

Rinker DL, Alm G and Olthof THA (1997) Use of nematodes to control mushroom flies on commercial mushroom farms in Ontario. *Mushroom News: Science and Technology* 45(4):6-11.

Rinker DL, Olthof THA, Dano J and Alm G (1995) Effects of entomopathogenic nematodes on control of a mushroom infesting sciarid fly and on mushroom production. *Biocontrol Science and Technology* 5:109-119.

Singh UA and Sharma K (2016) Pests of mushroom. *Advances in Crop Science and Technology* 4:1-6.

Yeates GW (2003) Nematodes as soil indicators: Functional and biodiversity aspects. *Biol. Fertil. Soils* 37:199-210.

Sustainable Agriculture and Biodiversity: Innovations and Insights, Pages: 345–380
Edited by: Bibhuti Bhusan Sahoo, Harvinder Kumar Singh, Niranjan Chinara and Amitabh Panda

CHAPTER - 15

SUSTAINABLE MILLET CROP IMPROVEMENT THROUGH MODERN BIOTECHNOLOGY

Rajeeb Lochan Moharana[1*], Chandana Behera[1], Srujani Behera[1], Simly Das[1], Nirakar Ranasingh[1], Sujit Majhi[1] and Kumar Avinash Biswal[2]

[1]*College of Agriculture, OUAT, Bhawanipatna, Kalahandi-766001, Odisha*
[2]*MS Swaminathan School of Agriculture, CUTM, Paralakhemundi-761200, Odisha*
**Corresponding email: rlmoharana@ouat.ac.in*

INTRODUCTION

In recent decades, both global and Indian agriculture have experienced significant growth in food production. These days, the industry is facing issues including malnutrition and climate change. Due to growing worries about the abuse of irrigated farm land, it is essential to prioritize dry and marginal lands due to the normally lower fertile soil in these areas, in an effort to lessen the effects of both current and projected climate change it is challenging to grow conventional grains like wheat, rice and millets are a viable alternative for planting in these areas due to their robust and climate-resilient qualities. In addition to their resistance, millets are a great way to fight malnutrition because they are full of vital vitamins and minerals. Millets, often referred to as nutricereals, are high in antioxidants, vitamins, minerals, vital fatty acids, and phytochemicals that can help combat hidden hunger. Health benefits include slower fat absorption, a low glycemic index, and a decreased risk of cardiovascular diseases, diabetes, as well as elevated blood pressure are also associated with their polyphenol level and bioactive chemicals. Proso millet (*Panicum miliaceum*), Finger millet (*Eleusine coracana*), foxtail millet (*Setaria italica*) & pearl millet (*Pennisetum glaucum*, sometimes known by synonyms such as *P. Americanum*, *P. typhoideum* & *P. typhoideum*) are the four main millets. Barnyard millet (*Echinochloa species*), Browntop millet (*Urochloa ramose/Brachiaria ramose/Panicum ramosum*) small millet (*Panicum sumatrense*), and kodo millet (*Paspalum scrobiculatum*) are examples of minor millets (Upadhyaya *et al.*, 2006). The management of genetic resources includes tasks including gathering, characterizing, and conserving the diversity of crop species for use in future crop enhancement initiatives (Singh *et al.,* 2014). Plant breeding lines, wild relatives, and cultivated cultivars can all be examples of this genetic variety. Plant Genetic Resources (PGR) are crucial for improving a plant species' genetic makeup, and they can be sourced from a range of natural variants like both native and exotic collections, advanced cultivars and new varieties, recombinant inbred lines (RILs), near-isogenic lines (NILs), mutants, genetic stocks, and relatives that are weedy and unpredictable (Singh *et al.,* 2014). Effective in management of these resources is vital for supporting both current and future research and crop development efforts, ensuring the preservation and use of genetic diversity for crop improvement.

Traditional enhancement of crops programs based on indicators of phenotype often progress more slowly than expected due to the fact that genotypic performance is not always accurately reflected by phenotypic traits. In addition, environmental factors influence phenotypic markers, which may only partially align with the desired traits. In contrast, molecular markers exhibit coinheritance with target traits, are easier to screen, and are unaffected by environmental conditions (Chhabra *et al.*, 2001). Among molecular markers, isozyme markers are among the oldest (Manivannan *et al.*, 2013).

The phrase "isozymes" was initially coined by Markert & Moller in 1959 to describe the various molecular forms of proteins that share identical enzymatic specificity (Shannon, 1968). Isozymes, or isoenzymes, are enzymes that catalyze the same chemical reaction but differ in their structural forms and catalytic efficiencies. These enzymes are often identified by their different electrophoretic mobilities (Eun, 1996). For species of plant breeding and genetic research, isomerases have proven to be reliable genetic markers due to their consistent expression under a range of environmental conditions (Heinz, 1987).

According to William and Mujeeb (1992), isozymes offer important information when comparing cultivars according to the existence or lack of other bands and the degree of intensity of common bands. According to Manivannan *et al.* (2013), isozyme analysis of patterns is also regarded as a crucial technique for identifying hybrids and evaluating the genetic links between individuals. Isozymes are gene expression products that serve as key tools in molecular characterization (Manivannan *et al.,* 2013). Such techniques have been used to characterize specific type of plant genotypes, breeds, and inbred lines, as well as to make chromosomal maps and evaluate variation in genetics in plant populations (Moore & Collins, 1983).

At the molecular level, isozymes are frequently used as genetic markers to track differences and inheritance across species or cultivars among a species (Manivannan *et al.*, 2013). Additionally, they contribute to the direct and indirect research of host-pathogen interactions (Mahatma *et al.*, 2011). Research on enzyme diversity in pearl millet has primarily focused on its relationship with disease resistance or susceptibility, as genetic markers for linkage map development, and to evaluate taxonomic or phylogenetic relationships within the species, as well as to study genetic variability in relation to geographic distribution and evolution.

ISOZYME MARKERS IN CROP IMPROVEMENT

Phenotypic marker-based crop improvement programs often face delays due to the inability of phenotypic assessments to accurately reflect genetic performance. Environmental influences can alter phenotypic markers, leading to weak correlations with the desired traits (Manivannan *et al.*, 2013). Isozymes, however, serve as reliable tools in genetic studies and are also valuable in understanding host-pathogen interactions (Mahatma *et al.*, 2011). The analysis of isozymes in pearl millet has been instrumental in evaluating disease resistance, constructing linkage maps, analyzing taxonomic relationships, and assessing genetic diversity based on geographic distribution and evolutionary trends. Mahatma *et al.* (2011) conducted a comprehensive study utilizing isozyme markers, total seed protein profiling, (ISSR) Inter-simple sequence repeat markers & (RAPD) randomly amplified polymorphic DNA to distinguish pearl millet genotypes based on resistance to downy mildew. Resistant genotypes displayed two unique proteins (~97 kDa and ~100 kDa) identified through seed protein analysis. ISSR analysis revealed six while these vulnerable genotypes were the only ones with UBC-857 (400 bp), UBC-825 (900 bp), UBC-827 (900 bp), & UBC-857 (1000 bp, 700 bp, 375 bp, & 200 bp) were separate markers in resistant genotypes. Combining these markers successfully categorized genotypes into resistant and susceptible groups, highlighting the potential of ISSR and protein profiling for germplasm screening against illness.

Anatala *et al.* (2015) investigated pearl millet's drought tolerance mechanisms, examining physiological, biochemical, and protein changes under water stress. Biochemical results showed elevated proline and glycine betaine levels under drought conditions. Isozyme profiling revealed consistent banding patterns across treatments, though the precise biochemical pathways for drought tolerance remained unclear. Manivannan *et al.* (2013) explored peroxidase isozyme diversity in 21 pearl millet genotypes, including hybrids, male parents, female parents, and open-pollinated varieties. They detected five bands with values between 0.5 and

0.64 for Rm/Rf, with one polymorphic band at Rm = 0.55 common across all genotypes. The study successfully distinguished only three genotypes (H 77/29-2, HMS 7A, and HHB 94). Similarity indices further revealed that hybrids and their female parents, such as HHB 146, HHB 50, HHB 67, and HHB 60 shared high genetic similarity (SI = 1.000), indicating that the female parent contributed more significantly to hybrid development. Peroxidase isozyme activity was further analyzed by Shivakumar *et al.* (2003) in relation to downy mildew resistance. Highly resistant seedlings exhibited 22 isozymes with pI values between 3.6 and 9.8, whereas there were only twelve isozymes visible in sensitive seedlings. Differences in peroxidase activity were observed across categories: seedlings that are extremely resistant, resistant, vulnerable, and highly susceptible recorded activity levels of 55.2–54.9, 40.9–44.8, 40.8–41.9, and 33.6–32.9 units, respectively. Following pathogen inoculation, enzyme activity increased in resistant genotypes but decreased in susceptible ones, indicating peroxidase's role in disease resistance. Superoxide dismutase (SOD) has also been linked to disease resistance in pearl millet. After being vaccinated against *S. graminicola,* Babitha *et al.* (2002) reported a notable increase in the activity of SOD in resistant genotypes., whereas susceptible genotypes showed a delayed and lower response. Similarly, Babitha *et al.* (2004) observed an increased lipoxygenase (LOX) expression in tolerant seedlings by 2.4 times, attributed to the emergence of a novel LOX isozyme, LOX-6, absent in susceptible genotypes. These findings emphasize the roles of SOD and LOX enzymes in downy mildew resistance.

IMPORTANCE OF ISOZYME MARKERS

Isozyme markers offer significant advantages over phenotypic markers due to their stability, consistent inheritance with target traits, and independence from environmental influences (Chhabra *et al.,* 2001). Markert and Moller were the first to use the phrase "isozyme" (1959) to describe proteins that catalyze the same biochemical reactions but differ in their structure and catalytic properties (Shannon, 1968). These enzymes are often identified by their distinct electrophoretic mobility (Eun, 1996). Isozymes have become trustworthy genetic markers in genetic research and breeding initiatives due to their stable expression across various environmental conditions (Heinz, 1987). According to William and Mujeeb (1992), isozyme analysis provides critical insights into cultivar differentiation by evaluating the intensity and presence of specific bands. This technique is beneficial for assessing genetic relationships, identifying hybrids, mapping chromosomes, and analyzing variability within species (Moore and Collins, 1983; Manivannan *et al.*, 2013). Isozymes are essential molecular tools for studying gene expression, inheritance patterns, and genetic characterization across diverse plant populations.

Kumar *et al.* (2007) conducted a study using seed esterase markers to identify the unique genotypes of 45 pearl millets, which included 14 hybrids together with their ancestry. The results showed that 36 genotypes were distinctly identifiable, while the remaining 11 genotypes were grouped into four clusters, with an extremely low probability (2.29×10^{-3}) of being identical by chance. The study further demonstrated that seed esterase markers were highly polymorphic, consistent across generations, untouched by environmental conditions, and not connected to any physical traits.

MILLET GENOMICS AND OMICS SCIENCE

In recent years, biofortified millet varieties have been developed using conventional breeding techniques. However, these methods alone may not suffice to address the nutritional and dietary needs of the increasing world's populations (Kudapa *et al.*, 2023). The introduction of advanced omics technologies offers a promising pathway to accelerate the development of biofortified millet and sorghum cultivars suited for future climate challenges (Varshney *et al.*, 2021). Integrating various omics methods, such as proteomics, metabolomics, transcriptomics, & ionomics is essential to creating nutritionally enriched crops for the future.

GENOMICS

Significant genomic developments have enabled the formation of a variety of marker systems, including SSRs (simple sequence repeats), SNPs (single nucleotide polymorphisms), and InDel's markers, and more recently, haplotype-based SNP markers (Kudapa *et al.*, 2023). One of the challenges in sorghum production lies in the limited bioavailability of micronutrients caused by complex compounds such as polyphenols and tannins. To address this, Phuke *et al.* (2017) developed two parent lines (296B × PVK 801) with varying amounts of iron (Fe) and zinc (Zn) in the grain to create the (RIL) recombinant inbred line population. Their results revealed significant differences in the quantities of Zn (10.2–58.7 mg kg^{-1}) & Fe (10.8–76.4 mg kg^{-1}) in the crop. In Grain, Fe & Zn concentration, notably, showed an inverse association with total yield but a significant correlation with 100-seed weight. Additionally, the study found a substantial genotype-by-environment (G×E) interaction, with environmental circumstances having a greater impact on Fe concentration than Zn. Shakoor *et al.* (2016) used (GWAS) genome-wide association mapping of variables associated with the sorghum seed ionome, which depicts the mineral, trace element, and nutritional composition of seeds, to look into the genetic factors that influence variance in mineral content (Salt *et al.*, 2008). The genes they found may be involved in controlling the buildup of cadmium, zn, magnesium, ca & nickel in the grains of sorghum.

In pearl millet, the observed genetic diversity in hybrids and inbred lines presents significant opportunities for improving grain micronutrient concentrations through selective breeding. Anuradha *et al.* (2017) carried out a GWAS on 130 diverse pearl millet lines and identified 16 areas of the genome linked to higher levels of iron & zinc in the grain. Additionally, using three million SNPs generated through genotyping-by-sequencing (GBS), the study pinpointed several important Fe and Zn content (MTAs) marker-trait associations. Six potential genes linked to Fe/Zn uptake are among these, were identified, with the YUCCA-11 gene highlighted for its role in enhancing zinc efficiency via auxin biosynthesis (Manwaring, 2018). When using finger millet, Puranik *et al.* (2020) applied mixed model and general linear model methods to identify 418 significant MTAs linked to variations in mineral content. Of these, 34 MTAs were considered particularly relevant, with 18 showing associations with genes related to metal ion transport, binding, and remobilization. These markers, once functionally validated, can be applied to finger millet grains to increase their nutritional value through genomic-assisted breeding techniques.

TRANSCRIPTOMICS

Generative sequencing technology developments have significantly transformed transcriptomics, allowing for a more profound understanding of the expression of genes in crops. This method has been particularly useful in identifying genes associated with the synthesis and accumulation of nutrients in various crop species, thanks to its broad coverage and cost-effectiveness (*Mishra et al., 2019*). for RNA sequencing is instrumental in developing high-throughput molecular markers and offers valuable data on a particular transcript's relative abundance under different conditions or stages of growth. For example, an investigation by Zhou *et al.* (2020) applied RNA-seq to analyze grains from three distinct sorghum cultivars, each with different grain colors, and found over 3,000 genes that are expressed differently. These genes were mainly connected to the metabolism of carbohydrates, amino acids, and flavonoids, which could influence the nutritional characteristics of the grains. Moreover, an alpha-amylase-3 gene was identified as a formidable contender linked to variations in the amount of fat and protein in sorghum grains, based on its expression patterns in relation to a GWAS QTL dataset (Rhodes *et al.*, 2017).

In another research, Satyavathi *et al.* (2022) conducted transcriptome analysis of pearl millet genotypes' stage-specific spikes, which differed Zinc (Zn) and iron

(Fe) composition. As Fe and Zn, respectively, 155 and 251 expressions have been demonstrated to be increased in the alleles with higher concentrations of these elements. Furthermore, the expression of transcripts 349 and 378 varied while the flowering & milking phases. According to RNA-seq analysis of spike tissues coming from two genotypes of finger millet, which has higher levels of calcium than other millet varieties, 24 calcium sensor genes, including CRKs, CaM, CBLs, CIPKs, and CaMLs, seemed significantly greater expressed within the high-calcium genotype (GP45) than in the low-calcium genotype (GP-1). These studies highlight how transcriptomics, particularly RNA-seq, can be a powerful tool for identifying and understanding genes that influence the nutritional composition of millets.

PROTEOMICS AND METABOLOMICS

Among the many "omics" approaches, proteomics & metabolomics are vital instruments for locating important biomolecules and trace elements with significant nutritional value. These approaches provide important insights into biological processes and offer potential to enhance the output of essential metabolic substances in millets, while also helping to understand the function of proteins. For instance, a study by Ramalingam *et al.* (2021) employed non-targeted metabolomics to analyze 61 different sorghum accessions, distinguishing between them by examining the building of phenolic acids, flavonoids, phytosterols, fatty acids, carotenoids, amino acids, & sugars temperate and tropical types were identified. New directions for creating biofortified sorghum cultivars with improved medicinal and nutraceutical properties were proposed. Even if millets are a major source of nourishment in impoverished nations, it is still important to investigate possible genes and their functions in the production and accumulation of nutrients by examining the proteome & metabolome at different phases of seed development.

IONOMIX

High-throughput profiling has become more and more important for enhancing the cereal grains' nutritional content as genotyping and sequencing technology increase. Ionomics has become well-known as a high-throughput technique in this regard. for "elemental profiling," which enables precise measurement of the mineral content in living organisms (Huang and Salt, 2016). From a nutritional perspective, ionomics is a valuable tool for studying. By determining the genes encoding transporters and comprehending their molecular functions, mineral transport systems in millets can be identified (Kumar *et al.*, 2014a). Even though ionomics research in crops remains in its infancy, it holds significant potential for evaluating the nutrient composition of grain crops.

Table 1: Genetic Variation in Millet Species

Sl. No.	Crop	Markers	Accession	Allele per Locus	PIC	Diversity Index	References
1	Pearl millet	SSR	86	3.7	0.68	0.61	Bougma *et al.* (2021)
2	Foxtail millet	SSR	250	20.9	0.84	0.86	Wang *et al.* (2012)
3	Finger millet	SSR	128	2.1	0.44	0.14	Ramakrishnan *et al.* (2016)
4	Proso millet	SSR	90	2.4	0.36	0.39	Hunt *et al.* (2011)
5	Pearl millet	SSR	214	0.77	0.82	-	Bashir *et al.* (2015)
6	Sorghum	SSR	3367	19.1	0.67	-	Billot *et al.* (2013)
7	Foxtail millet	SSR	128	0.72	0.75	-	Liu *et al.* (2011)
8	Finger millet	SSR	138	11.1	0.61	-	Lule *et al.* (2018)
9	Foxtail millet	SSR	26	10.6	0.62	0.65	Kim *et al.* (2012)
10	Sorghum	SSR	100	7.2	0.25	0.80	Mamo *et al.* (2023)
11	Proso millet	SSR	88	2.7	0.38	0.44	Liu *et al.* (2016)
12	Finger millet	RAPD	40	6.3	0.31	-	Joshi *et al.* (2020)
13	Finger millet	SSR	90	2.43	0.66	0.18	Pandian *et al.* (2018)
14	Sorghum	SSR	24	1.5	0.24	0.29	Enyew *et al.* (2022)
15	Foxtail millet	SSR	125	1.8	0.18	-	Kumari *et al.* (2011)
16	Proso millet	SSR	98	4.9	0.36	0.39	Hunt *et al.* (2011)
17	Finger millet	SSR	959	0.53	0.58	-	Bharathi (2011)
18	Pearl millet	SSR	404	8.4	0.48	-	Diack *et al.* (2017)

GENETIC VARIABILITY IN MILLETS: MOLECULAR MARKERS

Finger Millet (*Eleusine coracana* Gaertn)

Finger millet is an extremely nourishing grain, containing, 60.6% carbohydrates, 4.7% fat, 12.3% protein & 3.2% ash (Jayawardana *et al.*, 2019). The finger millet's genomic size is 1.5 Gb (*Vetriventhan et al., 2020*). To evaluate genetic diversity, SSR indicators have been shown to extensively utilized in finger millet experiment. For example, Panwar *et al.* (2010) utilized 10 SSR markers and found 70.19% polymorphism across 83 genotypes. Kumar *et al.* (2012) employed 24 SSR markers, detecting 66.6% polymorphism in 52 genotypes, and Nirgude *et al.* (2014) used 30 genic SSR markers to identify 68.23% polymorphism in 103 genotypes. A study by Ramakrishnan *et al.* (2016) identified only 72 of the 87 SSR markers (82.75%) in 128 genotypes were polymorphic, resulting in a 73.80% polymorphism rate. Furthermore, according to Krishna *et al.* (2018), the adaptability of SSR markers in foxtail and finger millet varied from 47.52% – 61.38% of finger millet & 30.76% to 69.23% of foxtail millet. Using five SSR and nine RAPD markers, 40 landraces from the Kaski and Dhading districts were also evaluated for genetic diversity. The findings indicated mild diversity, having (PIC) polymorphism information content values of 0.37 for SSR markers and 0.314 for RAPD markers. (Joshi *et al.,* 2020). In a separate study, three DNA marker techniques (RFLP, ISSR and RAPD) were utilized to examine 22 accessions of *Eleusine* species, which showed 14%, 10%, and 26% polymorphism, respectively, in African and Asian *E. coracana* accessions. The global germplasm diversity of finger millet was assessed using 314 accessions from ICRISAT, genotyped with the DArTseq technique. This study revealed 33,884 superior SNP markers, indicating high genetic diversity, with mean PIC, gene diversity, and corresponding Shannon index levels of 0.110, 0.114, & 0.194. The typical genetic distance between accessions was 0.301, with the highest genetic divergence (0.326) observed in the race elongata accessions (Backiyalakshmi *et al.*, 2021).

Pearl Millet (*Pennisetum glaucum* (L.) R. Br.)

An investigation into 214 pearl millet accessions from Sudan, alongside 11 additional accessions, used 30 SSR indicators for evaluating genomic diversity. The results showed high polymorphism with a PIC value of 0.77, gene diversity (GD) of about 0.82, and observed heterozygosity (Ho) of about 0.72Molecular variance analysis revealed that regional variation was greater than the variation between regions (Bashir *et al.*, 2015). These findings demonstrated the pearl millet accessions by way of Sudan exhibit significant genetic diversity, making them valuable for breeding programs aimed at improving pearl millet varieties throughout the region.

Foxtail Millet (*Setaria italica* (L.) Beauv.)

Foxtail millet, as stated by Vetriventhan *et al.* (2020) is renowned for having a modest genome size (423-510 Mb). Kumari *et al.* (2011) used RAPD & ISSR markers to assess the variety of genetics in 125 Indian foxtail millet accessions. According to the findings, Tamil Nadu, Karnataka, & Andhra Pradesh are states in southern India, had most alleles on the locus (1.78), trailed by the alpine northern states of Uttarakhand, Jammu, & Kashmir (1.61). The fewest alleles in every locus (1.5) were found in the middle-western area, which includes Rajasthan, MP & Maharashtra. In an analysis of acquisitions from China, Pakistan, and Korea have average gene diversity of 0.491, 0.692, and 0.652, respectively, respectively, with PIC values of 0.621, 0.653, and 0.438. The study concluded that Chinese accessions exhibited greater SSR diversity compared to those from Korea and Pakistan (Kim *et al.*, 2012).

Proso Millet (*Panicum miliaceum* L.)

Molecular diversity research on proso millet is limited, with few studies employing technology for next-generation sequencing (NGS). Most evaluations of proso millet's genetic diversity have traditionally used markers like RAPD, AFLP, and SSR (Boukail *et al.,* 2021). One study examined 90 proso millet genotypes, including both landraces and cultivars, using 100 SSR markers, which resulted in the identification of 1,287 alleles with sizes between 40 & 1,500 bp (Rajput *et al.*, 2016).

MARKER-LINKED TRAITS

To advance crop breeding, it is essential to collect beneficial alleles from various agro-ecological zones worldwide. Crop landraces and wild relatives often harbor valuable genetic diversity that can be leveraged to create cultivars resistant to environmental stress (Mbinda and Masaki, 2000). Molecular markers facilitate faster characterization and validation of germplasm, which in turn accelerates the development of improved varieties. The creation of multiparent populations will be critical for genomic selection and the identification of key traits in the future (Dida *et al.*, 2021). The initial genomic map of sorghum was built with RFLP markers, which was complemented by comparable maps of finger, foxtail, & pearl millet (Kiranmayee *et al.*, 2020).

Magnaporthe grisea-caused blast disease poses a serious danger to the finger millet industry. The use of comparative genomics is crucial for finding QTLs or genes connected to blast tolerance through SSR markers because finger millet genome sequence data is scarce (Table 2). For instance, the FMBLEST32 SSR

Table 2: Millets Molecular Markers Linked to Agronomic Features

Sl. No.	Trait	Marker Name	Marker Type	Millet Crop	Resistant Variety/Source	Reference
1	*Pyricularia* leaf spot	OP-D11700	RAPD	Pearl millet	THAK0285, THAK0201	Verma *et al.* (2021)
2	Ear length	TP351642	SNP	Finger millet crop	-	Sharma *et al.* (2018)
3	Maturity days	TP1600637	SNP	In Finger millet	-	Sharma *et al.* (2018)
4	50% flowering days	TP1071491	SNP	Finger millet	-	Sharma *et al.* (2018)
5	Flag leaf width	FMBLEST2	SSR	Finger millet	-	Babu *et al.* (2018)
6	Neck blast	FMBLEST35, FMBLEST36	SSR	Finger millet	-	Babu *et al.* (2018)
7	Basal tiller number	RM5963	SSR	Finger millet	-	Babu *et al.* (2018)
8	Blast of Finger	FMBLEST32	SSR	Finger millet	VHC3997, VHC3996, VHC3930, IE6082 (Nepal)	Babu *et al.* (2014)

marker's tolerant allele (710 bp) was primarily present in foreign genotypes, while it was only identified in a small number of Indian genotypes, mainly those from southern & northern India (Babu *et al.*, 2014).

GENOME SEQUENCING OF MILLET CROPS

Globally, approximately 164,447 millet accessions are preserved across different research organizations and gene banks. The majority of these collections consist of foxtail millet, proso millet, & finger millet while smaller millets have relatively lower representation (Muthamilarasan and Prasad, 2021). Advances in genome sequencing, particularly low-coverage (1x) sequencing, have enabled the identification of 972,863 bi-allelic single nucleotide polymorphisms based on 85 proso millet accessions originating through Twenty-three nations. These findings have been instrumental in analyzing population structures and phylogenetic relationships among the accessions.

Sorghum, a widely studied cereal, has its reference genome based on the inbred line "BTx623," known for its short growth period and early maturity (Smith and Frederiksen, 2000). The initial genome assembly for Sorghum bicolor comprised around 730 megabases (Mb) (Paterson *et al.*, 2009), later refined to 732.2 Mb, according to McCormick *et al.* (2018), including roughly 91.5% of the genetic code. Another assembly of the genome, derived from the "Rio" line, covered 729.4 Mb, representing 99.6% of the "BTx623" reference genome (Cooper *et al.*, 2019). A pan- DNA analysis involving 354 genetically diverse sorghum accessions identified several genes connected to agronomically crucial characteristics (Ruperao *et al.*, 2021).

Barnyard millet, part of the Echinochloa genus, includes both cultivated and wild species. Among the 35 known species that are grown are *E. frumentacea* & *E. utilis*, while *E. colona* (commonly known as "Jungle grass") is a wild species (Gomashe, 2017). Whole-genome sequencing of *E. crusgalli* using Illumina HiSeq 2000 technology revealed roughly 1.27 GB in size, with 108,771 protein-coding genes & 785 microRNAs predicted, & other non-coding RNAs (Guo *et al.*, 2017). Chloroplast genome sequencing of *E. frumentacea* identified a genome size of 139,593 base pairs, with 112 genes, of which 77 were protein-coding (Perumal *et al.,* 2016).

Kodo millet (*Paspalum scrobiculatum*), although significant in certain regions, lacks a comprehensive molecular map (Vetriventhan *et al.*, 2020). However, research on molecular markers has revealed phylogenetic groupings of genotypes between Indian and African origins (M'Ribu and Hilu, 1996). Based on panicle morphology, the crop is divided into three races: variabilis, irregularis & regaris. (de Wet *et al.*, 1983; Prasada *et al.*, 1993). Johnson *et al.* (2019) conducted the first genome-wide analysis of 165 kodo millet accessions, revealing that genetic diversity was not strongly aligned with traditional race classifications. The study emphasized the potential of genomic data to enhance the crop's core collections. Additionally, transcriptome analysis of the variety "CO_3" identified genes linked to drought resistance (Suresh *et al.*, 2022).

Little millet (*Panicum sumatrense*), an economically important cereal grown primarily in India, western Myanmar & Nepal is one of the least studied small millets (Baltensperger, 1996). Parani *et al.* (2001) used chloroplast DNA to identify seven species within 119 accessions of little millet. RFLP analysis, or restriction fragment length polymorphism, employing specific restricting enzymes like HaeIII and MspI, was effective for species identification. Complete chloroplast genome sequencing of *P. sumatrense*, conducted using the MiSeq platform, revealed a genome size of 139,384 base pairs, encompassing 125 genes, including thirty genes for tRNA and four rRNA, and ninety-one genes that code for proteins. According to Sebastian *et al.* (2018), the phylogenetic analysis revealed a tight link between witchgrass (*P. capillare*), common millet (*P. miliaceum*), & tiny millet (*P. sumatrense*). Genes implicated in both salt and drought tolerance have also been found by transcriptome analysis (Das *et al.*, 2020).

GENETIC ENGINEERING AND GENE EDITING IN MILLETS

Genetic engineering, also known as recombinant DNA technology, encompasses techniques to modify DNA from diverse sources, allowing the development of new genetic materials. These modifications enable organisms to inherit specific traits or produce proteins for targeted purposes. In millet crops like pearl millet and sorghum, genetic engineering has significantly advanced their improvement by introducing new traits through transgenic methods. One pioneering achievement in millet genetic engineering was reported by Casas *et al.* (1993), who successfully developed transgenic sorghum plants using particle bombardment. This method involved delivering DNA-coated particles into plant cells, enabling the introduction of beneficial genes. Later, *Zhao et al. (2000)* expanded these advancements by employing Agrobacterium-mediated transformation. This approach used the Agrobacterium's innate capacity to introduce genes within plant cells makes it a useful tool for precisely modifying sorghum's genetic makeup.

A remarkable application of genetic engineering is advancements in biofortified sorghum types. Zhao *et al.* (2019) engineered sorghum with enhanced and stable pro-vitamin A content, meeting a significant portion of the dietary requirement for

Table 3: Summary of Genome Sequencing in Millet Crops

Sl. No.	Crop	Ploidy Level	Genotype	Sequencing Platform	Key Findings	Reference
1	Sorghum	Diploid with 2n=2x=20	BTx623	Sequencing by Sanger	Approx 730 Mb genome size	Paterson *et al.* (2009)
2	Barnyard Millet	Hexaploid (2n=6x=54)	STB08	Illumina HiSeq 2000	Estimated genome size of ~1.27 GB	Guo *et al*. (2017)
3	Foxtail Millet	Diploid (2n=2x=18)	Zhang gu and A2	Illumina Genome Analyzer II, HiSeq 2000	423 Mb genome size anchored to 9 chromosomes; annotated 38,801 genes	Zhang *et a*l. (2012)
4	Pearl Millet	Diploid with (2n=2x=14)	Tift 23 D2B1-P1-P5	The Illumina Hi Seq 2000	Size of the genome ~1.79 GB, covering 90% of the estimated size; contains 38,579 predicted genes	Varshney *et al.* (2017)
5	Proso Millet	Tetraploid with 2n=4x=36	Landrace (ACC#00000390)	Illumina Short Read, PacBio RS II	Genome size of ~923 Mb, annotated 55,930 genes for protein-coding	Zou *et al*. (2019)
6	Finger Millet	Tetraploid (2n=4x=36)	PR 202	Illumina, NextSeq 500HiSeq 4000,	Genome assembly size of 1.19 GB (82.31% of the estimated 1.45 GB), predicted 85,243 genes	Hittalmani *et al*. (2017)

this nutrient, especially for children in areas where the staple grain is sorghum. This breakthrough has the potential to combat vitamin A deficiencies in undernourished populations. Further advancements in sorghum include increasing lysine content, an essential amino acid often deficient in sorghum-based diets. Zhao *et al.* (2003) enhanced lysine levels by introducing a gene encoding a high-lysine protein, significantly improving Sorghum's nutritional value. Similarly, genetic engineering in pearl millet has focused on developing herbicide-resistant varieties. Girgi *et al.* (2002) introduced a selective marker, the bar gene marker using a particle inflow gun, simplifying weed management and increasing yield potential in pearl millet cultivation.

Comprehensive reviews by Ceasar and Ignacimuthu (2009) and O'Kennedy *et al.* (2006) summarize the key achievements in millet genetic engineering, highlighting methods like biolistics and Agrobacterium-mediated transformation. These reviews also discuss genes of interest for improving many types of millet, including finger millet, foxtail millet, barnyard millet, & pearl millet. Of the two primary transformation methods, biolistics remains the more commonly employed for developing transgenic varieties (see Table 4). In addition to genetic engineering, gene editing has become a potent instrument for precise DNA modifications in plants. This technology allows researchers to insert, delete, or modify specific genes within the genome. However, successful gene editing requires a complete understanding of the organism's genetic structure. Currently Only a few numbers of millet have annotated genomic sequences available. species, with foxtail millet being a notable example. As a result, gene editing efforts in millets have primarily focused on foxtail millet (Ceasar, 2022).

The limited availability of annotated genomes for other millet species poses challenges in reducing unintended off-target effects of gene editing. Consequently, future advancements in millet gene editing will rely heavily on the development of complete and annotated genome sequences (Kudapa *et al.,* 2023). These advancements are expected to unlock new opportunities for precisely improving the agronomic and nutritional traits of millets, enhancing their contribution to global food security.

The introduction of CRISPR/Cas9 technology has brought about a major shift in agricultural research, particularly in plant genetics. Nowadays, CRISPR/Cas9 is regarded as the most efficient gene-editing tools available. In foxtail millet, the technology was first tested by disrupting the SiPDS gene in transient protoplast assays, although stable mutant lines were not produced in the initial study (Lin *et al.*, 2018). Despite this, researchers have continued to explore CRISPR/Cas9's potential in millet crops, for instance, Li *et al.* (2018) utilized CRISPR/

Table 4: Development of Crop Varieties through Biotechnological Methods

Sl. No.	Crop	Biotechnology Method	Plant Material	Varieties	Gene Used	Objective	References
1	**Teff**	*Agrobacterium*-mediated transformation	Callus	DZ-01-196	*PcGA2ox*	Induction of semi-dwarf traits	Gebre *et al.* (2013)
2	**Foxtail millet**	*Agrobacterium*-mediated transformation	Callus	Jigu11	*SiLEA14*	Enhanced drought and salinity tolerance	Wang *et al.* (2011)
3	**Foxtail millet**	*Agrobacterium*-mediated transformation	Callus	Jigu11	*SiARDP*	Enhanced resistance to drought	Li *et al.* (2014)
4	**Finger millet**	*Biolistic* method	Shoot tip and embryogenic callus	PGEC 2	*PcSrp*	Tolerance to salinity	Mahalakshmi *et al.* (2006)
5	**Finger millet**	*Agrobacterium*-mediated transformation	Embryogenic callus	Tropikanka, Yaroslav 8	*HvTUB1* and *TUAm1*	Herbicide resistance development	Bayer *et al.* (2014)
6	**Finger millet**	*Agrobacterium*-mediated transformation	Shoot apex	GPU 45	*Rice chitinase (chi11)*	Resistance to leaf blast	Ignacimuthu and Ceasar (2012)
7	**Barnyard millet**	*Biolistic* method	Callus	VL 29	Various promoter genes	Testing promoter efficiency	Gupta *et al.* (2001)
8	**Foxtail millet**	*Biolistic* method	Florets	3661	*SiPf40*	Characterization of *SiPf4* gene	Liu *et a*l. (2009)

Cas9 to target k1C genes in sorghum, aiming to reduce kafirin levels, improve protein quality, and enhance digestibility, while also increasing lysine content.

In more recent developments, Liang *et al.* (2022) succeeded in base editing foxtail millet using both to target the genes SiACC and SiALS, use Editors of Cytosine Base (CBE) & Adenine Base (ABE). Through the application of CBE to SiALS, they have been enabled to produce a herbicide-tolerant homozygous mutant. This illustrates the increasing accuracy of millet crop genome editing. Using CRISPR/Cas9 and base-editing, Liang *et al.* (2022) also presented a very efficient genome editing system that enables single-base changes and the deletion of one or more genes. This system offers a powerful tool for precisely modifying millet genomes, enabling the development of customized genetic traits. As this technology continues to improve and more genetic data becomes accessible, gene editing in millets is expected to significantly enhance both their agronomic and nutritional qualities.

MUTATION BREEDING

Mutation breeding is an essential method used in plant breeding that employs physical or chemical agents to generate genetic variations in plants. These induced changes help in the development of novel crop types. In India, mutation breeding has significantly contributed to millet crop improvement, with approximately 5% of millet varieties being developed using this approach (Patel and Vekariya, 2023). This technique has proven particularly effective in enhancing the genetic diversity of millets, leading to new traits and improved crop performance. The choice of mutagens plays an essential part in the mutation breeding process. Within crops like pearl millet and sorghum, different mutagens, such as ethidium bromide, streptomycin, and mitomycin, have been used to induce cytoplasmic mutations, especially for developing cytoplasmic male sterility. Gamma rays and thermal neutrons are effective in causing chromosomal breaks, which helps in promoting genetic diversity. Gamma rays, along with ethyl methane sulfonate and diethyl sulfate, have proven especially effective in producing a variety of mutations in sorghum and pearl millet (Hanna, 1982).

Research on proso millet & finger millet has highlighted the potential of gamma irradiation in generating novel variations with improved characteristics, such higher grain yield (Francis *et al.*, 2022; Msikita, 2002). In finger millet, mutation breeding has led to better grain yields, as observed by Ganapathy *et al.* (2021). Likewise, foxtail millet has benefited from mutation breeding through gamma irradiation, producing variants with diverse traits (Bolbhat and Ishte, 2020).

Kodo millet, recognized for its cleistogamous flowering nature, is another species where mutation breeding has been applied to generate genetic diversity and enhance agronomic traits (Nagaraja *et al.*, 2023). Similarly, barnyard millet has responded positively to mutagenic treatments such as EMS and gamma rays (Ramesh *et al.,* 2019). Mutation breeding continues to be an effective tool for developing new millet varieties that offer higher yields, disease resistance, and adaptability to environmental changes.

QUANTITATIVE TRAIT LOCI (QTL)

Only 14 (1.6%) of the 888 foxtail millet additions from around the globe shown strong resistance to the leaf blast in greenhouse experiments (Li *et al.*, 2020). It additionally noted that cultivars from Asian and European nations differ genetically in their ability to withstand leaf blast (Nakayama *et al.,* 2005). 155 accessions of foxtail millet retrieved from the gene bank maintained by the International Crops Research Institute for the Semi-Arid Tropics (ICRISAT) comprise the core collection. revealed several genotypes resistant to leaf blast (Sharma *et al.*, 2014). In research conducted in the cultivar Yugu 5, three QTLs promoting leaf blast resistance were discovered on chromosomes one, two, and eight by Tian *et al.* (2021). The population of recombinant inbred lines (RILs) 35,065 SNP markers were employed to create a very dense genetic linkage map. These QTLs were identified at the genomic regions of 1.62 Mb and 276.6 kb, & 1.75 Mb, respectively. They were the QLB-czas1 and QLB-czas2 and QLB-czas8. Common SNPs in the QLB-czas8 region were found using the majority RNA sequencing (BSR-Seq) and segregant analysis (BSA) techniques, validating its position. Within the QLB-czas2 and QLB-czas8 areas, many genes linked to disease resistance were discovered, offering valuable insights for improving cultivars and understanding the genetic regulation of blast resistance (Tian *et al.*, 2021).

An association study of 190 global genotype uses 46 SSR markers for finger millet and 15 agro-morphological traits revealed several noteworthy correlations between SSR alleles & characteristics like basal tiller number. One marker, UGEP81, located on the chromosome 6B, was strongly associated with basal tiller number. Another research by Bharathi *et al.* (2011) identified a QTL at the UGEP8 for the basal tiller number locus on chromosome 3B, with a P value of 0.003, explaining 4.2% of the phenotypic changes. These loci offer valuable targets for marker-assisted breeding, enabling the introduction of beneficial traits into locally adapted germplasm.

BIOFORTIFICATION

One important tactic for raising the nutritional value of crops is biofortification. It is possible to accomplish through agronomic practices, traditional breeding, as well as sophisticated biotechnological techniques like genome editing and genetic engineering. Millets, due to their economic advantages, have become an important target for biofortification, particularly in addressing micronutrient deficiencies. The Harvest Plus initiative has developed Millet made of high-iron pearl through conventional breeding in India to help combat iron (Fe) deficiency (Vinoth and Ravindhran, 2017). Additionally, biofortified sorghum, with increased levels of zinc and iron, has been released in both India and Africa.

In order to increase the acceptability of species like foxtail millet, proso millet, as well as barnyard millet as newborn meals, research has looked into the molecular foundation of their waxy starch. Moreover, genes and QTLs linked to finger millet protein quality have been found with the aid of comparative genomics. Although various millet varieties are nutritionally rich, the biofortification efforts have mainly focused on increasing iron content in pearl millet. Current biofortification strategies aim not only to raise the amount of nutrients in the grains but also to reduce antinutrients, improving the bioavailability of essential minerals (Vinoth and Ravindhran, 2017). However, it is it's crucial to remember that biofortified millets developed through biotechnology are not yet commercially available.

DNA BANK, DNA BARCODING, BIOINFORMATICS, AND TISSUE CULTURE IN MILLET RESEARCH

A DNA bank acts as a crucial tool for safeguarding the genetic component variety of millet species. This involves carefully isolating and preserving millet DNA under extremely low temperatures to maintain its viability over the long term. Such collections are invaluable for future breeding initiatives and scientific studies. For instance, Nepal's Genebank took a significant step in 2013 by creating a specialized DNA bank for finger millet. This initiative provides a reliable repository of genetic material, supporting researchers and breeders in studying critical genetic attributes like disease resistance, nutritional qualities, and environmental adaptability. DNA banks are essential to the advancement of research, enhancing crop breeding programs, and bolstering food security in areas where millets are an essential dietary staple.

DNA barcoding is an innovative technique for identifying plant species using their DNA sequences. By analyzing multiple DNA regions simultaneously, this method enables precise species identification. In millets, DNA barcoding has been a cornerstone for conserving safeguarding farmers' rights, traditional knowledge, and genetic resources. For example, Ragupathy *et al.* (2016) used the ITS2 DNA barcode to achieve a 100% accuracy rate in identifying 32 landraces and six millet species. This approach is particularly valuable for safeguarding the genetic integrity of plants in countries like Nepal and India. Studies like those by Newmaster *et al.* (2013) further underscore the usefulness of DNA barcoding, as specific genomic regions such as ITS and trnH-psbA have proven effective in assessing genetic differences among millet species and their landraces. The field of bioinformatics has become indispensable for managing and analyzing complex biological data, offering significant insights into crop genetics. Genomic databases have emerged as key resources, housing extensive collections of complete genome sequences, intergenic regions, open reading frames (ORFs), and DNA sequences of vital crops, such as millets. For species like finger millet an foxtail millet, databases like Gramene, millet genome repositories, & Phytozome offer essential genetic information that enables researchers to examine and enhance these crops. Foxtail millet, in particular, benefits from dedicated bioinformatics tools, including databases for transcription factors and microRNAs, which shed light on the genetic regulatory mechanisms within the species. Marker-based and transposable-element databases are also instrumental for breeding and genetic studies. Han *et al.* (2014) identified 271 microRNAs spanning 44 families in foxtail millet, highlighting its intricate regulatory systems. Additionally, Millet, a web-based multi-omics platform created by Sun *et al.* (2023), offers extensive data on six millet species, including genomic and stress-related multi-omics information. By providing a user-friendly interface, Millet enables researchers to better understand millet genomes and their responses to environmental pressures. Publicly available nuclear genome sequences for several kinds of millet, including foxtail millet, proso millet and kodo millet, among others, further enhance research potential, contributing to advancements in millet genetics and breeding (Thulasinathan *et al.*, 2022).

TISSUE CULTURE: UNLOCKING NEW POSSIBILITIES FOR MILLETS

Tissue culture, though not widely adopted for millet production in Nepal, presents immense potential for improving millet crops. Traditionally, millets are propagated through seeds, but tissue culture offers opportunities for breakthroughs in crop improvement. Applications include regenerating plants from various explants, rescuing embryos in hybridization processes, developing mutants, conducting cellular hybridization, and regenerating transgenic organs.

Various explants, such as immature embryos, seeds, inflorescences, seedling bases, and roots, have been successfully used in tissue culture studies. Key advancements in this field include:

Regenerative Callus Cultures in Finger Millet: Callus cultures derived from immature embryos of finger millet have shown promising results for regeneration.

Embryogenic Callus Cultures in Kodo Millet: These cultures establish a basis for in vitro propagation & modifications about genetic makeup as kodo millet.

Advancements in Pearl Millet Tissue Culture: Researchers have successfully developed embryogenic callus in pearl millet, paving the way for genetic enhancement and crop improvement.

Somatic Embryogenesis Protocols in Finger Millet: Effective methods for somatic embryogenesis and plant regeneration have been developed, allowing the growth of genetically modified finger millet plants (Ngetech *et al.*, 2018).

Mass Production Using Mature Seeds in Finger Millet: Regeneration techniques using mature seeds as explants have proven valuable for large-scale cultivation (Kashyap *et al.*, 2018).

CONCLUSION

Biotechnology presents immense opportunities for transforming millet cultivation, particularly in overcoming the challenges posed by their diminutive flowers. Innovative biotechnological approaches have been successfully employed across various millet species, facilitating advancements in genetic research, crop improvement, genetic variation, and the safeguarding of genetic diversity. Techniques such as tissue culture, mutagenesis, genetic diversity analysis, marker-assisted selection, DNA barcoding, genomic databases, biofortification, and genetic engineering each play a critical role in elevating millet agriculture to new heights. DNA markers have been essential for investigating the millet species' genetic diversity, offering crucial insights that inform crop enhancement strategies. Open-access genetic resources hold the potential to accelerate progress in millet research, enabling improvements in yield, resilience, and disease resistance. By fostering continued innovation and collaboration, the integration of biotechnology into millet farming can address key agricultural challenges, contributing to sustainable food systems and bolstering food security, where millets serve as a dietary cornerstone.

REFERENCES

Ajeesh Krishna TP, T Maharajan and SA Ceasar (2022) Improvement of millets in the post-genomic era. Physiol. *Mol. Biol. Plants* 28(3): 669-685. DOI: https://doi.org/10.1007/s12298-022-01158-8

Anatala TJ, MK Mandavia, RA Dave, W Kothari, HP Gajera, HR Ramani and B Golakiya (2015) Protein profile and isoenzymes in response to water stress in pearl millet. *J. Cell Tissue Res.* 15: 5255-5266.

Anuradha N, CT Satyavathi, C Bharadwaj, T Nepolean, SM Sankar, SP Singh *et al.* (2017) Deciphering genomic regions for high grain iron and zinc content using association mapping in pearl millet. *Front. Plant Sci.* 8: 412. DOI: https://doi.org/10.3389/fpls.2017.00412

Babitha MP, HS Prakash and HS Shetty (2004) Purification and properties of lipoxygenase induced in downy mildew resistant pearl millet seedlings due to infection with Sclerospora graminicola. *Plant Sci.* 166(1): 31-39.

Babitha MP, SG Bhat, HS Prakash, and HS Shetty (2002) Differential induction of superoxide dismutase in downy mildew resistant and susceptible genotypes of pearl millet. *Plant Pathology* 51(4): 480-48

Babu BK, P Dinesh, PK Agrawal, S Sood, C Chandrashekara, JC Bhatt and A Kumar (2014) Comparative genomics and association mapping approaches for blast resistant genes in finger millet using SSRs. *PLoS ONE* 9(6) DOI: https://doi.org/10.1371/journal.pone.0099182

Babu BK, S Sood, C Chandrashekara, A Pattanayak and L Kant (2018) Mapping quantitative trait loci for important agronomic traits in finger millet (Eleusine coracana) mini core collection with genomic and genic SSR markers. *J. Plant Biochem. Biotechnol.* 27: 401–414. DOI: https://doi.org/10.1007/s13562-018-0449-7

Backiyalakshmi C, M Vetriventhan, S Deshpande, C Babu, V Allan, D Naresh, R Gupta and VCR Azevedo (2021) Genomewide assessment of population structure and genetic diversity of the global finger millet germplasm panel conserved at the ICRISAT genebank. *Front Plant Sci.* 12:692463. DOI: https://doi.org/10.3389/fpls.2021.692463

Bashir EMA, AM Ali, ETI Mohamed, AE Melchinger, HK Parzies and BIG Haussmann (2015) Genetic diversity of Sudanese pearl millet (Pennisetum glaucum (L.) R. Br.) landraces as revealed by SSR markers, and relationship between genetic and agro-morphological diversity. *Gen. Res. Crop Evol.* 62(4): 579–591. https://doi.org/10.1007/s10722-014-0183-5

Bayer GY, AI Yemets and YB Blume (2014) Obtaining the transgenic lines of finger millet *Eleusine coracana* (L.) with dinitroaniline resistance. *Cytol Genet.* 48:139–44.

Bharathi A (2011) Phenotypic and genotypic diversity of global finger millet (*Eleusine coracana* (L.) Gaertn.) composite collection. PhD Thesis. Tamil Nadu Agricultural University, Coimbatore, India.

Billot C, P Ramu, S Bouchet, J Chantereau, M Deu, L Gardes, JL Noyer, JF Rami, R Rivallan, Y Li, P Lu, T Wang, RT Folkertsma, E Arnaud, HD Upadhyaya, JC Glaszmann and CT Hash. Massive sorghum collection genotyped with SSR markers to enhance use of global genetic resources. *PLoS One*, 8(4): DOI: https://doi.org/10.1371/journal.pone.0059714

Bolbhat SN and K Ishte Dipali (2020) Mutagenic effect on growth parameters of foxtail millet [Setaria italica (L.) Beauv.] in M1 generation. *Int. J. Creative Research Thoughts* 8(5): 1111-1115.

Bougma LA, MH Ouedraogo, A Ouoba, AA Zoure, N Sawadogo and M Sawadogo (2011) Genetic differentiation for gene diversity among pearl millet (*Pennisetum glaucum* (L.) R. Br.) landraces as revealed by SSR markers. *International Journal of Agronomy*, 2021. DOI: https://doi.org/10.1155/2021/6160903

Boukail S, M Macharia, M Miculan, A Masoni, A Calamai, E Palchetti and M Dell'Acqua (2021) Genome wide association study of agronomic and seed traits in a world collection of proso millet (*Panicum miliaceum* L.) *BMC Plant Biology* 21(1) https://doi.org/10.1186/s12870-021-03111-5

Casas AM, AK Kononowicz, UB Zehr, DT Tomes, JD Axtell, LG Butler, RA Bressan and PM Hasegawa (1993) Transgenic sorghum plants via microprojectile bombardment. *Proc. Natl. Acad. Sci.* USA 90: 11212-11216.

Ceasar SA and S Ignacimuthu (2009) Genetic engineering of millets: current status and future prospects. *Biotechnol. Lett.* 6: 779-88. DOI: https://doi.org/10.1007/s10529-009-9933-4

Ceasar SA and T Maharajan (2022) The role of millets in attaining United Nation's sustainable developmental goals. *Plants People Planet* 4(4): 345–349. https://doi.org/10.1002/ppp3.10254

Ceasar SA (2019) Genome wide identification and in silico analysis of PHT1 family genes and proteins in Setariaviridis: The best model to study nutrient transport in millets. *The plant genome* 12(1), 180019.

Ceasar SA (2022) Genome-editing in millets: current knowledge and future perspectives. *Mol. Biol. Rep.* 49: 773–781. DOI: https://doi.org/10.1007/s11033-021-06975-w

Chhabra AK, RK Behl and El Basam (2001) Role of isozymes in pearl-millet improvement (*Pennisetum glaucum*) *Landbauforschung Volkenrode* 4(51): 165-173.

Chowdari KV, AP Davierwala, VS Gupta, PK Ranjekar and OP Govila (1998) Genotype identification and assessment of genetic relationships in pearl millet [*Pennisetum glaucum* (L.) R. Br] using microsatellites and RAPDs. *Theor. Appl. Genet.* 97: 154–162.

Cooper EA, ZW Brenton, BS Flinn, J Jenkins, S Shu, D Flowers, F Luo, Y Wang, P Xia, K Barry, C Daum, A Lipzen, Y Yoshinaga, J Schmutz, C Saski, W Vermerris and S Kresovich (2019) A new reference genome for Sorghum bicolor reveals high levels of sequence similarity between sweet and grain genotypes: implications for the genetics of sugar metabolism. *BMC Genomics* 20, 420. https://doi.org/10.1186/s12864-019-5734-x

Das RR, S Pradhan and A Parida (2020) De-novo transcriptome analysis unveils differentially expressed genes regulating drought and salt stress response in Panicum sumatrense. *Scientific Reports* 10(1): 21251. DOI: https://doi.org/10.1038/s41598-020-78118-3

De Wet J, KE Rao Prasada, MH Mengesha and DE Brink (1983) Domestication of mawa millet (*Echinochloa colona*) *Economic Botany* 37(3): 283-291.

Desai, H., Hamid, R., Ghorbanzadeh, Z., Bhut, N., Padhiyar, S. M., Kheni, J., & Tomar, R. S (2021) Genic microsatellite marker characterization and development in little millet (*Panicum sumatrense*) using transcriptome sequencing. *Scientific Reports* 11(1) DOI: https://doi.org/10.1038/s41598-021-00100-4

Diack O, NA Kane, CB Salazar, MC Gueye, BM Diop, A Fofana, O Sy, H Tall, L Zekraoui, M Piquet, M Couderc, Y Vigouroux, D Diouf and A Barnaud (2017) New genetic insights into pearl millet diversity as revealed by characterization of early- and late-flowering landraces from Senegal. *Front. Plant Sci.* 8: 818. DOI: https://doi.org/10.3389/fpls.2017.00818

Dida M, CA Oduori, SJ Manthi, MO Avosa, EO Mikwa, HF Ojulong and DA Odeny (2021) Novel sources of resistance to blast disease in finger millet. *Crop Science* 61(1): 250–262.DOI: https://doi.org/10.1002/csc2.20378

Enyew M, T Feyissa, AS Carlsson, K Tesfaye, C Hammenhag and M Geleta (2022) Genetic diversity and population structure of sorghum [Sorghum Bicolor (L.) Moench] accessions as revealed by single nucleotide polymorphism markers. *Front. Plant Sci.*, 12-2021. DOI: https://doi.org/10.3389/fpls.2021.799482

Eun HM (1996) Enzymology primer for recombinant DNA technology. Elsevier

Francis N, R Rajasekaran, I Krishnamoorthy, R Muthurajan, C Thiyagarajan and S Alagarswamy (2022) Gamma irradiation to induce beneficial mutants in proso millet (*Panicum miliaceum* L.): an underutilized food crop.

International Journal of Radiation Biology 98(7): 1277-1288. DOI: https://doi.org/10.1080/09553002.2022.2024292

Ganapathy KN, TSSK Patro, KB Palanna, IK Das, M Elangovan, B Prashant, SC Venkatesha, N Anuradha, TE Nagaraja and VA Tonapi (2021) Development of improved mutants for grain yield and related traits in finger millet (*Eleusine coracana* L Gaertn.) through gamma mutagenesis. *Int. J. Plant and Soil Science* 33: 225-235.

Gebre E, L Gugsa, U Schlüter and K Kunert (2013) Transformation of tef (Eragrostis tef) by Agrobacterium through immature embryo regeneration system for inducing semi-dwarfism. *S. Afr. J. Bot.* 87: 9–17.

Genebank (n.d.) Millets in Nepal: For human, livestock and environment. https://www.researchgate.net/publication/348049968_Conservation_and_Utilization_of_Agricultural_PlaGirgi M, MM O'Kennedy, A Morgenstern, G Mayer, H Lorz and KH Oldach (2002) Transgenic and herbicide resistant pearl millet (*Pennisetum glaucum* L.) R. Br. via microprojectile bombardment of scutellar tissue. *Mol. Breed.* 10: 243–252.

Guo L, J Qiu, C Ye, G Jin, L Ma *et al.* (2017) Echinochloa crus-galli genome analysis provides insight into its adaptation and invasiveness as a weed. *Nat. Com.* 8: 1031. DOI: https://doi.org/10.1038/s41467-017-01067-5

Gupta P, S Raghuvanshi and AK Tyagi (2001) Assessment of the efficiency of various gene promoters via biolistics in leaf and regenerating seed callus of millets, Eleusine coracana and Echinochloa crusgalli. *Plant Biotech.* 18: 275–282

Han J, H Xie, Q Sun, J Wang, M Lu, W Wang, E Guo and J Pan (2014) Bioinformatic identification and experimental validation of miRNAs from foxtail millet (*Setaria italica*) *Gene* 546(2): 367-77. DOI: https://doi.org/10.1016/j.gene.2014.05.050

Hanna WW (1982) Mutation breeding of pearl millet and sorghum. IAEA.

Hash CT, A Sharma, MA Kolesnikova-Allen, SD Sing, RP Thakur, AG Bhasker Raj, MNV Ratnaji Rao, DC Nijhawan, CR Beniwal, P Sagar, HP Yadav, YP Yadav, Shrikant, SK Bhatanagar, IS Khairwal, CJ Howarth, GP Cavan, MD Ghale, C Liu, KM Devos, WA Breese and JR Witcombe (2006) Teamwork delivers biotechnology products to Indian smallholder crop-livestock producers: Pearl millet hybrid "HHB 67 Improved" enters seed delivery pipeline. *J SAT Agri Res.* 2: 1–4

Hatakeyama M, S Aluri, MT Balachadran, SR Sivarajan, A Patrignani, S Grüter, L Poveda, R Shimizu-Inatsugi, J Baeten, KJ Francoijs, KN Nataraja, YAN Reddy, S Phadnis, RL Ravikumar, R Schlapbach, SM Sreeman and KK

Shimizu (2018) Multiple hybrid de novo genome assembly of finger millet, an orphan allotetraploid crop. *DNA Research* 25(1): 39-47.

He Q, S Tang, H Zhi, J Chen, J Zhang *et al.* (2023) A graph-based genome and pan-genome variation of the model plant Setaria. *Nat. Genet.* 55: 1232–1242. DOI: https://doi.org/10.1038/s41588-023-01423-w

Heinz DJ (1987) Sugarcane Improvement through Breeding, Elsevier, Amsterdam 273-311.

Hema R, RS Vemanna, S Sreeramulu, CP Reddy, M Senthil-kumar and M Udayakumar (2014) Stable expression of mtlD gene imparts multiple stress tolerance in finger millet. *PLoS ONE* 9: e99110

Hittalmani S, H Mahesh, MD Shirke, H Biradar, G Uday, Y Aruna, H Lohithaswa and A Mohanrao (2017) Genome and transcriptome sequence of finger millet (*Eleusine coracana* (L.) Gaertn.) provides insights into drought tolerance and nutraceutical properties. *BMC Genomic*s 18: 1–16.

Huang XY and DE Salt (2016) Plant ionomics: From elemental profiling to environmental adaptation. *Mol. Plant* 9: 787–797. DOI: https://doi.org/10.1016/j.molp.2016.05.003

Hunt HV, MG Campana, MC Lawes, YJ Park, MA Bower, CJ Howe and MK Jones (2011) Genetic diversity and phylogeography of broomcorn millet (*Panicum miliaceum* L.) across Eurasia. *Molecular Ecology* 20: 4756–4771. DOI: https://doi.org/10.1111/j.1365-294X.2011.05318.x

Ignacimuthu S, Ceasar SA (2012) Development of transgenic finger millet (*Eleusine coracana* (L.) Gaertn.) resistant to leaf blast disease. *J Biosci* 37: 135–47.

Jayawardana SAS, JKRR Samarasekera, GHCM Hettiarachchi, J Gooneratne, SD Mazumdar and R Banerjee (2019) Dietary fibers, starch fractions and nutritional composition of finger millet varieties cultivated in Sri Lanka. J. F. Com. Anal. 82(363): 103249. DOI: https://doi.org/10.1016/j.jfca.2019.103249

Joshi BK, D Joshi and SK Ghimire (2020) Genetic diversity in finger millet landraces revealed by RAPD and SSR markers. *Nepal J. Biotechnol.* 8(1): 1-11 DOI: https://doi.org/10.3126/njb.v8i1.30204

Joshi BK, KH Ghimire, RP Mainali, SK Shrestha and M Bhattarai (2023) Pictorial dictionary of millets and their wild relatives in Nepal. NAGRC, Khumaltar, Kathmandu. https://www.researchgate.net/publication/371984042

Kashyap A, S Madanrao Penak, A Saha and BR Singh (2018) In Vitro plant development of Eleusine Coracana via indirect organogenesis and somatic embryogenesis using mature seeds as explants. *Current Science* 115(1): 91–98. https://www.jstor.org/stable/26978152.

Kim EJ, KJ Sa, KC Park and JK Lee (2012) Study of genetic diversity and relationships among accessions of foxtail millet [*Setaria italica* (L.) p. beauv.] in Korea, China and Pakistan using SSR markers. *Genes and Genomics* 34(5): 529–538. DOI: https://doi.org/10.1007/s13258-012-0074-0

Kiranmayee KNSU, Hash CT, Sivasubramani S, Ramu P, Amindala BP, Rathore, A, Kishor PBK, Gupta R and Deshpande SP (2020) Fine-mapping of sorghum stay-green QTL onchromosome10 revealed genes associated with delayed senescence. *Genes* (Basel), 11, 1026. DOI: https://doi.org/10.3390/genes11091026

Kothari SL, S Kumar, RK Vishnoi, A Kothari and KN Watanabe (2005) Applications of biotechnology for improvement of millet crops: Review of progress and future prospects. *Plant Biotechnol.* 22(2): 81-88. DOI: https://doi.org/10.5511/plantbiotechnology.22.81

Krishna TPA, T Maharajan, RH Antony David, M Ramakrishnan, SA Ceasar, V Duraipandiyan, GV Roch and S Ignacimuthu (2018) Microsatellite markers of finger millet (*Eleusine coracana* (L.) Gaertn) and foxtail millet (*Setaria italica* (L.) Beauv) provide resources for cross-genome transferability and genetic diversity analyses in other millets. *Biocatalysis and Agricultural Biotechnology* 16: 493–501. DOI: https://doi.org/10.1016/j.bcab.2018.09.009

Kudapa H, R Barmukh, H Vemuri, S Gorthy, R Pinnamaneni, M Vetriventhan, RK Srivastava, P Joshi, E Habyarimana, SKGupta and M Govindaraj (2023) Genetic and genomic interventions in crop biofortification: Examples in millets. *Front. Plant Sci.* 14: 1123655. DOI: https://doi.org/10.3389/fpls.2023.1123655

Kumar A, N Mirza, T Charan, N Sharma and VS Gaur (2014b) Isolation, characterization and immunolocalization of a seed dominant CaM from finger millet (*Eleusine coracana* L. Gartn.) for studying its functional role in differential accumulation of calcium in developing grains. *Applied biochemistry and biotechnology*, 172, 2955-2973.

Kumar A, N Sharma N, P Panwar and AK Gupta (2012) Use of SSR, RAPD markers and protein profiles-based analysis to differentiate Eleusine coracana genotypes differing in their protein content. *Mol. Bio. Rep.* 39: 4949-4960.

Kumar A, RK Chowdhury, RL Kapoor and OS Dahiya (1995) Identification of pearl millet hybrids and their lines using seed and seedling characters, chemical tests and gel electrophoresis. *Seed Sci. Technol*. 23: 21–32.

Kumar A, VS Gaur, A Goel and AK Gupta (2014a) De novo assembly and characterization of developing spikes transcriptome of finger millet (*Eleusine coracana*): a minor crop having nutraceutical properties. *Plant Mol. Biol. Rep.* 33: 905–922. DOI: https://doi.org/10.1007/s11105-014-0802-5

Kumar MA, RJ Sherry, A Varier and M Dadlani (2007) Suitability of seed esterases for establishing distinctness, uniformity and stability of pearl millet genotypes. *Current Sci.* 951-956.

Kumari R, N Dikshit, D Sharma and KV Bhat (2011) Analysis of molecular genetic diversity in a representative collection of foxtail millet [*Setaria italica* (L.) P. Beauv.] from different agro-ecological regions of India. *Phy. Mol. Biol. Plants* 17(4): 363–374. DOI: https://doi.org/10.1007/s12298-011-0085-3

KW Puranik S, PP Sahu, S Beynon, RK Srivastava, D Sehgal, H Ojulong *et al.* (2020) Genome-wide association mapping and comparative genomics identifies genomic regions governing grain nutritional traits in finger millet (*Eleusine coracana* L. Gaertn.) *Plants People Planet* 2: 649–662. DOI: https://doi.org/10.1002/ppp3.10120

Lata C, S Gupta and M Prasad (2013) Foxtail millet: a model crop for genetic and genomic studies in bioenergy grasses. *Critical reviews in biotechnology* 33(3): 328-43.

Li A, S Jia, A Yobi, Z Ge, SJ Sato, C Zhang *et al.* (2018) Editing of an alphakafirin gene family increases, digestibility and protein quality in sorghum. *Plant Physiol.* 177: 1425–1438. DOI: https://doi.org/10.1104/pp.18.00200

Li C, J Yue, X Wu, C Xu and J Yu (2014) An ABA-responsive DRE-binding protein gene from Setaria italica, SiARDP, the target gene of SiAREB, plays a critical role under drought stress. *J Exp Bot.* 65: 5415–27.

Li ZH, GQ Jia, Y LX, Li YC, H Zhi, S Tang, JF Ma, S Zhang, YD Li, ZL Shang and XM Diao (2021) Identification of blastresistance loci through genome-wide association analysis in foxtail millet (*Setaria italica* (L.) Beauv.) *Journal of Integrative Agriculture*, 20(8): 2056–2064. DOI: https://doi.org/10.1016/S2095-3119(20)63196-3

Liang Z, Y Wu, L Ma, Y Guo and Y Ran (2022) Efficient genome editing in *Setaria italica* using CRISPR/Cas9 and base editors. *Front. Plant Sci.* 12: 815946. DOI: https://doi.org/10.3389/fpls.2021.815946

Lin CS, CT Hsu, LH Yang, LY Lee, JY Fu, QW Cheng *et al.* (2018) Application of protoplast technology to CRISPR/Cas9 mutagenesis: from single-cell mutation detection to mutant plant regeneration. *Plant Biotechnol.* J. 16, 1295–1310. DOI: https://doi.org/10.1111/pbi.12870

Liu M, Y Xu, J He, S Zhang, Y Wang and P Lu (2016) Genetic diversity and population structure of broomcorn millet (*Panicum miliaceum* L.) cultivars and landraces in China based on microsatellite markers. *Int. J. Mol. Sci.*, 17, 370. DOI: https://doi.org/10.3390/ijms17030370

Liu Z, G Bai, D Zhang, C Zhu, X Xia, R Cheng and Z Shi (2011) Genetic diversity and population structure of elite foxtail millet [*Setaria italica* (L.) P. Beauv.] germplasm in China. *Crop Science* (Plant Genetic Resource), 51(4): 1655–1663. DOI: https://doi.org/10.2135/cropsci2010.11.0643

Liu Z, T Zhang, C Li and G Bai (2014) Genetic diversity and classification of cytoplasm of Chinese elite foxtail millet [*Setaria italica* (L.) P. Beauv.] germplasm. *Crop Sci.* 54: 659–66.

Lule D, S De Villiers, M Fetene, DA Odeny, A Rathore, RR Das and K Tesfaye (2018) Genetic diversity and association mapping of Ethiopian and exotic finger millet accessions. *Crop and Pasture Science*, 69(9), 879–891. DOI: https://doi.org/10.1071/CP18175

Mahalakshmi S, GSB Christopher, TP Reddy, KV Rao and VD Reddy (2006) Isolation of a cDNA clone (PcSrp) encoding serine-rich protein from Porteresia coarctata T. and its expression in yeast and finger millet (*Eleusine coracana* L.) affording salt tolerance. *Planta* 224: 347–59

Mahatma MK, R Bhatnagar, RK Solanki, GK Mittal and RR Shah (2011) Characterization of downy mildew resistant and susceptible pearl millet (*Pennisetum glaucum* (L.) R. Br.) genotypes using isozyme, protein, randomly amplified polymorphic DNA and inter-simple sequence repeat markers. *Archives of Phytopathology and Plant Protection* 44(20): 1985-1998.

Mamo W, M Enyew, T Mekonnen, K Tesfaye and T Feyissa (2023) Genetic diversity and population structure of sorghum [*Sorghum bicolor* (L.) Moench] genotypes in Ethiopia as revealed by microsatellite markers. *Heliyon*, 9(1): DOI: https://doi.org/10.1016/j.heliyon.2023.e12830

Manivannan A, AK Chhabra and S Nimbal (2013) Characterization of pearl millet [*Pennisetum glaucum* (L.) R. BR.] genotypes by using superoxide dismutase (SOD) isozyme. *Indian J. Agri. Res.* 47(6): 471-479.

Manwaring HR (2018) Accessing and dissecting useful genetic variations for iron and zinc content in pearl millet (*Pennisetum glaucum*) using genome wide association studies. Doctoral dissertation, Aberystwyth University.

Markert CL and F Moller (1959) Multiple forms of enzymes: tissue, ontogenetic and species-specific patterns. *Proc. Nat. Acad. Sci.* 45: 753-763.

Mbinda WM and HI Masaki (2021) Breeding strategies and challenges in the improvement of blast disease resistance in finger millet. *A Current Review. Front. Plant Sci.* 11.

McCormick RF, SK Truong, A Sreedasyam, J Jenkins, S Shu, D Sims *et al.* (2018) The Sorghum bicolor reference genome: improved assembly, gene annotations, a transcriptome atlas, and signatures of genome organization. *The Plant Journal* 93(2): 338-354.

Mirza, N, G Taj, S Arora, and A Kumar (2014) Transcriptional expression analysis of genes involved in regulation of calcium translocation and storage in finger millet (*Eleusine coracana* L. Gartn.) Gene 550(2): 171-179.

Mishra VK, S Gupta, R Chand, PS Yadav, SK Singh, AK Joshi *et al.* (2019) Comparative transcriptomic profiling of highand low- grain zinc and iron containing Indian wheat genotypes. *Curr. Plant Biol.* 18: 100105. DOI: https://doi.org/10.1016/j.cpb.2019.100105

Mondini L, A Noorani and MA Pagnotta (2009) Assessing plant genetic diversity by molecular tools. *Diversity* 1(1): 19–35. DOI: https://doi.org/10.3390/d1010019

Moore GA and GB Collins (1983) New challenges confronting plant breeders. In: Isozymes in plant genetics and breeding (SD Tanksley and TJ Orton eds) Elsevier Science Publishers. pp. 25–58.

M'Ribu HK and KW Hilu (1996) Application of random amplified polymorphic DNA to study genetic diversity in *Paspalum scrobiculatum* L (Kodo millet, Poaceae) *Genetic Resources and Crop Evolution* 43: 203-210.

Msikita RN (2002) Use of mutation breeding technique in improving finger millet in northern Zambia. Proceedingsof the 8 Biennial Scientific Conference Dem and -Driven Agricultural Research and Development for Sustainable

Resource Management and Increased Production, (p. 99) Kenya: Kenya Agricultural Research Institute, Nairobi. Muthamilarasan M and M Prasad (2021) Small Millets for enduring food security amidst pandemics. *Trends in plant science* 26(1): 33–40. DOI: https://doi.org/10.1016/j.tplants.2020.08.008

Nagaraja TE, C Nandini, S Bhat, S Parveen and Gazala (2023) Artificial hybridization techniques in small millets—A review. *Front Plant Science* 14. DOI: https://doi.org/10.3389/fpls.2023.1112117

Nakayama H, T Nagamine and N Hayashi (2005) Genetic variation of blast resistance in foxtail millet (*Setaria italica* (L.) P. Beauv.) and its geographic distribution. *Genet Resour. Crop Evol.* 52: 863–868

Newmaster SG, S Ragupathy, D Shanmughanandhan, CJ Jijo, R Sathishkumar and K Patel (2013) Genomic valorization of the fine scale classification of small millet landraces in southern India. *Genome* 56: 123–127

Ngetich A, C Mweu, M Ngugi, A Mukami, H Ojulong and W Mbinda (2018) Efficient plant regeneration protocol for finger millet [*Eleusine coracana* (L.) Gaertn.] via somatic embryogenesis. *Afr. J. Biotechnol.* 17: 660–667. DOI: https://doi.org/10.5897/AJB2018.16452

Nirgude, M, BK Babu, Y Shambhavi, UM Singh, HD Upadhyaya and A Kumar (2014) Development and molecular characterization of genic molecular markers for grain protein and calcium content in finger millet (*Eleusine coracana* (L.) Gaertn.) *Mol. Bio. Rep.* 41: 1189-1200.

O'Kennedy MM, A Grootboom and PR Shewry (2006) Harnessing sorghum and millet biotechnology for food and health. *Journal of Cereal Science* 44(3): 224-235. DOI: https://doi.org/10.1016/j.jcs.2006.08.001

Oh BJ, RA Frederiksen and CW Magill (1994) Identification of molecular markers linked to head smut resistance gene (shs) in sorghum by RFLP and RAPD analysis. *Phytopathol.* 84:830–833.

Pandian S, L Satish, R Rameshkumar, P Muthuramalingam, AS Rency, P Rathinapriya and M Ramesh (2018) Analysis of population structure and genetic diversity in an exotic germplasm collection of *Eleusine coracana* (L.) Gaertn. using genic-SSR markers. *Gene*, 653, 80–90. DOI: https://doi.org/ 10.1016/j.gene.2018. 02.018

Panwar P, M Nath M, VK Yadav and A Kumar (2010) Comparative evaluation of genetic diversity using RAPD, SSR and cytochrome P450 gene-based markers with respect to calcium content in finger millet (*Eleusine coracana* L. Gaertn.) *J Gen.* 89: 121–133

Parani M, K Rajesh, M Lakshmi, L Parducci, AE Szmidt and A Parida (2001) Species identification in seven small millet species using polymerase chain reaction—restriction fragment length polymorphism of trnS-psbC gene region. *Genome* 44: 495–499

Patel AA and RD Vekariya (2023) Advanced breeding approaches for small millet crops to accelerate improvements for a new green revolution. *Just Agricultlure* 3(10): 16-23

Paterson AH, JE Bowers, R Bruggmann, I Dubchak, J Grimwood, H Gundlach *et al.* (2009) The sorghum bicolor genome and the diversification of grasses. *Nature* 457(7229): 551-556.

Phuke RM, K Anuradha, K Radhika, F Jabeen, G Anuradha, T Ramesh *et al.* (2017) Genetic variability, genotype × environment interaction, correlation, and GGE biplot analysis for grain iron and zinc concentration and other agronomic traits in RIL population of sorghum (*Sorghum bicolor* L. Moench) Front. Plant Sci. 8: 712. DOI: https://doi.org/10.3389/fpls.2017.00712

Prasada Rao K, J De Wet, VG Reddy, M Mengesha (1993) Diversity in the small millets collection at ICRISAT.

Ragupathy S, S Dhivya, K Patel, A Sritharan, K Sambandan, H Gartaula, R Sathiskumar, K Khadka, BC Nirmala, A Nirmala Kumari and SG Bigmaster (2016) DNA record of some traditional small millet landraces in India and Nepal. *Biotech* 6 (2): 1–19. DOI: https://doi.org/10.1007/s13205-016-0450-6

Rajput SG, DK Santra and J Schnable (2016) Mapping QTLs for morpho-agronomic traits in proso millet (*Panicum miliaceum* L.) *Molecular Breeding* 36: 1-8.

Ramakrishnan M, AS Ceasar, V Duraipandiyan, NA Al-Dhabi and S Ignacimuthu (2016) Assessment of genetic diversity, population structure and relationships in Indian and non-Indian genotypes of finger millet (*Eleusine coracana* (L.) Gaertn) using genomic SSR markers. *Springer Plus* 5(1): 1–11. DOI: https://doi.org/10.1186/s40064-015-1626-y

Ramalingam AP, W Mohanavel, A Premnath, R Muthurajan, PVV Prasad and R Perumal (2021) Large-Scale nontargeted metabolomics reveals antioxidant, nutraceutical and therapeutic potentials of sorghum. *Antioxidants* 10: 1511. DOI: https://doi.org/10.3390/antiox10101511

Ramakrishnan M, SA Ceasar, V Duraipandiyan, NA Al-Dhabi and S Ignacimuthu (2016) Assessment of genetic diversity, population structure and relationships in Indian and non-Indian genotypes of finger millet [*Eleusine coracana* (L.) Gaertn] using genomic SSR markers. *Springer Plus*, 5(1), 1–11. DOI: https://doi.org/10.1186/s40064-015-1626-y

Ramesh M, C Vanniarajan, R Ravikesavan, K Eraivan Arutkani Aiyanathan and PP Mahendran (2019) Mutagenic effectiveness and efficiency in barnyard millet Echinochloa frumentacea using physical, chemical and combination of mutagens. *Electronic Journal of Plant Breeding* 10(2): 949-956.

Rangan TS (1974) Morphogenic investigations on tissue cultures of *Panicum miliaceum. Z. Pflanzenphysiol.* 72(5): 456–459

Rangan TS (1976) Growth and plantlet regeneration in tissue cultures of some Indian millets: Paspalum scrobiculatumL., Eleusine coracana Gaertn. and *Pennisetum typhoideum Pers. Z. Pflanzenphysiol.* 78(3): 208–216

Repellin A, m Båga, PP Jauhar and RN Chibbar (2001) Genetic enrichment of cereal crops via alien gene transfer: New challenges. *Plant Cell Tiss. Org.* Cult. 64: 159–183

Rhodes DH, L Jr. Hoffmann, WL Rooney, TJ Herald, S Bean, R Boyles *et al.* (2017) Genetic architecture of kernel composition in global sorghum germplasm. *BMC Genomics* 18: 15. DOI: https://doi.org/10.1186/s12864-016-3403-x

Riley S.C. Gupta A. Seetharam J.N. Mushonga, editors, Advances in small millets. Oxford and IBH Publishing Co. Pvt. Ltd., New Delhi, India.

Ruperao P, N Thirunavukkarasu, P Gandham, S Selvanayagam, M Govindaraj, B Nebie, E Manyasa, R Gupta, RR Das, DA Odeny, H Gandhi, D Edwards, SP Deshpande and A Rathore (2021) Sorghum pan-genome explores the functional utility for genomic-assisted breeding to accelerate the genetic gain. *Frontier Plant Science* 12: 666342. DOI: https://doi.org/10.3389/fpls.2021.666342

Saha D, MC Gowda, L Arya, M Verma and KC Bansal (2016) Genetic and genomic resources of small millets. *Critical Reviews in Plant Sciences* 35(1): 56-79.

Saleem S, N Mushtaq, W Shah, A Rasool, K Hakeem and R Rehman (2021) Morpho-physiological, biochemical and molecular adaptation of millets to abiotic stresses: A review. *Phyton* 90: 1363–1385. DOI: https://doi.org/10.32604/phyton.2021.014826

Salimath SS, AC de Oliveira, ID Godwin and JL Bennetzen (1995) Assessment of genome origins and genetic diversity in the genus Eleusine with DNA markers. *Genome* 38: 757–763

Salson M. J Orjuela, C Mariac, L Zekraouï, M Couderc, S Arribat, N Rodde, A Faye, N Kane, C Tranchant-Dubreuil, Y Vigouroux, and C Berthouly-Salazar (2023) An improved assembly of the pearl millet reference genome using Oxford Nanopore long reads and optical mapping. G3 Genes|Genomes|*Genetics* 13(5), jkad051. DOI: https://doi.org/10.1093/g3journal/jkad051

Salt DE, I Baxter and B Lahner (2008) Ionomics and the study of the plant ionome. *Annu. Rev. Plant Biol.* 59: 709–733. DOI: https://doi.org/10.1146/annurev.arplant.59.032607.092942

Satyavathi CT, RS Tomar, S Ambawat, J Kheni, SM Padhiyar, H Desai *et al.* (2022) Stage specific comparative transcriptomic analysis to reveal gene networks regulating iron and zinc content in pearl millet [*Pennisetum glaucum* (L.) R. Br.]. *Sci. Rep.* 12(1): 276. DOI: https://doi.org/10.1038/s41598-021-04388-0

Sebastin R, GA Lee, KH Lee, MJ Shin,BGT Cho, JR Lee, KH Ma and JW Chung (2018) The complete chloroplast genome sequences of little millet (*Panicum sumatrense* Roth ex Roem. and Schult.) (Poaceae) *Mitochondrial DNA Part B*, 3(2): 719–720. DOI: https://doi.org/10.1080/23802359.2018.1483771

Shakoor N, G Ziegler, BP Dilkes, Z Brenton, R Boyles, EL Connolly *et al.* (2016) Integration of experiments across diverse environments identifies the genetic determinants of variation in Sorghum bicolor seed element composition. *Plant Physiol.* 170: 1989–1998. DOI: https://doi.org/10.1104/pp.15.01971

Shannon LM (1968) Plant isoenzymes. *Ann. Rev. Plant physiol.* 19(1): 187-210.

Sharma D, A Tiwari, S Sood, G Jamra, NK Singh, PK Meher and A Kumar (2018) Genome wide association mapping of agro-morphological traits among a diverse collection of finger millet (*Eleusine coracana* L.) genotypes using SNP markers. *PLoS ONE* 13(8): e0199444. DOI: https://doi.org/10.1371/journal.pone.0199444

Sharma R, AG Girish, HD Upadhyaya, P Humayun, TK Babu, VP Rao and RP Thakur (2014) Identification of blast resistance in a core collection of foxtail millet germplasm. *Plant Dis.* 98: 519–524

Shivakumar PD, HM Geetha and HS Shetty (2003) Peroxidase activity and isozyme analysis of pearl millet seedlings and their implications in downy mildew disease resistance. *Plant Sci.* 164(1): 85-93.

Singh UM, M Chandra, SC Shankhdhar and A Kumar (2014) Transcriptome wide identification and validation of calcium sensor gene family in the developing spikes of finger millet genotypes for elucidating its role in grain calcium accumulation. *PloS One* 9(8) e103963. DOI: https://doi.org/10.1371/journal.pone.0103963

Smith CW and RA Frederiksen (2000) Sorghum: Origin, history, technology, and production. New York, NY: John Wiley and SonsSun M, H Yan, A Zhang, Y jin, C Lin, L Luo, B Wu, Y Fan, S Tian, X Cao, Z Wang, J Luo, Y Yang, J Jia, P Zhou,Q Tang, CS Jones, RK Varshney, RK Srivastava, M He, Z Xie, X Wang, G Feng, G Nie, D Huang, X Zhang, F Zhu and L Huang (2023) Milletdb: a multi-omics database to accelerate the research of functional genomics and molecular breeding of millets. *Plant Biotechnol.* J. 21: 2348-2357. DOI: https://doi.org/10.1111/pbi.14136

Suresh BV, P Choudhary, PR Aggarwal, S Rana, RK Singh, R Ravikcsavan, M Prasad, and M Muthamilarasan (2022) De novo transcriptome analysis identifies key genes involved in dehydration stress response in kodo millet (*Paspalum scrobiculatum* L.) Genomics 114(3): 110347. DOI: https://doi.org/https://doi.org/10.1016/j.ygeno.2022.110347

Taylor MG and IK Vasil (1991) Histology of, and physical factors affecting, transient GUS expression in pearl millet (*Pennisetum glaucum* (L.) R.B.) embryos following microprojectile bombardment. *Plant Cell Rep.* 10: 120-125.

Teama S (2018) DNA Polymorphisms: DNA-Based molecular markers and their application in medicine. In: Genetic diversity and disease susceptibility (Y Liu, ed) In Tech. DOI: https://doi.org/10.5772/intechopen.79517

Thulasinathan T, P Jain, AK Yadav, V Kumar, AM Sevanthi and AU Solanke (2022) Current status of bioinformatics resources of small millets. In: Omics of Climate Resilient Small Millets (RN Pudake, AU Solanke, AM Sevanthi and P Rajendrakumar, eds), pp. 227-234. Springer, Singapore. DOI: https://doi.org/10.1007/978-981-19-3907-5_11

Titan B, L Zhang, Y Liu, P Wu, W Wang, Y Zhang and H Li (2021) Identification of QTL for resistance to leaf blast in foxtail millet by genome re-sequencing analysis. *Theoretical and Applied Genetics* 134(2): 743–754. DOI: https://doi.org/10.1007/s00122-020-03730-w

Upadhyaya HD, Wang YH, Sharma R and Sharma S (2013) Identification of genetic markers linked to anthracnose resistance in sorghum using association analysis. *Theoretical and Applied Genetics* 126: 1649–1657. DOI: https://doi.org/10.1007/s00122-013-2081-1

Varshney RK, A Bohra, M Roorkiwal, R Barmukh, WA Cowling, A Chitikineni *et al.* (2021) Fast-forward breeding for a food-secure world. Trends Genet. 37: 1124–1136. DOI: https://doi.org/10.1016/j.tig.2021.08.002

Varshney RK, C Shi, M Thudi, C Mariac, J Wallace *et al.* (2017) Pearl millet genome sequence provides a resource to improve agronomic traits in arid environments. *Nat. Biotechnol.* 35: 969–976. DOI: https://doi.org/10.1038/nbt.3943

Vasil IK (1987) Developing cell and tissue culture systems for the improvement of cereal and grass crops. *J. Plant Physiol.* 128: 193–197

Verma R, MK Tripathi, S Tiwari, N Tripathi, K Pandya and P Parihar (2021) Screening of pearl millet [*Pennisetum glaucum* (L.) R. Br.] genotypes against blast disease on the basis of disease indexing and gene specific SSR markers. *International Journal of Current Microbiology and Applied Sciences* 10(2): 2218–2227. DOI: https://doi.org/10.20546/ijcmas.2021.1002.264

Vetriventhan M, VCR Azevedo, HD Upadhyaya, A Nirmalakumari, Joanna Kane-Potaka, S Anitha, SA Ceasar, M Muthamilarasan, BV Bhat, K Hariprasanna, A Bellungdagi, D Cheruku, C Backiyalakshmi, D Santra, C Vanniarajan and VA Tonapi (2020) Genetic and genomic resources, and breeding for accelerating improvement of small millets: current status and future interventions. *The Nucleus* 63(3): 217–239. DOI: https://doi.org/10.1007/s13237-020-00322-3

Vinoth A and R Ravindhran (2017) Biofortification in millets: A sustainable approach for nutritional security. *Front. Plant Sci.* 8: 29. DOI: https://doi.org/10.3389/fpls.2017.00029

Wakizuka T and T Yamaguchi (1987) The induction of enlarged apical domes in vitro and multi shoot formation from finger millet (*Eleusine coracana*) Ann. Bot. 60: 331–336

Wang C, G Jia, H Zhi, Z Niu, Y Chai, W Li, Y Wang, H Li, P Lu, B Zhao and X Diao (2012) Genetic diversity and population structure of Chinese foxtail millet [*Setaria italic*a (L.) Beauv.] landraces. *Gene Genomes Genetics* 2, 769. DOI: https://doi.org/10.1534/g3.112.002907

Wang M, Y Pan, C Li, C Liu, Q Zhao and G Ao (2011) Culturing of immature inflorescences and Agrobacterium-mediated transformation of foxtail millet (*Setaria italica*) *Afr J Biotechnol.* 10: 16466–79.

William DHM and KA Mujeeb (1992) Isozyme and cytological markers of some Psathyrostachys juncea accessions. *Theor. Appl. Genet.* 84(5- 8): 53.

Yadav S, S Jain, V Jain and RK Jain (2007) Genetic analysis of CMS, restorer, hybrid and open pollinated genotypes of Indian pearl millet [*Pennisetum glaucum* (L.) R. Br.] using ISSR markers. *Indian J. Biotechnolnol.* 6: 205–209.

Yang HB, WH Kang, SH Nahm and BC Kang (2015) Methods for developing molecular markers. In: Current technologies in plant molecular breeding (HJ Koh, SY Kwon, M Thomson, eds) Springer Netherlands; pp. 15-50. DOI: https://doi.org/10.1007/978-94-017-9996-6_2

Yang HB, Kang WH, Nahm, SH, and Kang BC (2015) Methods for developing molecular markers. In: Current Technologies in Plant Molecular Breeding (pp. 15–50) Springer Netherlands. DOI: https://doi.org/10.1007/978-94-017-9996-6_2

Zhao ZU, T Cai, L Tagliani, M Miller, N Wang, H Pang, M Rudert, S Schroeder, D Hondred, J Seltzer, and D Pierce (2000) Agrobacterium-mediated sorghum transformation. *Plant Mol. Bio.* 44: 789–798.

Zhao ZY, K Glassman, V Sewalt, N Wang, M Miller, S Chang *et al.* (2003) Nutritionally improved transgenic sorghum. In: Plant biotechnology 2002 and beyond (I. K. Vasil, ed) Dordrecht: Kluwer Academic Publishers, pp. 413–416.

Zhao ZY, P Che, K Glassman and M Albertsen (2019) Nutritionally enhanced sorghum for the arid and semiarid tropical areas of Africa. *Methods Mol.* Biol. 1931: 197–207. DOI: https://doi.org/10.1007/978-1-4939-9039-9_14

Zhou W, Y Zhou, Z Wang, Y Li, H Liu, F Cui *et al.* (2020) Comparative transcriptome analysis in three sorghum (*Sorghum bicolor*) cultivars reveal genomic basis of differential seed quality. *Plant Biosyst.* 156: 232–241. DOI: https://doi.org/10.1080/11263504.2020.1851790

Zou C, L Li, D Miki, D Li, Q Tang, L Xiao, S Rajput, P Deng, Li Peng, W Jia, R Huang, M Zhang, Y Sun, J Hu, X Fu, PS Schnable, Y Chang, F Li, H Zhang, B Feng, X Zhu, R Liu, JC Schnable, JK Zhu and H Zhang (2019) The genome of broomcorn millet. *Nature Communication* 10: 436. DOI: https:/ /doi.org/10.1038/s41467-019-08409-5.

Fig 22.1 PC : Asia Financial (Pg. 117)

Fig 23. 1 PC :SP (Pg. 121)

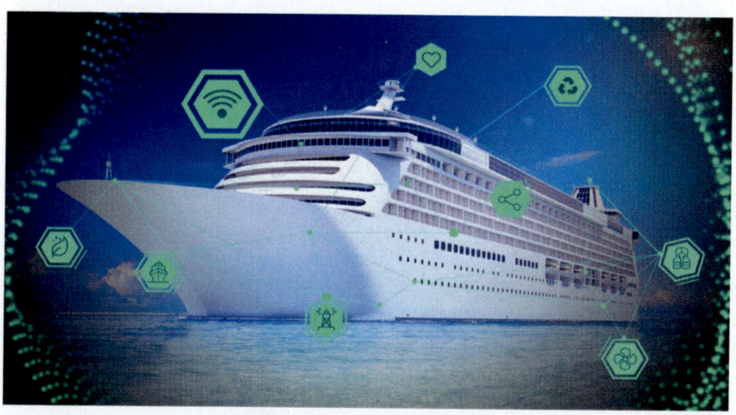

Fig 26.1 PC: Siemens (Pg. 134)

Fig 27.1 PC: CNET (Pg. 138)

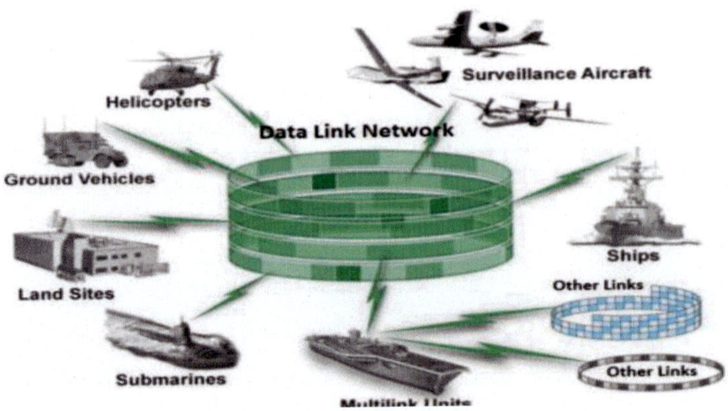

Fig 28. 1 PC: FICCI Events (Pg. 145)

Fig 29.1 PC: George Town security studies review (Pg. 149)